Machine Learning from Weak Supervision

Machine Learning from Weak Supervision

An Empirical Risk Minimization Approach

Masashi Sugiyama, Han Bao, Takashi Ishida,
Nan Lu, Tomoya Sakai, and Gang Niu

The MIT Press
Cambridge, Massachusetts
London, England

The MIT Press would like to thank the anonymous peer reviewers who provided comments on drafts of this book. The generous work of academic experts is essential for establishing the authority and quality of our publications. We acknowledge with gratitude the contributions of these otherwise uncredited readers.

This book was set in Times New Roman by Westchester Publishing Services. Printed and bound in the United States of America.

Library of Congress Cataloging-in-Publication Data

Names: Sugiyama, Masashi, 1974– author.
Title: Machine learning from weak supervision : an empirical risk
    minimization approach / Masashi Sugiyama, Han Bao, Takashi Ishida,
    Nan Lu, Tomoya Sakayi and Gang Niu.
Description: Cambridge, Massachusetts : The MIT Press, [2022] |
    Series: Adaptive computation and machine learning series |
    Includes bibliographical references and index.
Identifiers: LCCN 2021045984 | ISBN 9780262047074 (hardcover)
Subjects: LCSH: Supervised learning (Machine learning)
Classification: LCC Q325.75 .S84 2022 | DDC 006.3/1—dc23/eng/20211119
LC record available at https://lccn.loc.gov/2021045984

10 9 8 7 6 5 4 3 2 1

Contents

IV   ADVANCED TOPICS AND PERSPECTIVES

Preface

The term *artificial intelligence* (AI) has been coined at the celebrated Dartmouth Workshop in 1956. Since then, AI has been a topic of central interest for many researchers and engineers. After more than half a century, AI has finally become a genuine technology to enrich our daily life. *Machine learning* (ML) is a sub-field of AI aimed at investigating computer algorithms that improve themselves automatically through experience. ML is one of the most evolved and deeply researched topics in science and technology in the early twenty-first century, and the technology has boosted the use of AI in the real world. Nowadays, ML-based AI systems have been deployed in pioneering new businesses as well as in advanced scientific research and technological development.

So far, success of ML has mainly been in the virtual world, in e-commerce, social networks, and gaming. ML-based AI systems need to be trained with big data containing rich, supervised information. When we try to apply ML to real-world problems in our physical world, such as medical diagnosis, natural disaster, and education, it is extremely difficult or even impossible to collect such a huge amount of fully supervised data. Thus, data collection is one of the critical bottlenecks for AI to overcome before it can further penetrate our society. The difficulty of collecting data in the physical world has become even more pronounced during the COVID-19 pandemic, and thus there is an urgent need to develop a novel theory and algorithms that allow us to train ML-based AI systems with limited supervision.

The purpose of this book is to give readers practical algorithms of *weakly supervised classification*, as well as basic explanations and the advanced mathematical theories behind them. By weakly supervised classification, we do not mean that we try to train a classifier from small training data, which is mathematically not possible without imposing strong assumptions on data and models. Instead, we aim to train a classifier with a large amount of data that can be easily collected. Easily collectible data usually contains weaker supervised information than expensive, fully supervised data labeled by experts, but we will show that it is possible to train a classifier from such weakly supervised data as if we have fully supervised data.

Authors' photo taken during a video conference in March 2021.

The authors' research team started working on weakly supervised classification around 2010 in Japan, at a laboratory at Tokyo Institute of Technology. Since then, we have been developing various weakly supervised classification methods and theoretically elucidating their mechanisms and mathematical properties. The laboratory was then moved to the University of Tokyo in 2014, and an additional research team was also built at the RIKEN Center for Advanced Intelligence Project in 2016.

In 2017, we started to think that it was time to summarize and publish an account of our weakly supervised classification research, and we submitted a book proposal to the MIT Press. Luckily, in September 2017, the first author (M. Sugiyama) had a chance to visit Paris and meet Dr. Francis Bach at Ècole Normale Supérieure, who is the editor of the MIT Press Adaptive Computation and Machine Learning series. We had a fruitful discussion on the contents of the book and also had precious feedback from two reviewers.

The contents of this book are mostly based on our own research. The authors would like to thank all the coworkers and students involved in this research project. In particular, our special thanks go to Dr. Marthinus Christoffel du Plessis, who was one of the first and most active members in our project. Without his contribution, we could not initiate this weakly supervised classification project. In addition, we want to express our appreciation to (in alphabetical order) Bo An, Daniele Calandriello, Nontawat Charoenphakdee, Yu-Ting Chou, Bo Dai, Lei Feng, Xin Geng, Yoav Goldberg, Mingming Gong, Hirotaka Hachiya, Bo Han, Yu-Guan Hsieh, Weihua Hu, Alon Jacovi, Wittawat Jitkrittum, Takuo Kaneko, Ryuichi Kiryo, Hsuan-Tien Lin, Song Liu, Tongliang Liu, Jiaqi Lv, Yao Ma, Aditya Krishna Menon, Takuya Shimada, Dacheng Tao, Zhenguo Wu, Ning Xie, Miao Xu, Makoto Yamada, Ikko Yamane, Yu Yao, Tianyi Zhang, and Tingting Zhao for their fruitful and valuable discussion and feedback on our research. While writing the book, we severely suffered the COVID-19 pandemic and our writing fell behind schedule. Nevertheless, our MIT Press editors, Marie Lufkin Lee, Stephanie Cohen, Elizabeth Swayze, and Alex Hoopes, were steadfast in their support, and we would like to express our gratitude. Some illustrations used in this book are taken from *Irasutoya*, a popular website of royalty-free illustrations in Japan: https://www.irasutoya.com/. We thank Takashi Mifune, the host of the website, for making his great illustrations available.

As we are writing this preface, we are still suffering the effects of the COVID-19 pandemic, and vaccination has started in several countries. We sincerely hope that we all overcome COVID-19's challenges and uncertainties and can return to our previous activities and routines (the good old days!) and thrive in the future.

Finally, our research has been supported by the following funding sources: KAKENHI 20680007, 23120001, 23300069, 23120004, 25700022, 26280054, 17H01760, 17H00757, 20H04206, 19J21094, 20J11937; JST CREST JPMJCR1403, JPMJCR18A2; JST AIP Acceleration Research JPMJCR20U3; JST ACT-X PMJAX2005; the International Research Center for Neurointelligence (WPI-IRCN) at the University of Tokyo Institutes for Advanced Study; the Institute for AI and Beyond at the University of Tokyo; Google AI Focused Research; and Microsoft Research CORE Project.

Masashi Sugiyama, Han Bao, Takashi Ishida,
Nan Lu, Tomoya Sakai, and Gang Niu

March 2021
Tokyo, Japan

I MACHINE LEARNING FROM WEAK SUPERVISION

1 Introduction

Together with the advancement of information technologies such as programming, computer architecture, mathematics for information processing, and devices for data collection, *artificial intelligence* has become one of the most active fields of scientific research. In particular, its core technology called *machine learning* has been gathering a great deal of attention. In this chapter, we provide a brief overview of the field of machine learning and describe our focus on machine learning from weak supervision.

1.1 Machine Learning

As human beings, we can improve our knowledge and behavior by learning from experience. The objective of machine learning is to let a computer learn like a human. Machine learning has been studied in a variety of research communities such as computer science, statistics, mathematics, physics, and neuroscience.

Studies of machine learning can be categorized into three types, depending on information available to learning machines: *supervised learning*, *unsupervised learning*, and *reinforcement learning*.

1.1.1 Supervised Learning

Supervised learning (Vapnik, 1998; Bishop, 2006; Sugiyama, 2015a) is the most standard formulation of machine learning, where pairs of input and output samples called *training data* are available. "Input" is also called "query," "question," "independent variable," "feature," "pattern," "measurement," and "covariate," while "output" is also referred to as "supervision," "answer," "dependent variable," "response," and "label." Supervised learning can be regarded as a natural realization of our daily learning style, where a student (a learning machine or computer) learns by asking questions to a teacher (an expert who has knowledge).

The objective of supervised learning is to estimate the input-output relationship from input-output paired training data. This allows a learning machine to estimate output of unseen input, i.e., the learning machine generalizes to unseen data. With this *generalization capability*, machine learning resembles "human-like" learning.

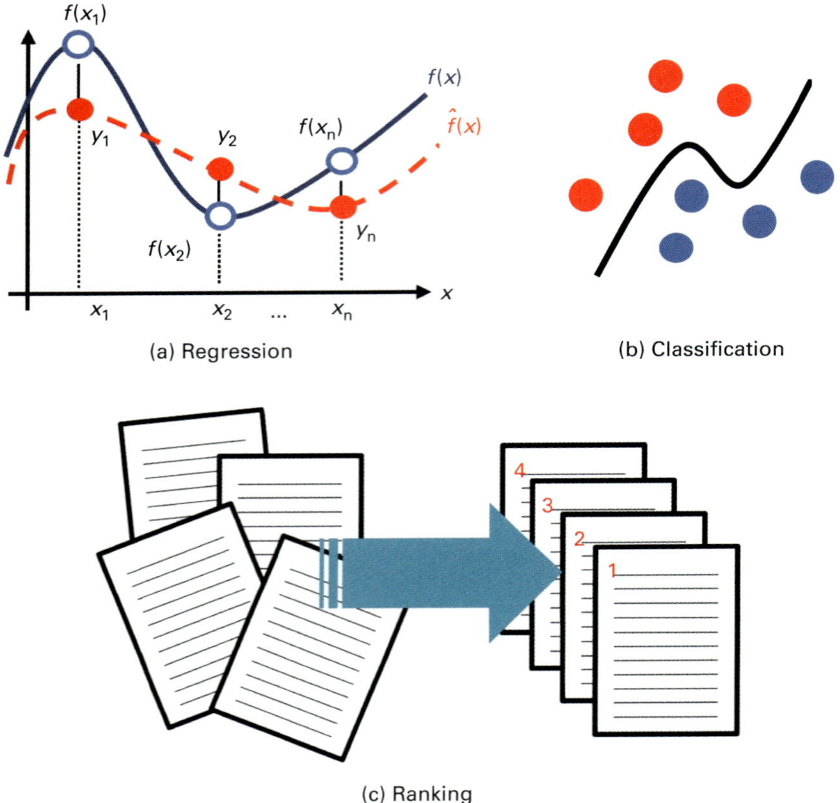

Figure 1.1
Tasks of supervised learning.

Typical tasks of supervised learning include *regression* (figure 1.1a), *classification* (figure 1.1b), and *ranking* (figure 1.1c). Regression is the problem of estimating a *real-valued* output value for an input point, while classification considers a *categorical* output value (or a class label). Classification is also referred to as *pattern recognition* since a pattern is automatically recognized by a learning machine. Ranking is the problem of estimating a *relative rank* of an input point, which has been extensively studied in the context of information retrieval (Li, 2014).

Supervised *feature extraction* (or *dimension reduction*) is a data pre-processing method aimed at finding a low-dimensional compact representation of a high-dimensional (redundant) input vector for enhancing the generalization capability. If a sub-vector consisting of a subset of variables of the original high-dimensional vector is found, the task is called *feature selection* (or *variable selection*), and it is used for interpreting the behavior of learning machines.

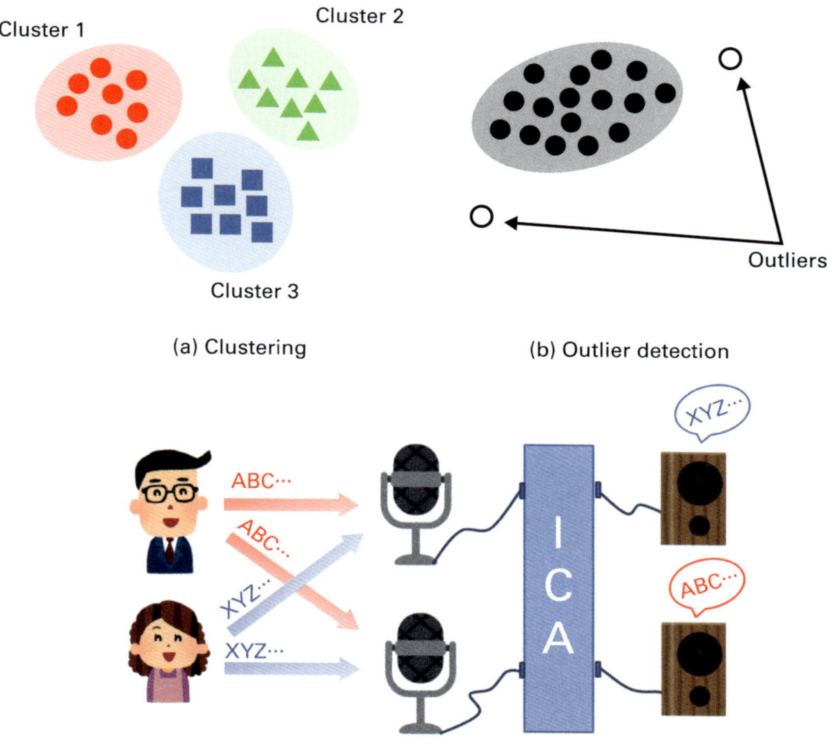

Figure 1.2
Tasks of unsupervised learning.

1.1.2　Unsupervised Learning

As noted previously, supervised learning uses input-output paired data for learning. On the other hand, in unsupervised learning, input-only data is provided and corresponding output is not available. Unsupervised learning may be regarded as a situation where a student learns alone by reading books.

While supervised learning is mathematically rigorously formulated as the problem of optimizing the generalization capability, unsupervised learning is often only vaguely and subjectively defined as the problem of extracting "informative" structure from data. A typical task of unsupervised learning is *clustering* (or *unsupervised classification*), which is aimed at grouping input samples based on their similarity (figure 1.2a). The problem of *outlier detection* (which is also referred to as *anomaly detection* or *novelty detection*) is also an important unsupervised learning task aimed at finding "irregular" samples in the set of given samples (figure 1.2b). *Density estimation* is also a representative unsupervised

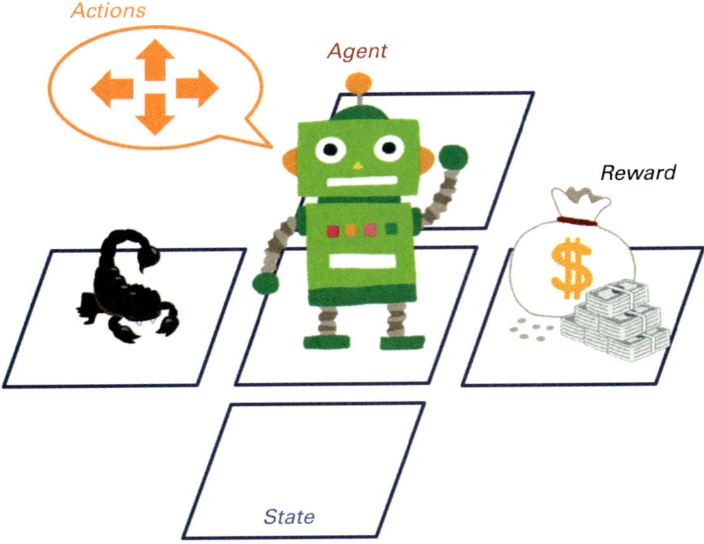

Figure 1.3
Reinforcement learning.

learning task, where the underlying probability density function behind data is estimated. Density estimation may also be used for clustering and outlier detection. Another interesting unsupervised learning topic is *independent component analysis* (Hyvärinen et al., 2001), which is aimed at separating a mixture of independent components into original independent sources (figure 1.2c). Feature extraction and feature selection can also be performed in an unsupervised fashion based on some measure of informativeness.

1.1.3   Reinforcement Learning

Reinforcement learning (Sutton and Barto, 1998; Sugiyama, 2015b) is a framework of letting an agent optimize its decision-making policy through interaction with an unknown environment. Reinforcement learning is formulated in a quite different way from supervised learning and unsupervised learning since it involves *sequential* decision making. Whereas standard supervised and unsupervised learning formulations handle a given set of training data in a *batch* manner, reinforcement learning is carried out in an *online* manner.[1] Nevertheless, various supervised and unsupervised learning techniques such as regression, feature extraction, feature selection, clustering, and outlier detection can be used to efficiently solve reinforcement learning problems.

In reinforcement learning, an agent located at some *state* takes an *action*, by which the agent moves to a next state and receives a *reward* (figure 1.3). The objective of reinforcement learning is to find a best decision-making policy that maximizes the sum of future rewards. A difficulty of reinforcement learning is that the best action to take at the current state is not explicitly supervised—the sum of future rewards has to be maximized, but an agent can receive only a single-step reward that corresponds to immediate evaluation of a taken action. This long-term nature makes reinforcement learning practically more attractive but technically more challenging than supervised learning.

Reinforcement learning may also be regarded as a form of *learning from weak supervision*, since immediate rewards do not include explicit "answers" but only implicit "hints" for maximizing the long-term return. However, because the framework of reinforcement learning is mathematically so different from standard supervised learning and unsupervised learning, we do not discuss reinforcement learning in this book. It would be an interesting future direction to discuss how the "learning from weak supervision" approaches we explore can be used for improving reinforcement learning.

1.2   Elements of Classification

Clearly, a wide variety of learning tasks are involved in machine learning. In this book, we discuss the problem of classification in supervised learning, unsupervised learning, and their intermediate situations called *weakly supervised classification*.

The process of classification consists of the *training* and *test* phases. In the training phase, a classifier is trained based on training data. Then, in the test phase, given a test input sample, the class to which the test input sample belongs is predicted based on the trained classifier. When we train a classifier, we need to take into account various components such as *training data, classifiers, learning criteria*, and *optimization algorithms*. As training data, we typically consider labeled data and unlabeled data. In discussing weakly supervised classification, we further consider other types of data such as pairwise similarity and complementary labels. First, though, we briefly discuss classifiers, learning criteria, and optimization algorithms.

1.2.1   Classifiers

A classifier is essentially a function that maps an input pattern to a class to which the pattern belongs. When designing a classification algorithm, a function class from which a final classification function is selected needs to be determined. Such a function class is referred to as a *model*. At a glance, a very expressive model that contains a wide variety of functions seems to be preferable since it is capable of learning a very complex classification function. However, since we need to train a classifier (i.e., selecting the best function from the function class) based on a finite amount of training data, a too big model can result in *overfitting*—a trained classifier perfectly fits given training data, but it does not generalize well to unseen test data. On the other hand, if the model is small, a nearly optimal function in the model may be chosen from a limited amount of training data. However, due to the low representation capability of the model, even the best function in the small model may exhibit a poor generalization capability. Such a situation is referred to as *underfitting*—the trained classifier cannot even fit given training data well, and therefore it does not generalize well to unseen test data either.

1.2.2   Learning Criteria

Given data, we train a classifier from data. How to train a classifier—i.e., how to obtain a good classification function from a model—is the role of a learning criterion. The most fundamental learning criterion is the *training error* (also called the *empirical error*), which evaluates the goodness of fit to given training data (or empirical samples). However, as mentioned, simply minimizing the training error can result in overfitting if the model is flexible. Therefore, a *regularizer* is usually included in the learning criterion, which penalizes overfitting by typically imposing the classification function to be smooth.

The main focus of this book is to develop learning criteria in the scenario of classification from weak supervision. Consequently, various learning criteria that allow a learning machine to generalize well only from weak supervision are explored.

1.2.3   Optimization Algorithms

Once the learning criterion is fixed, finally its minimizer is obtained by an optimization algorithm.

In the operations research community, various optimization algorithms have been developed so far, depending on the type of objective functions to be minimized. For a *convex* optimization problem, the global minimizer may be found efficiently; otherwise, a local minimizer is sought for. Among various optimization algorithms, *gradient descent* would be the most versatile and popular algorithm used in machine learning whose objective function is differentiable. The basic idea of gradient descent is to go down the slope of the objective function from some initial point until a zero-gradient point is found (figure 1.4).

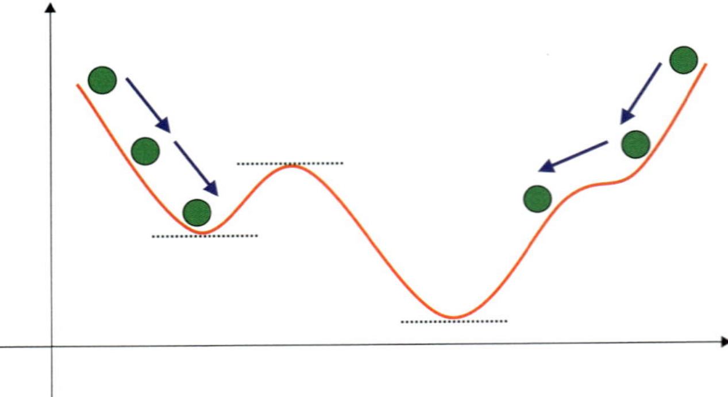

Figure 1.4
Optimization by gradient descent.

When the number of training samples is large, the gradient is often computed not by all samples but by a mini-batch (subset) of samples. Such a mini-batch-based gradient descent is called *stochastic gradient descent.*

In this gradient descent process, the choice of the *step size* is critical for better convergence: If the step size is too large, going down the slope may be fast, but overshooting may cause instability of the optimization. On the other hand, if the step size is too small, going down the slope takes too long to get to a stationary point. Gradually shrinking the step size is a standard strategy in practice, but the best choice of the step size is still an ongoing research issue.

1.3 Aspects of Machine Learning

Machine learning has a long history, and it has been investigated from many different aspects. In this section, we review and compare typical approaches to machine learning and explain our choice.

1.3.1 Logical Learning, Biologically Inspired Learning, and Statistical Learning

Machine learning research has been conducted in different communities, resulted in different schools of studies—logical learning, biologically inspired learning, and statistical learning (figure 1.5).

Logical learning expresses a classification rule based on *if-then* or other logical rules (Russell and Norvig, 2010). Logical learning can give a clear decision rule that is easily interpretable by humans. However, since real-world data is usually noisy, it is difficult and even impossible to derive clear logical rules that explain given data. Furthermore, noisy data often induces a large number of logical rules that can easily result in severe overfitting. Therefore, properly handling ambiguous and even contradictory real-world data as well as *pruning* irrelevant and redundant rules is important to make logical learning more useful in practice. However, for highly complex real-world problems such as computer vision and natural language processing, deriving meaningful logical rules from low-level features such as pixel values and a sequence of words is still extremely challenging.

Biologically inspired learning tries to mimic the learning behavior of organisms (typically, human beings). To do so, it is necessary to elucidate the behavior of organisms first, which is a very active area of research such as *brain science.* However, understanding the learning behavior of organisms is still extremely challenging, and only superficial concepts have been brought to the machine learning community, such as *reinforcement learning* (Doya, 2007), *memory* (Weston et al., 2015), and *visual data processing* (Fukushima, 1980). Making biologically inspired learning more practical in the context of modern machine learning is an important future challenge.

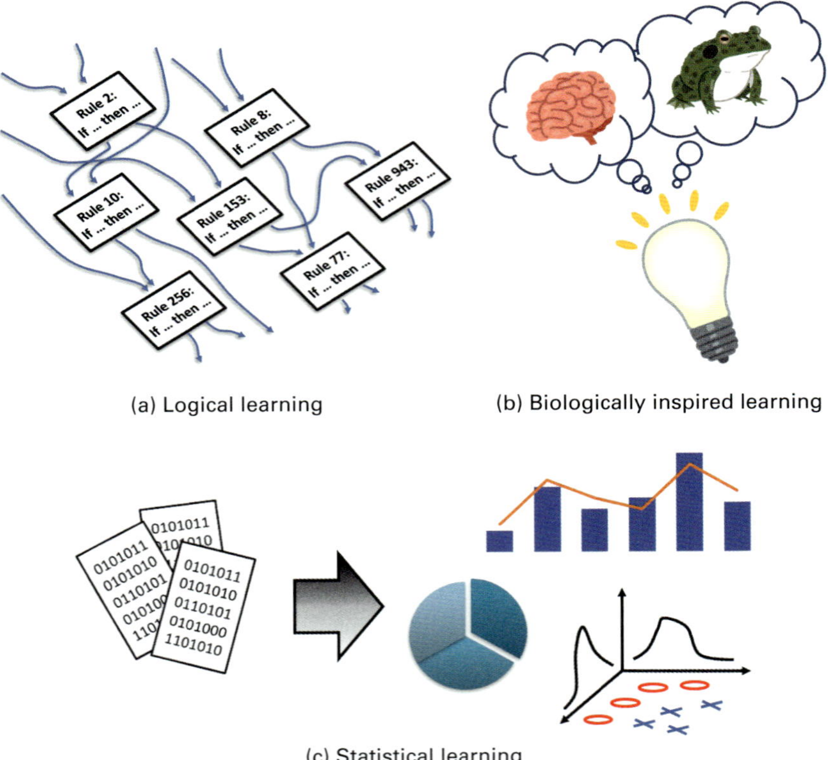

(a) Logical learning   (b) Biologically inspired learning

(c) Statistical learning

Figure 1.5
Logical learning, biologically inspired learning, and statistical learning.

Statistical learning handles ambiguous and contradictory information in a systematic fashion through *probability theory*. The essence of statistical learning is not to directly analyze given training data but to consider and estimate the probability distribution behind the data, allowing us to discuss the generalization capability in a mathematically rigorous way. Together with the development of information technologies, statistical learning has been dramatically advanced in the last decades, and it is playing a central role in modern research and practice of artificial intelligence.

Statistical learning is the most practical approach to machine learning at this moment. For this reason, in this book, we discuss *classification from weak supervision* in the framework of statistical learning. Combining statistical learning with logical and biologically inspired learning will be a promising future direction of research to further advance machine learning.

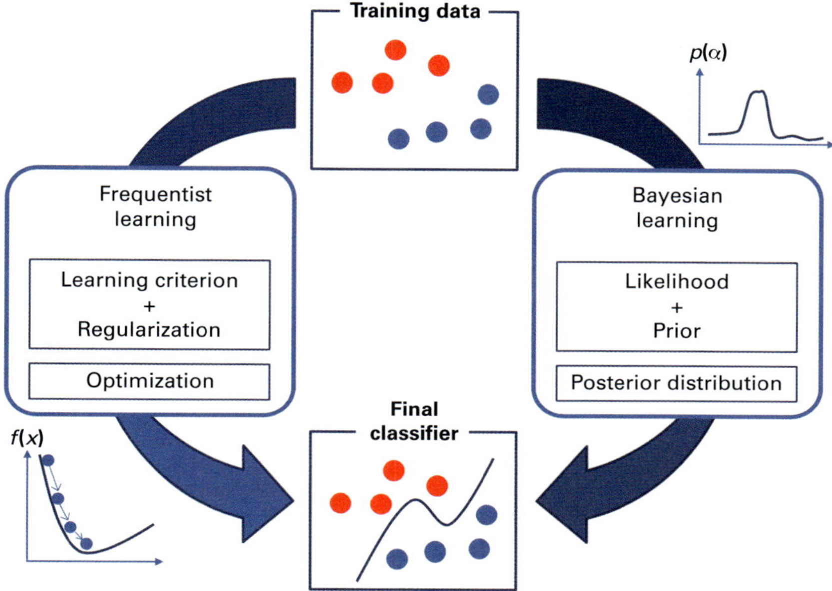

Figure 1.6
Frequentist learning and Bayesian learning.

1.3.2   Frequentist Learning and Bayesian Learning

Frequentist learning and Bayesian learning are two different approaches in machine learning
(figure 1.6).

In the framework of frequentist learning, the single best solution that optimizes a certain
learning criterion is adopted as the final classifier. This approach would be the most straightforward and natural one to use when training a classifier. *Regularization* is a key technique
to avoid overfitting in frequentist learning. Statistics and *mathematical optimization* (Boyd
and Vandenberghe, 2004) play important roles when developing frequentist-type machine
learning algorithms.

Information-theoretic learning is a special instance of frequentist learning that trains
a classifier by optimizing an information measure such as *mutual information* (Shannon,
1948) and its variants (Sugiyama et al., 2012). Information-theoretic learning is particularly
useful in *unsupervised learning* scenarios, where "informativeness" of unlabeled data is
evaluated based on the information measure. Information theory (MacKay, 2003; Cover
and Thomas, 2006) offers central mathematical tools for developing information-theoretic
machine learning algorithms.

On the other hand, in the framework of Bayesian learning, functions in a model are
regarded as *probabilistic*, and a *prior* probability distribution of functions in the model is

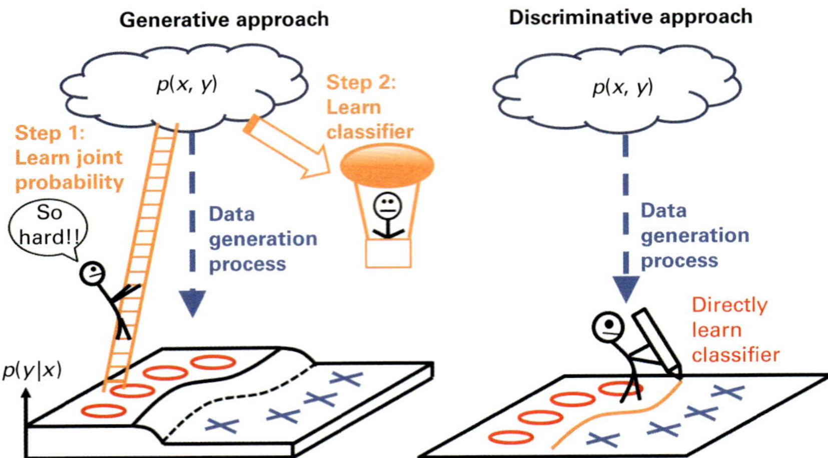

Figure 1.7
Generative classification and discriminative classification.

considered. Then, based on the *Bayes rule*, a *posterior* probability distribution of functions
is inferred based on given training data. Finally, a classification function is obtained either
as the maximizer of the posterior distribution (*maximum a posteriori classification*) or as
the mean of the posterior distribution (*posterior mean classification*); sometimes a single
classification function is randomly selected following the posterior distribution (*Gibbs classification*). Note that the posterior distribution can be constructed also for each test input
point, which is called the *posterior predictive distribution*.

In the Bayesian framework, a prior distribution plays a role as a regularizer and thus
overfitting can be naturally avoided. However, how to set the prior distribution is not
straightforward in practice; often it is chosen from a computationally tractable family of
distributions (called a *conjugate prior*) and tuning parameters included in the prior (called
hyper-parameters) are set either by considering a prior distribution for hyper-parameters
(called a *hyper-prior*) or by maximizing the *marginal likelihood* (called *empirical Bayes*).
Statistics and probability theory are key mathematical tools for developing Bayesian
machine learning algorithms (Bishop, 2006; Murphy, 2012; Nakajima et al., 2019).

1.3.3   Generative Classification and Discriminative Classification

On top of the choice of frequentist learning and Bayesian learning, there are two approaches
to classification: generative and discriminative (figure 1.7).

The generative approach first learns the data generation rule, which is the joint probability
distribution over input and output in the framework of statistical learning. Actually, the generative approach is *universal*, since the complete knowledge of the data generation process
allows us to solve *all* machine learning tasks—once the data generation process is perfectly

identified, we can generate as much data as we want and thus we can know all about the data.[2] However, accurately learning the data generation rule (more specifically, estimating the joint probability distribution over input and output) without strong prior knowledge is highly challenging. For this reason, just naively taking the generative approach for training a classifier does not necessarily result in practically useful solutions.

While the generative classification is the most general approach to machine learning, the discriminative approach is completely opposite and tries to be as task-specific as possible. The ultimate goal of classification is to classify test patterns into correct classes. To this end, all we need is the classification rule that can be applied to test patterns. Following this idea, the discriminative approach tries to learn the classification rule *directly*, without estimating the data generating probability distribution. Since discriminative classification algorithms are specifically designed for classification, they are expected to perform better than generative methods that can also be used for other purposes. For this reason, we focus on the discriminative approach in this book.

However, the price we have to pay for the specificity of the discriminative approach is high costs for research and development, since a novel machine learning algorithm needs to be developed for each task that can cleverly avoid estimating the data generation rule. To mitigate this problem, considering an *intermediate* approach between the generative and discriminative approaches is useful. For example, estimation of the *ratio* of probability density functions allows us to solve a wide variety of machine learning tasks, including classification, outlier detection, clustering, feature extraction, feature selection, and independent component analysis, while probability density functions themselves need not be estimated (Sugiyama et al., 2012). Extending our discriminative methods of *classification from weak supervision* to such an intermediate framework would be an important future direction.

1.3.4   Induction, Deduction, and Transduction

In a usual classification scenario, a classifier is trained from training data (*induction*) and then used to predict class labels of given input patterns (*deduction*). On the other hand, if test unlabeled patterns (without class labels) are available in the training phase, we may directly predict class labels of given test unlabeled patterns without using an inductive classifier. Such a direct approach is called *transduction* (Vapnik, 1998), and it is expected to achieve better prediction accuracy than the two-step induction-deduction approach (figure 1.8). However, transduction does not provide a classifier, and thus class labels of *unseen* unlabeled patterns cannot be predicted in a deductive way.

1.4   **Improving Data Collection and Weakly Supervised Learning**

A current major trend in the research and practice of classification is to use an enormous amount of labeled data (often called *big data*) for training a highly complex classifier (Goodfellow et al., 2016). However, if the amount of labeled training data is limited, even

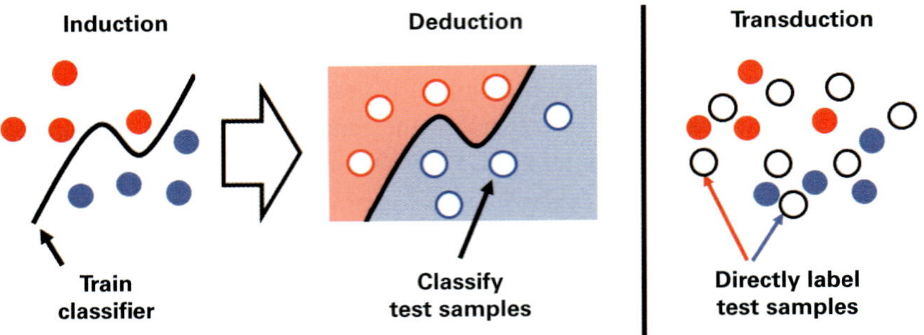

Figure 1.8
Induction, deduction, and transduction.

state-of-the-art supervised learning methods do not perform well in general. Therefore, it is essential to collect a large amount of labeled data in supervised learning.

In practical classification scenarios, input data is often labeled *manually* by human experts, so collecting a large amount of labeled data is laborious and costly. To cope with this problem, the data collection process may be improved. In this section, we review some of such improvements.

When we are allowed to design the location of input points and label those selected points, we want to find the best set of input points that maximally enhances the generalization capability. Such a problem is called *active learning* (or *experiment design*) and has been studied extensively in statistics and machine learning (Fedorov, 1972; Pukelsheim, 1993; Settles, 2009; Sugiyama and Kawanabe, 2012).

While active learning tries to sample the best training data given fixed labeling costs, we may consider using training data that can be collected in an inexpensive way. *Crowd-sourcing* is an active topic of research in applied machine learning and data mining communities (Law and von Ahn, 2011) to annotate unlabeled data in a cost-efficient way. In crowd-sourcing, unlabeled data is distributed over the web and, with small allowance, a crowd of (nonexpert) workers is asked to annotate the unlabeled data. Compared with asking experts to annotate unlabeled data, collecting labels through crowd-sourcing is inexpensive. However, such crowd-sourced labels are often of lower quality than labels collected from experts.

Another possibility to obtain labeled data in an inexpensive way is to borrow data from other similar classification tasks. This line of research is called *transfer learning* (or *domain adaptation*), which has been studied extensively (Pan and Yang, 2010; Sugiyama and Kawanabe, 2012). A bidirectional variant of transfer learning where multiple related tasks are solved simultaneously by mutually sharing information is called *multitask learning* (Caruana, 1997).

The aforementioned approaches try to improve the data collection process, which are highly useful in many real-world machine learning problems. On the other hand, the approach we have chosen to explore is *weakly supervised learning*: A classifier is trained based on data that contains "weaker" information than fully supervised data. A typical example of such weakly supervised learning scenarios is *semi-supervised learning* (Chapelle et al., 2006), where unlabeled data is used in addition to labeled data for training a classifier. Unlabeled data contains no label information, and thus it is "weaker" than labeled data. Nevertheless, in semi-supervised learning, we still want to obtain a classifier that generalizes as if we train it on fully labeled data. In subsequent chapters, we will explore various scenarios of classification from weak supervision, design learning criteria, provide practical algorithms, elucidate their theoretical properties, and investigate their numerical behavior.

Note that the term *weakly supervised learning* has been used rather broadly. For example, Ratner et al. (2017) presented a weakly supervised learning system called *Snorkel* that allows users to train various learning models only by writing labeling functions that express arbitrary heuristics. Zhou (2018) considered three types of weak supervision called *incomplete supervision* (only a subset of training data is labeled), *inexact supervision* (only coarse-grained labels are given), and *inaccurate supervision* (only noisy labels are given). Among various approaches and features of weakly supervised learning, we focus on *discriminative classification that follows the principle of empirical risk minimization*. See chapter 3 for the details of empirical risk minimization.

1.5   Organization of This Book

The organization of this book is illustrated in figure 1.9. Before going into the details of weakly supervised learning, we mathematically formulate classification problems and define common notations in chapter 2. Then, in chapter 3, we review various algorithms for supervised binary and multi-class classification and investigate their theoretical properties. We often refer to supervised binary classification as *positive-negative (PN) classification*, since it uses PN samples for training a classifier.

1.5.1   Weakly Supervised Learning for Binary Classification

In part II, we explore various problems of *binary* weakly supervised classification. In chapter 4, we consider *positive-unlabeled (PU) classification*, which is the most standard form of binary weakly supervised classification. In PU classification, negative (N) data is not available and only PU data is used for training a classifier. We introduce an algorithm for PU classification that compensates for a lack of N data based on the fact that the distribution of U data is a mixture of the distributions of P and N data. Theoretically, we show that even without any single N sample, infinitely many PU samples can yield the optimal classifier. We further show that the PU classification method can achieve better generalization performance than PN classification if we have a large amount of U data compared with N data. This is a surprising result since PU classification is thought of as a compromised version of PN classification where N data is replaced with U data—but instead of relying on a small

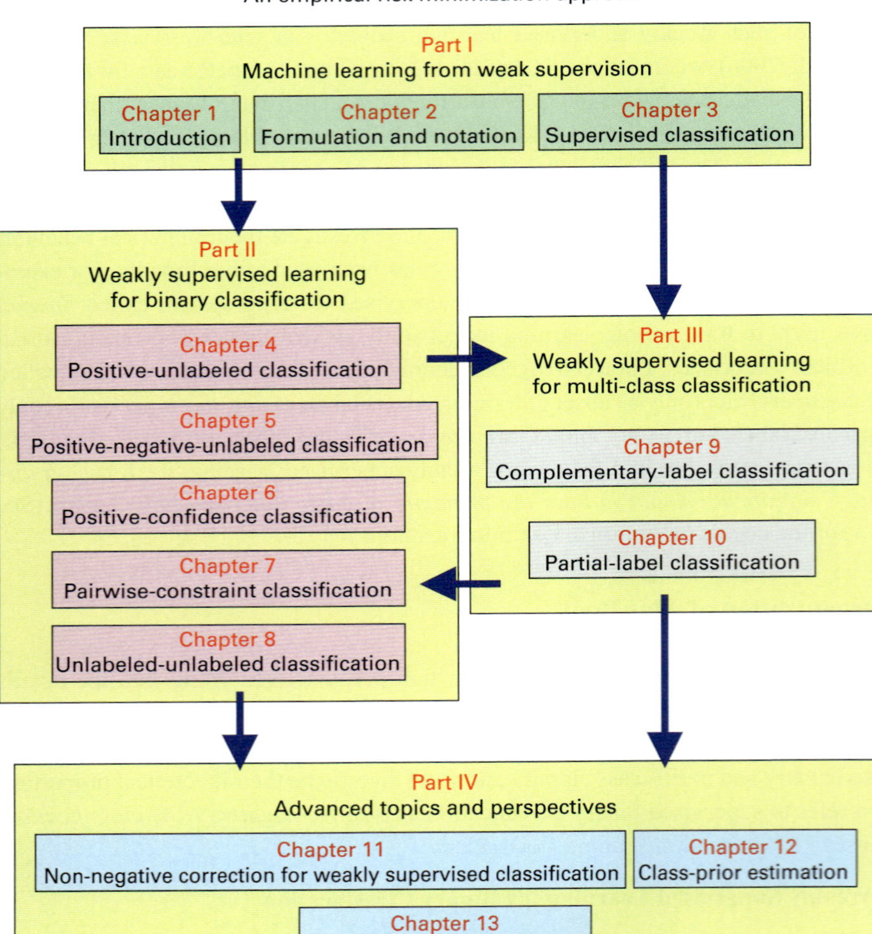

Figure 1.9
Organization of this book.

number of N samples, using a large number of U samples can achieve better generalization performance.

In chapter 5, we extend PU classification to *semi-supervised classification, which we* specially refer to as *positive-negative-unlabeled (PNU) classification* since the method introduced here is based on PU classification with additional N data. As explained previously, PU classification can achieve better generalization performance than PN classification if we have more U samples than N samples. However, further utilizing the N samples in PU classification should further improve the generalization performance. The method introduced here follows this idea and combines the methods of PU classification, PN classification, and negative-unlabeled (NU) classification into PNU classification. We theoretically show that, without imposing strong assumptions on U data, the PNU classification method can perform well with many U samples. The superiority of the PNU classification is also demonstrated through experiments.

In chapter 6, we try to remove the necessity of U data from PU classification, since U data is sometimes hard to collect because of privacy concerns or legal restrictions. However, only-P classification is essentially equivalent to unsupervised classification, so additional supervision is needed to obtain a meaningful solution to supervised classification. Here, we consider *positive-confidence (Pconf) classification*, which uses P data equipped with *confidence* for training a classifier (in the absence of U and N data). The Pconf classification method introduced here allows us to train a classifier only from Pconf data, which is based on the idea that a training sample with Pconf α $(0 \leq \alpha \leq 1)$ is also a training sample with negative-confidence (Nconf) $1 - \alpha$. We theoretically and experimentally show the usefulness of the Pconf classification method.

In chapter 7, we try to explore another form of weakly supervised classification that does not require any single explicit class labels, which we call *pairwise-constraint classification*. In this scenario, we consider pairs of unlabeled data annotated as *similar (S)* or *dissimilar (D)*. We show that obtaining the optimal boundary between the P and N classes[3] is possible only from SU data, which we call *similar-unlabeled (SU) classification*. Furthermore, we can similarly show that *similar-dissimilar (SD) classification* and *dissimilar-unlabeled (DU) classification* are also possible; this allows us to develop the *similar-dissimilar-unlabeled (SDU) classification* method, derived in the same way as PNU classification discussed in chapter 5. We also show the usefulness of the pairwise-constraint classification methods through theoretical analysis and experimental evaluation.

Finally, in chapter 8, we consider the weakest form of weakly supervised classification, which we call *unlabeled-unlabeled (UU) classification*. In UU classification, we use two sets of U data with *unknown* label proportion for training a classifier. UU classification is actually a very general framework of weakly supervised learning, and it includes PN, PU, NU, SD, SU, and DU classification as special cases. We show that it is possible to obtain the optimal boundary between the P and N classes from UU data. This surprising result is based on the fact that one of the U datasets is regarded as a corrupted P dataset and the other U dataset is regarded as a corrupted N dataset, and the corruption is compensated for by the knowledge of class proportions.

1.5.2   Weakly Supervised Learning for Multi-Class Classification

In part III, we turn our focus toward multi-class classification. In chapter 9, we introduce *complementary-label (CL) classification*, where training patterns are accompanied with *incorrectly specified* class labels. Annotating unlabeled samples is significantly more challenging in multi-class classification than the binary case since choosing the correct label from a long list of candidate labels is highly time-consuming. Here, the idea is not to ask labelers to give a correct label (e.g., "cat" and "person" in the case of image classification) but to ask them to select one of the incorrect labels (e.g., "not dog" and "not car," for instance), which can be much easier to provide. We show that it is possible to train a classifier only from complementarily labeled data. Based on this result, we then design a practical algorithm that can use both ordinarily and complementarily labeled data. When collecting labels, instead of requesting multi-class labeling, we simply ask a yes/no question whether this pattern belongs to that class or not; the positive/negative answer yields an ordinary/complementary label and we can use both of them without loss of information. Experiments demonstrate the usefulness of this approach.

In chapter 10, we consider *partial-label (PL) classification*, which generalizes CL classification. A PL is a subset of labels in which one of them is correct (e.g., "either a dog or cat"). Thus, labelers do not have to narrow down the choice of labels to the uniquely correct one, which can significantly save the labeling costs. CL is the complementary of a PL with $c - 1$ candidates in c-class problems (e.g., among the "dog," "cat," and "bird" classes, "not a dog" is equivalent to "either a cat or bird"). We will give practical solutions to PL classification with solid theoretical guarantee.

1.5.3   Advanced Topics and Perspectives

In part IV, we address advanced issues in weakly supervised learning and prospects. In chapter 11, to improve the generalization performance of weakly supervised learning, a family of *correction methods* is introduced. In standard PN classification, the training error is minimized, which is non-negative when the loss function is non-negative. However, in many of the weakly supervised classification settings introduced in this book, the training error can become negative even when the loss function is non-negative; this tendency becomes more significant if a more complex model such as a deep network is used. Here, we discuss the idea of correcting the training error to be kept non-negative. Specifically, we will theoretically and experimentally show that this correction idea can contribute to improving the generalization performance of various weakly supervised classification methods.

In chapter 12, we discuss the problem of *class-prior estimation*. Some of the *binary* weakly supervised classification methods introduced in part II actually rely on the *class-prior probabilities*, which need to be estimated in practice. We first focus on PNU classification (explored in chapter 5) and introduce the method of class-prior estimation by *full distribution matching*. Then, we turn our focus to PU classification (explored in chapter 4) and discuss two major approaches to class-prior estimation: *mixture proportion estimation* and *partial distribution matching*. We further introduce the *regrouping* technique (to improve mixture proportion estimation) and a practical implementation of partial distribution matching based on a *penalized* divergence measure. A class-prior estimation method that is specific to classification from pairwise data (explored in chapter 7) is also discussed.

Finally, in chapter 13, we conclude by recapitulating the messages that we wanted to convey in this book and discuss future prospects of machine learning research and beyond.

2 Formulation and Notation

In this chapter, we formulate classification problems and define notations that are commonly used in the chapters that follow.

2.1 Binary Classification

First, we consider a two-class classification problem.

2.1.1 Formulation

Let us consider a classification problem where a binary class label

$$y \in \mathcal{Y} := \{+1, -1\}$$

is assigned to each d-dimensional real-valued pattern

$$\boldsymbol{x} \in \mathcal{X} \subset \mathbb{R}^d.$$

We treat pattern $\boldsymbol{x}$ and its label y as random variables, which are assumed to be equipped with unknown joint probability density

$$p(\boldsymbol{x}, y).$$

Let

$$g \colon \mathcal{X} \to \mathbb{R}$$

be a binary *classifier*, which is a real-valued function. Classifier g predicts class $\widehat{y}$ to which input pattern $\boldsymbol{x}$ belongs based on its sign as

$$\widehat{y} = \mathrm{sign}\big(g(\boldsymbol{x})\big) = \begin{cases} +1 & (g(\boldsymbol{x}) > 0), \\ -1 & (g(\boldsymbol{x}) \le 0). \end{cases}$$

Ultimately, we would like to obtain a classifier that minimizes the *classification error*:

$$I(g) := \mathbb{E}_{p(\boldsymbol{x},y)}[\ell_{0\text{-}1}(g(\boldsymbol{x}), y)], \tag{2.1}$$

where $\mathbb{E}_{p(x,y)}$ denotes the expectation over $p(x,y)$, i.e.,

$$\mathbb{E}_{p(x,y)}[\;\cdot\;] := \sum_{y=\pm 1} \int \;\cdot\; p(x,y)\mathrm{d}x,$$

and $\ell_{0\text{-}1} : \mathbb{R} \times \mathcal{Y} \to \{0, 1\}$ is the *zero-one loss* defined as

$$\ell_{0\text{-}1}(\widehat{y}, y) := \begin{cases} 0 & (y\widehat{y} > 0), \\ 1 & (y\widehat{y} \le 0). \end{cases} \tag{2.2}$$

The product of true and estimated output values

$$m = yg(x)$$

is called the *margin* of classifier g for sample (x, y). If the margin is positive, i.e., y and $g(x)$ take the same sign—the zero-one loss takes zero; otherwise, the zero-one loss takes one.

2.1.2   Classification Models

When training classifier g, it is convenient to have its explicit form called a *model*. Here we describe popular classification models.

2.1.2.1   Linear-in-input model

This most fundamental model for classification is defined as

$$g(x) = \alpha^\top x,$$

where

$$\alpha = (\alpha_1, \dots, \alpha_d)^\top \in \mathbb{R}^d$$

is a parameter vector and $^\top$ denotes the transpose. The linear-in-input model has linearity both in parameter α and input x, and thus it is very easy to handle mathematically. However, it can represent only a linear function with respect to input x, which is highly restrictive as a classifier in practice.

2.1.2.2   Linear-in-parameter model

A slight generalization of the linear-in-input model, this model is represented as

$$g(x) = \sum_{j=1}^{b} \alpha_j \phi_j(x) = \alpha^\top \phi(x), \tag{2.3}$$

where b denotes the number of parameters, $\alpha \in \mathbb{R}^b$ is a parameter vector, and

$$\phi(x) := (\phi_1(x), \dots, \phi_b(x))^\top \in \mathbb{R}^b$$

is a vector of *basis functions*. By properly designing basis functions (e.g., with polynomial functions and sinusoidal functions, for example), a linear-in-parameter model can represent a complicated nonlinear function, and thus it is more expressive than the linear-in-input model. Although linear-in-parameter models are easy to handle mathematically, as the name indicates, they are still linear with respect to parameter vector $\boldsymbol{\alpha}$. Note that the linear-in-parameter model is reduced to the linear-in-input model if basis functions are chosen to be linear, i.e.,

$$\boldsymbol{\phi}(\boldsymbol{x}) = \boldsymbol{x} \text{ for } b = d.$$

2.1.2.3 Kernel model
The model is defined by

$$g(\boldsymbol{x}) = \sum_{i=1}^{n} \alpha_i K(\boldsymbol{x}, \boldsymbol{x}_i),$$

where $K \colon \mathcal{X} \times \mathcal{X} \to \mathbb{R}$ is a *kernel function* and $\{\boldsymbol{x}_i\}_{i=1}^{n}$ are given input points on $\mathcal{X}$. An advantage of the kernel model is that the number of basis functions grows with the number n of given input points, therefore the model complexity can be adaptively determined based on the amount of data we are given.

Kernel models are still linear in parameter vector $\boldsymbol{\alpha}$, so they are easy to handle in terms of optimization. However, kernel models have intrinsically different asymptotic properties than the linear-in-parameter model (2.3) because the model complexity grows as the number n of data points is increased. Due to this different behavior, in statistics, kernel models are categorized as *nonparametric models* (Härdle, 1990) while the linear-in-input model and the linear-in-parameter model are called *parametric models*, which have fixed model complexity without regard to the number n of training samples.

As a kernel function, the *Gaussian kernel* is a popular choice (figure 2.1a), represented as

$$K(\boldsymbol{x}, \boldsymbol{x}') = \exp\left(-\frac{\|\boldsymbol{x} - \boldsymbol{x}'\|^2}{2\sigma^2}\right),$$

where $\sigma > 0$ is called the *Gaussian bandwidth* and $\boldsymbol{x}'$ is called the *Gaussian center*. The Gaussian kernel model puts smooth Gaussian functions at the location of given input points. For this reason, smooth function approximation is possible *locally* around given input points (figure 2.1b). This local nature of the Gaussian kernel model is highly useful, particularly when the input dimensionality d is high and input points are distributed on a low-dimensional *manifold* (figure 2.1c)—without any special treatment, function approximation can be performed automatically on the low-dimensional manifold and thus better approximation can be expected when the dimensionality of the manifold is relatively low.

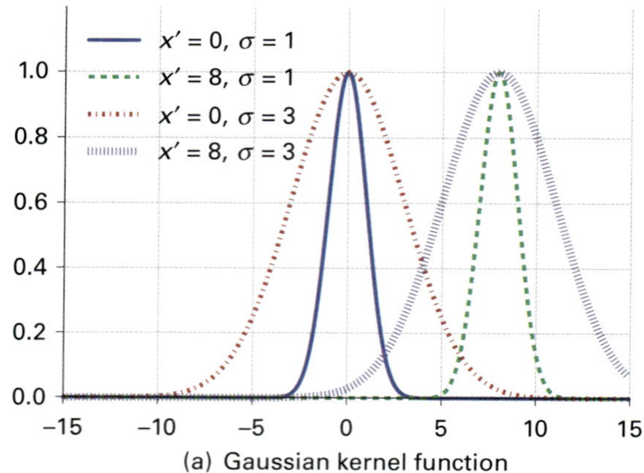

(a) Gaussian kernel function

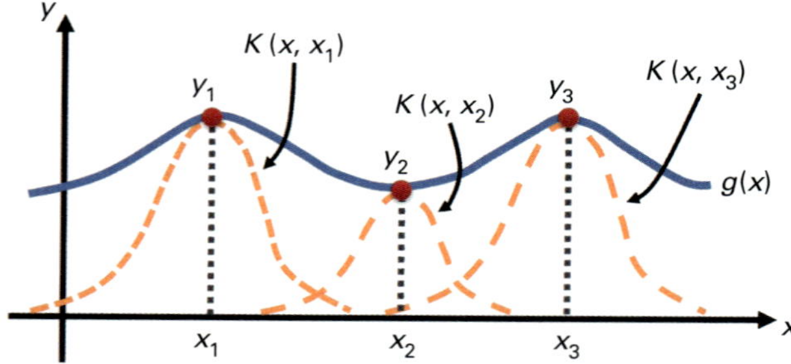

(b) One-dimensional Gaussian kernel model

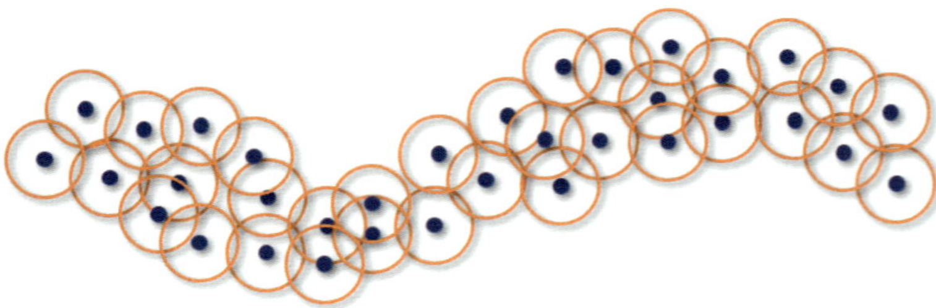

(c) Two-dimensional Gaussian kernel model

Figure 2.1
Gaussian kernel models.

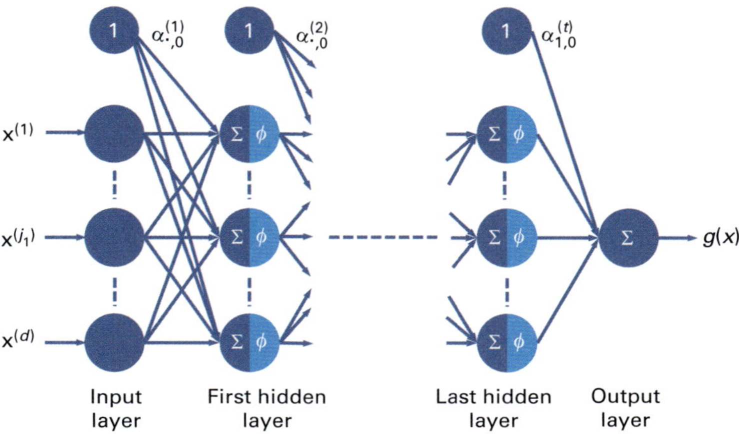

Figure 2.2
Neural network model.

2.1.2.4   Neural network model

In the kernel model, the Gaussian center c_j is fixed to x_j. If c_j is regarded as a free parameter, the model is no longer linear with respect to all parameters $(\alpha^\top, c_1^\top, \ldots, c_b^\top)^\top$.

Among such nonlinear models, hierarchical models called *neural networks* have recently been explored very actively (Rumelhart et al., 1986; Bishop, 1995; Orr and Müller, 1998; Goodfellow et al., 2016). The most basic form of the neural network model is given as follows (figure 2.2):

$$g(x) = \sum_{j_t=1}^{b_t} \alpha_{1,j_t}^{(t)} \phi\left(\sum_{j_{t-1}=1}^{b_{t-1}} \alpha_{j_t,j_{t-1}}^{(t-1)} \cdots \phi\left(\sum_{j_2=1}^{b_2} \alpha_{j_3,j_2}^{(2)} \right. \right.$$

$$\left. \left. \times \phi\left(\sum_{j_1=1}^{d} \alpha_{j_2,j_1}^{(1)} x^{(j_1)} + \alpha_{j_2,0}^{(1)} \right) + \alpha_{j_3,0}^{(2)} \right) \cdots + \alpha_{j_t,0}^{(t-1)} \right) + \alpha_{1,0}^{(t)},$$

where t denotes the number of layers, $x = (x^{(1)}, \ldots, x^{(d)})^\top$, and ϕ is called an *activation function*.

Popular activation functions are as follows (figure 2.3):

$$\text{Sigmoid function: } \phi(x) = \frac{1}{1 + \exp(-x)},$$

$$\text{Rectified linear unit: } \phi(x) = \max\{0, x\},$$

$$\text{Softplus function: } \phi(x) = \ln\left(1 + \exp(x)\right),$$

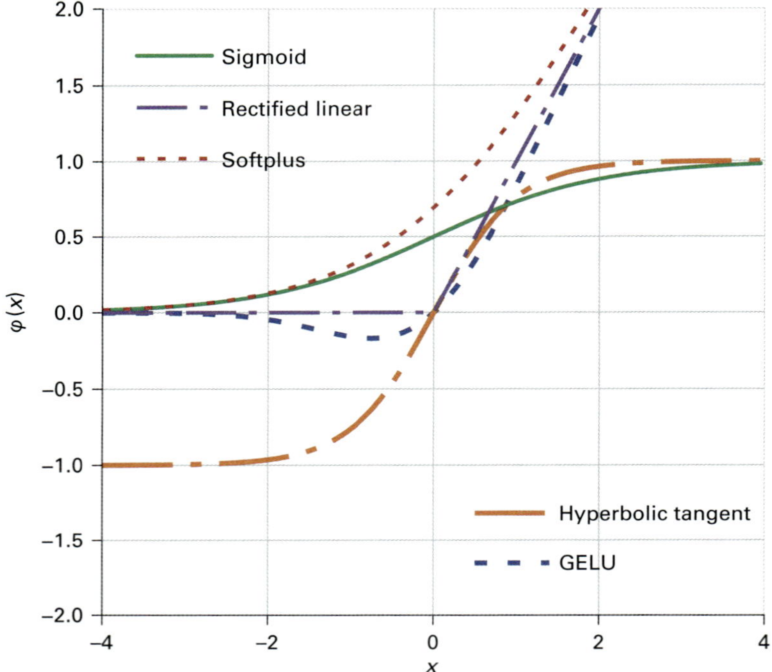

Figure 2.3
Activation functions in neural networks.

$$\text{Hyperbolic tangent function: } \phi(x) = \tanh(x),$$

$$\text{Gaussian error linear unit (GELU): } \phi(x) = x \cdot \frac{1}{2}(1 + \mathrm{erf}(x/\sqrt{2})),$$

where $\mathrm{erf}(x)$ is the Gauss error function.

When the number t of layers is large, neural networks have "deep" architectures and thus they are called *deep models*. Machine learning with deep models is called *deep learning*. Note that even though deep models are highly nonlinear in model parameters, their complexities are still independent of the number of training data and thus they are still categorized as parametric models in the statistics taxonomy.

2.1.3   Surrogate Losses

Given an explicit parametric form for classifier g, we want to obtain a classifier that minimizes the classification error $I(g)$ defined by (2.1). However, this is generally not possible because of the discrete nature of the zero-one loss—the zero-one loss has no gradient and thus minimization under the zero-one loss is essentially combinatorial optimization.

Indeed, it is known that minimization under the zero-one loss is *NP-hard* (Feldman et al., 2012), i.e., the computation time to find a global minimizer is expected to grow exponentially with respect to the number of variables to be optimized. Therefore, it is regarded as computationally intractable in practice.

To cope with this problem, we use a *surrogate loss* $\ell\colon \mathbb{R} \times \mathcal{Y} \to \mathbb{R}$ such that $\ell(g(\boldsymbol{x}), y)$ tends to take a small value for large margin $m = yg(\boldsymbol{x})$. Thus, a larger margin is more preferred under the surrogate loss formulation. The classification *risk* under surrogate loss ℓ is defined as

$$R(g) := \mathbb{E}_{p(\boldsymbol{x},y)}[\ell(g(\boldsymbol{x}), y)]. \tag{2.4}$$

If the loss function depends only on margin $m = yg(\boldsymbol{x})$, not on y and $g(\boldsymbol{x})$ separately, it is said to be *margin-based*. With a slight abuse of notation, a margin-based surrogate loss is denoted by $\ell(m)$.

The following losses, which are all margin-based, are popularly used in machine learning (figure 2.4):

$$\text{Linear loss: } \ell(m) = -m,$$

$$\text{Squared loss: } \ell(m) = (1 - m)^2,$$

$$\text{Logistic loss: } \ell(m) = \ln(1 + \exp(-m)),$$

$$\text{Hinge loss: } \ell(m) = \max\{0, 1 - m\},$$

$$\text{Squared-hinge loss: } \ell(m) = \big(\max\{0, 1 - m\}\big)^2,$$

$$\text{Huber-hinge loss: } \ell(m) = \begin{cases} \frac{1}{4}\big(\max\{0, 1 - m\}\big)^2 & (m \geq -1), \\ -m & (m < -1), \end{cases}$$

$$\text{Double-hinge loss: } \ell(m) = \max\big\{-2m, \max\{0, 1 - m\}\big\},$$

$$\text{Ramp loss: } \ell(m) = \frac{1}{2}\min\big\{2, \max\{0, 1 - m\}\big\},$$

$$\text{Sigmoid loss: } \ell(m) = \frac{1}{1 + \exp(m)}.$$

Note that all of the above losses except the linear loss are bounded from below, while the ramp loss and the sigmoid loss are also bounded from above. All losses except the ramp loss and the sigmoid loss are *convex* and allow efficient optimization when they are combined with linear-in-parameter models (see section 2.1.2). Note also that the zero-one loss defined by (2.2) is also a margin-based loss:

$$\text{Zero-one loss: } \ell_{0\text{-}1}(m) = \begin{cases} 1 & (m \leq 0), \\ 0 & (m > 0). \end{cases} \tag{2.5}$$

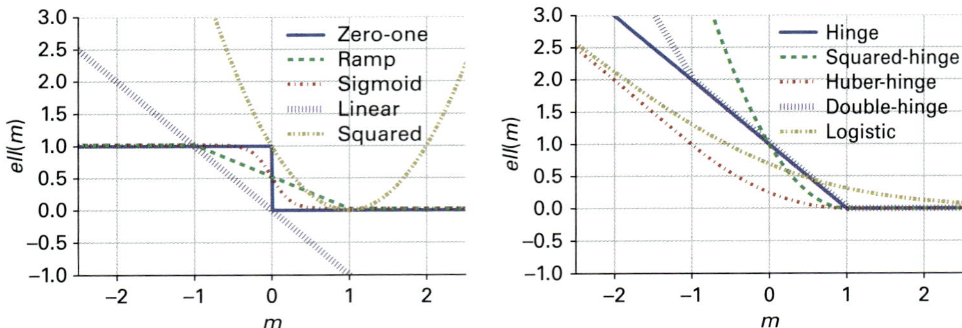

Figure 2.4
Example of losses.

2.1.4   Training Samples

Even if the zero-one loss is replaced with a surrogate loss, we cannot still directly obtain a minimizer of the risk $R(g)$ defined by (2.4) since it includes unknown $p(x, y)$.

Here, we assume that we are given training samples and use them for estimating the expectation over unknown $p(x, y)$. In the following chapters, we consider three sets of training samples called the *positive* (P), *negative* (N), and *unlabeled* (U) samples, respectively:

$$\mathcal{X}_{\mathrm{P}} := \{x_i^{\mathrm{P}}\}_{i=1}^{n_{\mathrm{P}}} \overset{\text{i.i.d.}}{\sim} p_{\mathrm{P}}(x) := p(x \mid y = +1),$$

$$\mathcal{X}_{\mathrm{N}} := \{x_i^{\mathrm{N}}\}_{i=1}^{n_{\mathrm{N}}} \overset{\text{i.i.d.}}{\sim} p_{\mathrm{N}}(x) := p(x \mid y = -1),$$

$$\mathcal{X}_{\mathrm{U}} := \{x_i^{\mathrm{U}}\}_{i=1}^{n_{\mathrm{U}}} \overset{\text{i.i.d.}}{\sim} p_{\mathrm{U}}(x) := \pi_{\mathrm{P}} p_{\mathrm{P}}(x) + \pi_{\mathrm{N}} p_{\mathrm{N}}(x),$$

where "i.i.d." stands for "*independent and identically distributed*" and

$$\pi_{\mathrm{P}} := p(y = +1),$$

$$\pi_{\mathrm{N}} := p(y = -1),$$

are the *class-prior probabilities* for the positive and negative classes such that

$$\pi_{\mathrm{P}} + \pi_{\mathrm{N}} = 1.$$

Note that $p_{\mathrm{U}}(x)$ agrees with the marginal density $p(x)$ derived from the joint density $p(x, y)$:

$$p_{\mathrm{U}}(x) = p(x) = \sum_{y=\pm 1} p(x, y).$$

Let $R_\mathrm{P}(g)$ and $R_\mathrm{N}(g)$ be the risks of classifier g under loss ℓ for positive and negative samples:

$$R_\mathrm{P}(g) := \mathbb{E}_\mathrm{P}[\ell(g(\boldsymbol{x}), +1)],$$

$$R_\mathrm{N}(g) := \mathbb{E}_\mathrm{N}[\ell(g(\boldsymbol{x}), -1)],$$

where $\mathbb{E}_\mathrm{P}$ and $\mathbb{E}_\mathrm{N}$ denote the expectations over $p_\mathrm{P}(\boldsymbol{x})$ and $p_\mathrm{N}(\boldsymbol{x})$, respectively:

$$\mathbb{E}_\mathrm{P}[\,\cdot\,] := \int \cdot\, p_\mathrm{P}(\boldsymbol{x})\mathrm{d}\boldsymbol{x},$$

$$\mathbb{E}_\mathrm{N}[\,\cdot\,] := \int \cdot\, p_\mathrm{N}(\boldsymbol{x})\mathrm{d}\boldsymbol{x}.$$

Similarly, let $R_{\mathrm{U},\mathrm{P}}(g)$ and $R_{\mathrm{U},\mathrm{N}}(g)$ be the risks of classifier g under loss ℓ when unlabeled samples are regarded as positive and negative samples:[1]

$$R_{\mathrm{U},\mathrm{P}}(g) := \mathbb{E}_\mathrm{U}[\ell(g(\boldsymbol{x}), +1)],$$

$$R_{\mathrm{U},\mathrm{N}}(g) := \mathbb{E}_\mathrm{U}[\ell(g(\boldsymbol{x}), -1)],$$

where $\mathbb{E}_\mathrm{U}$ denotes the expectation over $p_\mathrm{U}(\boldsymbol{x})$:

$$\mathbb{E}_\mathrm{U}[\,\cdot\,] := \int \cdot\, p_\mathrm{U}(\boldsymbol{x})\mathrm{d}\boldsymbol{x}.$$

The above risks can be approximated in an unbiased fashion by replacing the expectations with the corresponding sample averages as

$$\widehat{R}_\mathrm{P}(g) := \frac{1}{n_\mathrm{P}} \sum_{i=1}^{n_\mathrm{P}} \ell(g(\boldsymbol{x}_i^\mathrm{P}), +1),$$

$$\widehat{R}_\mathrm{N}(g) := \frac{1}{n_\mathrm{N}} \sum_{i=1}^{n_\mathrm{N}} \ell(g(\boldsymbol{x}_i^\mathrm{N}), -1),$$

$$\widehat{R}_{\mathrm{U},\mathrm{P}}(g) := \frac{1}{n_\mathrm{U}} \sum_{i=1}^{n_\mathrm{U}} \ell(g(\boldsymbol{x}_i^\mathrm{U}), +1),$$

$$\widehat{R}_{\mathrm{U},\mathrm{N}}(g) := \frac{1}{n_\mathrm{U}} \sum_{i=1}^{n_\mathrm{U}} \ell(g(\boldsymbol{x}_i^\mathrm{U}), -1).$$

They are called *empirical risks*.

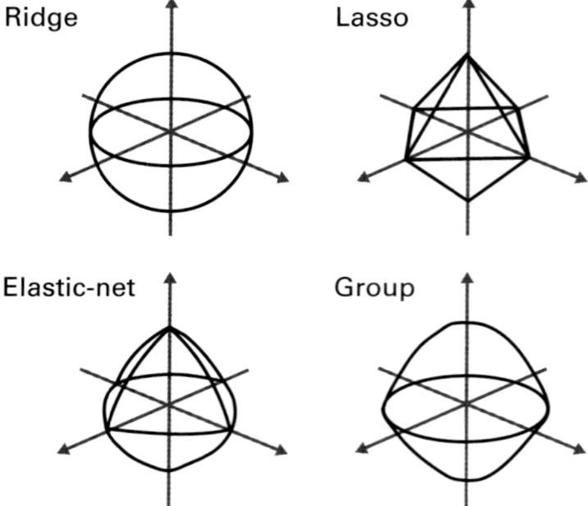

Figure 2.5
Example of regularizers.

2.1.5   Regularization

As discussed in section 1.2, it is important to include a *regularizer* to avoid overfitting. That is, classifier g is trained to minimize a weighted sum of an empirical risk $\widehat{R}$ and a regularizer Ω over a prespecified function class (model) $\mathscr{G}$:

$$\min_{g \in \mathscr{G}} \widehat{R}(g) + \lambda \Omega(g),$$

where $\lambda \geq 0$ is called the *regularization parameter*.

For parameter vector $\boldsymbol{\alpha} = (\alpha_1, \ldots, \alpha_b)^\top$ and its ℓ_p-norm $\|\cdot\|_p$ defined for $p > 0$ as

$$\|\boldsymbol{\alpha}\|_p = \left(\sum_{j=1}^{b} |\alpha_j|^p \right)^{\frac{1}{p}},$$

the following regularizers are popularly used (figure 2.5):

$$\text{Ridge regularizer: } \Omega(\boldsymbol{\alpha}) = \|\boldsymbol{\alpha}\|_2^2,$$

$$\text{Weighted ridge regularizer: } \Omega(\boldsymbol{\alpha}) = \boldsymbol{\alpha}^\top \boldsymbol{R} \boldsymbol{\alpha},$$

$$\text{Lasso regularizer: } \Omega(\boldsymbol{\alpha}) = \|\boldsymbol{\alpha}\|_1,$$

$$\text{Elastic-net regularizer: } \Omega(\boldsymbol{\alpha}) = \eta \|\boldsymbol{\alpha}\|_1 + (1 - \eta) \|\boldsymbol{\alpha}\|_2^2,$$

$$\text{Group regularizer: } \Omega(\boldsymbol{\alpha}) = \sum_{j=1}^{k} \|\boldsymbol{\alpha}_j\|_2,$$

where $\boldsymbol{R}$ is a positive semi-definite matrix called the *regularization matrix*, $0 \leq \eta \leq 1$, and $\boldsymbol{\alpha}_j$ is a sub-vector of $\boldsymbol{\alpha}$ such that $\boldsymbol{\alpha} = (\boldsymbol{\alpha}_1^\top, \ldots, \boldsymbol{\alpha}_k^\top)^\top$. The ridge regularizer *shrinks* the elements of $\boldsymbol{\alpha}$ toward zero (i.e., they are encouraged to take values close to zero) while the lasso regularizer *sparsifies* the elements of $\boldsymbol{\alpha}$ (i.e., they are encouraged to take exactly zero). The elastic-net regularizer combines the ridge and lasso regularizers, and the group regularizer encourages group-wise sparsity.

2.2 Multi-Class Classification

Next, we extend the binary classification problem to more than two classes.

2.2.1 Formulation

For integer c larger than 1, let us consider a c-class classification problem from pattern

$$\boldsymbol{x} \in \mathscr{X} \subset \mathbb{R}^d$$

to one of the c classes:

$$y \in \mathscr{Y} := \{1, \ldots, c\}.$$

In the same way as the binary case, we treat pattern $\boldsymbol{x}$ and label y as random variables and assume that they are equipped with unknown joint probability density $p(\boldsymbol{x}, y)$.

Let

$$\boldsymbol{g}(\boldsymbol{x}) := (g_1(\boldsymbol{x}), \ldots, g_c(\boldsymbol{x}))^\top$$

be a c-class classifier, where

$$g_y : \mathscr{X} \to \mathbb{R}$$

is a "score" for class y. In multi-class classification, the classes are competing with each other, and thus classifier $\boldsymbol{g}$ predicts class $\widehat{y}$ to which input pattern $\boldsymbol{x}$ belongs based on its maximizing index as

$$\widehat{y} = \underset{y \in \mathscr{Y}}{\operatorname{argmax}}\, g_y(\boldsymbol{x}).$$

Intuitively, $g_y(\boldsymbol{x})$ can be regarded as a binary classifier for class y versus the remaining classes $y' \neq y$; then any classification model explained in section 2.1.2 may be used as g_y (when the neural network model is considered, all different g_y would usually have many shared layers; for instance, all the layers except the last one would be shared by all g_y). Ultimately, we would like to obtain a classifier that minimizes the multi-class *classification error* defined as

$$I(\boldsymbol{g}) := \mathbb{E}_{p(\boldsymbol{x},y)}[\mathscr{L}_{\text{0-1}}(\boldsymbol{g}(\boldsymbol{x}), y)], \tag{2.6}$$

where $\mathbb{E}_{p(\boldsymbol{x},y)}$ denotes the expectation over $p(\boldsymbol{x},y)$ and $\mathscr{L}_{0\text{-}1}: \mathbb{R}^c \times \mathscr{Y} \to \mathbb{R}$ is the multi-class *zero-one loss* defined as

$$\mathscr{L}_{0\text{-}1}(\boldsymbol{g}(\boldsymbol{x}),y) := \begin{cases} 0 & (\mathrm{argmax}_{y' \in \mathscr{Y}}\, g_{y'}(\boldsymbol{x}) = y), \\ 1 & (\mathrm{argmax}_{y' \in \mathscr{Y}}\, g_{y'}(\boldsymbol{x}) \neq y). \end{cases}$$

2.2.2   Surrogate Losses

Similar to binary classification, it is computationally intractable to find a minimizer of classification error $I(\boldsymbol{g})$ defined by (2.6). To avoid this problem, we use multi-class *surrogate loss* $\mathscr{L}: \mathbb{R}^c \times \mathscr{Y} \to \mathbb{R}$ instead of the zero-one loss. $\mathscr{L}(\boldsymbol{g}(\boldsymbol{x}),y)$ basically encourages $g_y(\boldsymbol{x})$ to take a large (small) value if y is correct (incorrect). The multi-class classification *risk* under surrogate loss $\mathscr{L}$ is defined as

$$R(\boldsymbol{g}) := \mathbb{E}_{p(\boldsymbol{x},y)}\left[\mathscr{L}(\boldsymbol{g}(\boldsymbol{x}),y)\right]. \tag{2.7}$$

The following multi-class losses are often used as a surrogate (Zhang, 2004):[2]

$$\text{One-versus-the-rest: } \mathscr{L}(\boldsymbol{g}(\boldsymbol{x}),y) = \ell(g_y(\boldsymbol{x}),+1) + \frac{1}{c-1}\sum_{y' \neq y}\ell(g_{y'}(\boldsymbol{x}),-1),$$

$$\text{Pairwise comparison: } \mathscr{L}(\boldsymbol{g}(\boldsymbol{x}),y) = \sum_{y' \neq y}\ell(g_y(\boldsymbol{x}) - g_{y'}(\boldsymbol{x})),$$

$$\text{Softmax cross-entropy: } \mathscr{L}(\boldsymbol{g}(\boldsymbol{x}),y) = -g_y(\boldsymbol{x}) + \ln\left(\sum_{y' \in \mathscr{Y}}\exp(g_{y'}(\boldsymbol{x}))\right),$$

where $\ell: \mathbb{R} \times \mathscr{Y} \to \mathbb{R}$ (or $\ell: \mathbb{R} \to \mathbb{R}$) is a (margin-based) binary surrogate loss (see section 2.1.3). Originally, the softmax cross-entropy loss is defined as

$$\mathscr{L}(\boldsymbol{g}(\boldsymbol{x}),y) = -\sum_{y' \in \mathscr{Y}} e^{(y')} \ln q^{(y')},$$

where $e^{(y')}$ is the *one-hot* expression of y:

$$e^{(y')} := \begin{cases} 1 & (y' = y), \\ 0 & (y' \neq y). \end{cases}$$

In addition, $q^{(y')}$ is the *softmax* expression of $g_{y'}(\boldsymbol{x})$:

$$q^{(y')} = \frac{\exp(g_{y'}(\boldsymbol{x}))}{\sum_{y \in \mathscr{Y}}\exp(g_y(\boldsymbol{x}))},$$

which corresponds to the probability of making prediction as y'.

2.2.3   Training Samples

As with the binary classification case, training samples are used for estimating the expectation over unknown $p(\boldsymbol{x}, y)$ included in the risk $R(\boldsymbol{g})$ defined by (2.7):

$$\mathcal{X}_1 := \{\boldsymbol{x}_i^{(1)}\}_{i=1}^{n_1} \overset{\text{i.i.d.}}{\sim} p_1(\boldsymbol{x}) := p(\boldsymbol{x} \mid y = 1),$$

$$\vdots$$

$$\mathcal{X}_c := \{\boldsymbol{x}_i^{(c)}\}_{i=1}^{n_c} \overset{\text{i.i.d.}}{\sim} p_c(\boldsymbol{x}) := p(\boldsymbol{x} \mid y = c),$$

$$\mathcal{X}_{\mathrm{U}} := \{\boldsymbol{x}_i^{\mathrm{U}}\}_{i=1}^{n_{\mathrm{U}}} \overset{\text{i.i.d.}}{\sim} p_{\mathrm{U}}(\boldsymbol{x}) := \pi_1 p_1(\boldsymbol{x}) + \cdots + \pi_c p_c(\boldsymbol{x}),$$

where

$$\pi_1 := p(y = 1),$$

$$\vdots$$

$$\pi_c := p(y = c),$$

are the class-prior probabilities for each class such that

$$\pi_1 + \cdots + \pi_c = 1.$$

For $y = 1, \ldots, c$, let $R_y(\boldsymbol{g})$ be the risk of classifier $\boldsymbol{g}$ under loss $\mathcal{L}$ for samples in class y:

$$R_y(\boldsymbol{g}) := \mathbb{E}_{p_y(\boldsymbol{x})}[\mathcal{L}(\boldsymbol{g}(\boldsymbol{x}), y)],$$

where $\mathbb{E}_{p_y(\boldsymbol{x})}$ denotes the expectation over $p_y(\boldsymbol{x})$. R_y can be approximated in an unbiased fashion by replacing the expectation by the corresponding sample average as

$$\widehat{R}_y(\boldsymbol{g}) := \frac{1}{n_y} \sum_{i=1}^{n_y} \mathcal{L}(\boldsymbol{g}(\boldsymbol{x}_i^{(y)}), y),$$

which is called an *empirical risk*.

3 Supervised Classification

In this chapter, we discuss the problem of supervised binary/multi-class classification. Supervised classification forms the basis of the weakly supervised classification methods that will be introduced in the following chapters.

3.1 Positive-Negative (PN) Classification

In this section, we consider the supervised binary classification problem. We give its empirical risk estimators and analyze the corresponding minimizers.

3.1.1 Formulation

There are two different cases of the formulation, depending on whether positive (P) and negative (N) data are sampled together or separately.

3.1.1.1 One-sample case

The first case uses a set of labeled data in the form of (pattern, label) pairs:

$$\mathscr{X}_{\mathrm{PN}} := \{(\boldsymbol{x}_i, y_i)\}_{i=1}^{n} \overset{\text{i.i.d.}}{\sim} p(\boldsymbol{x}, y).$$

We can see that the P and N data are sampled together. Under this setting, the training data shares the same distribution as test data, and thus it is easy to develop classification algorithms and analyze them.

The goal of binary classification[1] is to obtain a classifier that minimizes the classification risk:

$$R(g) := \mathbb{E}_{p(\boldsymbol{x},y)}[\ell(g(\boldsymbol{x}), y)],$$

where $\ell \colon \mathbb{R} \times \mathscr{Y} \to \mathbb{R}$ is a properly chosen loss function (see section 2.1.3), and $g \colon \mathscr{X} \to \mathbb{R}$ is a decision function in a properly chosen function class $\mathscr{G}$ (see section 2.1.2).

Subsequently, $R(g)$ can be approximated using $\mathscr{X}_{\mathrm{PN}}$ as

$$\widehat{R}(g) := \frac{1}{n} \sum_{i=1}^{n} \ell(g(\boldsymbol{x}_i), y_i). \tag{3.1}$$

In practice, the empirical risk $\widehat{R}(g)$ is regularized by a properly chosen regularizer Ω (see section 2.1.5), and a minimizer of the regularized risk over the function class $\mathscr{G}$ is obtained as

$$\widehat{g} := \underset{g \in \mathscr{G}}{\operatorname{argmin}} \left[\widehat{R}(g) + \lambda \Omega(g) \right],$$

where $\lambda \geq 0$ is the regularization parameter to trade off the risk and the regularizer. The loss function ℓ, the function class $\mathscr{G}$, and the regularizer Ω can be selected either according to the domain knowledge from domain experts or with the help of an additional set of validation data sampled from $p(\boldsymbol{x}, y)$.

3.1.1.2   Two-sample case

The second case uses a set of P data and a set of N data:

$$\mathscr{X}_{\mathrm{P}} := \{\boldsymbol{x}_i^{\mathrm{P}}\}_{i=1}^{n_{\mathrm{P}}} \overset{\text{i.i.d.}}{\sim} p_{\mathrm{P}}(\boldsymbol{x}) := p(\boldsymbol{x} \mid y = +1),$$

$$\mathscr{X}_{\mathrm{N}} := \{\boldsymbol{x}_i^{\mathrm{N}}\}_{i=1}^{n_{\mathrm{N}}} \overset{\text{i.i.d.}}{\sim} p_{\mathrm{N}}(\boldsymbol{x}) := p(\boldsymbol{x} \mid y = -1).$$

We can see that the P and N data are sampled separately, such that the training data $\mathscr{X}_{\mathrm{P}}$ and $\mathscr{X}_{\mathrm{N}}$ are two independent sets, and the sample sizes n_{P} and n_{N} are two independent variables. For this reason, we refer to this supervised binary classification also as *positive-negative (PN) classification*. Under this setting, the P and N data can be collected in different ways from different sources, and thus it allows more flexibility to develop powerful classification algorithms.

The original classification risk $R(g)$ can be decomposed into

$$R(g) = \pi_{\mathrm{P}} R_{\mathrm{P}}(g) + \pi_{\mathrm{N}} R_{\mathrm{N}}(g),$$

where

$$\pi_{\mathrm{P}} := p(y = +1),$$

$$\pi_{\mathrm{N}} := p(y = -1),$$

are the class-prior probabilities for the P and N classes and

$$R_{\mathrm{P}}(g) := \mathbb{E}_{\mathrm{P}}[\ell(g(\boldsymbol{x}), +1)],$$

$$R_{\mathrm{N}}(g) := \mathbb{E}_{\mathrm{N}}[\ell(g(\boldsymbol{x}), -1)],$$

are the partial risks for the P and N data. Subsequently, $R_P(g)$ and $R_N(g)$ can be approximated using $\mathscr{X}_P$ and $\mathscr{X}_N$ as

$$\widehat{R}_P(g) := \frac{1}{n_P} \sum_{i=1}^{n_P} \ell(g(\boldsymbol{x}_i^P), +1),$$

$$\widehat{R}_N(g) := \frac{1}{n_N} \sum_{i=1}^{n_N} \ell(g(\boldsymbol{x}_i^N), -1),$$

and $R(g)$ can be empirically approximated as

$$\widehat{R}_{PN}(g) := \pi_P \widehat{R}_P(g) + \pi_N \widehat{R}_N(g). \tag{3.2}$$

The estimator $\widehat{R}_{PN}(g)$ of the classification risk is known as the *empirical risk estimator* or simply the *empirical risk*. Finally, a minimizer of the empirical risk $\widehat{R}_{PN}(g)$ over the function class $\mathscr{G}$ is obtained as

$$\widehat{g}_{PN} := \operatorname*{argmin}_{g \in \mathscr{G}} \widehat{R}_{PN}(g),$$

where any of the regularizers is omitted for simplicity. The minimizer $\widehat{g}_{PN}$ of the empirical risk estimator is known as the *empirical risk minimizer*.

3.1.1.3 Comparison

Let us compare the one-sample estimator (3.1) and the two-sample estimator (3.2). In equation (3.1), the labels $\{y_1, \ldots, y_n\}$ are present as they are explicitly given in $\mathscr{X}_{PN}$, while in (3.2), the labels are absent as they are implicitly implied by the membership of $\mathscr{X}_P$ or $\mathscr{X}_N$. Moreover, in (3.2), we need to know π_P and π_N. It is not difficult to know them, for example, by asking domain experts or estimating them from a set of validation data sampled from $p(\boldsymbol{x}, y)$. It is also possible to estimate them with the help of an additional set of unlabeled data sampled from $p(\boldsymbol{x})$, which is known as *class-prior estimation* (Saerens et al., 2002; du Plessis and Sugiyama, 2014b; Sugiyama et al., 2013; Iyer et al., 2014) and will be discussed in detail in chapter 12.

In fact, the two-sample estimator (3.2) is slightly more general than the one-sample estimator (3.1). Indeed, we cannot approximate the classification risk by (3.1) in the two-sample case, while we can still approximate the classification risk by (3.2) in the one-sample case. To do so, we simply need to split $\mathscr{X}_{PN}$ into $\mathscr{X}_P$ and $\mathscr{X}_N$ and replace the unknown π_P and π_N with their empirical estimates

$$\widehat{\pi}_P = \frac{n_P}{n},$$

$$\widehat{\pi}_N = \frac{n_N}{n}.$$

After the split and replacement operations, (3.2) is reduced to (3.1). Of course, this is exactly as expected—the data are the same before and after the operations, and thus there is no reason to expect to obtain different empirical risk estimators. We will focus on the two-sample case because of its generality.

Last but not least, we would like to give a remark on the names *one-sample* and *two-sample* estimators. They are actually statistical jargons where a "sample" means a set of data. Nevertheless, we are using this meaning only in this context. Hence, in other contexts of this book, a "sample" means a single data point and "samples" do not indicate many sets of data.

3.1.2   Theoretical Analysis

This subsection establishes the theoretical guarantee of PN classification, i.e., convergence properties of the empirical risk minimizer $\widehat{g}_{\mathrm{PN}}$. It also serves as a paradigm for some theoretical analyses in the following chapters. Note, however, that the contents in this section are mathematically highly advanced. Those who are interested only in conceptual and algorithmic issues of PN classification may skip ahead to section 3.2.

3.1.2.1   Targets of convergence

In order to establish the convergence properties of $\widehat{g}_{\mathrm{PN}}$, let us think about certain optimal classifiers that could serve as the target to which $\widehat{g}_{\mathrm{PN}}$ converges. We have three choices.

First, we may consider the minimizer of the true risk $R(g)$ over the function class $\mathscr{G}$ as

$$g^* := \operatorname*{argmin}_{g \in \mathscr{G}} R(g),$$

which is known as the *true risk minimizer*. Compared with $\widehat{g}_{\mathrm{PN}}$, this g^* is built on access to the underlying joint probability density $p(x, y)$ or, in terms of data, access to uncountably infinite training data.

Second, the target may be the minimizer of $R(g)$ over the set of all measurable functions, i.e.,

$$g_R^{\star} := \operatorname*{argmin}_{g} R(g).$$

This $g_R^{\star}$ is known as the *Bayes optimal classifier* under loss ℓ, corresponding to the *Bayes risk* defined by

$$R^{\star} := \inf_{g} R(g) = R(g_R^{\star}).$$

Compared with g^*, this $g_R^{\star}$ is built on a very large function class. Indeed, this implicit function class (i.e., the set of all measurable functions) is the largest one we can use when minimizing $R(g)$ and obtaining $g_R^{\star}$ even with access to uncountably infinite training data.

Third, recall that ℓ is a surrogate loss for the zero-one loss $\ell_{0\text{-}1}$ and $R(g)$ is a surrogate learning criterion for the classification error $I(g)$ defined by

$$I(g) := \mathbb{E}_{p(x,y)}[\ell_{0\text{-}1}(g(\boldsymbol{x}), y)].$$

Then, the target may also be the minimizer of $I(g)$ over the set of all measurable functions, i.e.,

$$g_I^\star := \underset{g}{\operatorname{argmin}}\, I(g).$$

This $g_I^\star$ is known as the Bayes optimal classifier under loss $\ell_{0\text{-}1}$, corresponding to the *Bayes error* defined by

$$I^\star := \inf_g I(g) = I(g_I^\star).$$

Compared with $g_R^\star$, this $g_I^\star$ comes from a non-smooth, non-convex optimization and thus may not be unique.

In principle, we should always make use of *classification-calibrated* losses. A surrogate loss ℓ is classification-calibrated if and only if there is a transformation $\psi_\ell : [0, 1] \rightarrow [0, +\infty)$ satisfying

- ψ_ℓ is convex, invertible, and non-decreasing,

- $\psi_\ell(0) = 0$,

such that the following condition holds (Bartlett et al., 2006):

$$\psi_\ell(I(g) - I^\star) \leq R(g) - R^\star,$$

or equivalently,

$$I(g) - I^\star \leq \psi_\ell^{-1}(R(g) - R^\star).$$

There is no need to worry about the classification calibration of ℓ, because all popular surrogate losses are indeed classification-calibrated.

Since the above inequality holds for every measurable function g including $g_R^\star$, we can know that

$$0 \leq I(g_R^\star) - I^\star \leq \psi_\ell^{-1}(R(g_R^\star) - R^\star) = \psi_\ell^{-1}(0) = 0,$$

which indicates $g_R^\star$ is also one of the Bayes optimal classifiers under loss $\ell_{0\text{-}1}$. As a consequence, we will not distinguish Bayes optimal classifiers under different losses. We will simply call it the Bayes optimal classifier and denote it by $g^\star$ from now on.

In summary, we end up with two choices for the target of convergence: the true risk minimizer g^* and the Bayes optimal classifier $g^\star$.

3.1.2.2   Measures of convergence

Next, let us think about how to measure the convergence, for instance, from $\widehat{g}_{PN}$ to $g^\star$. In mathematics and statistics, we usually regard the set of all measurable functions as a metric space, equipped with a distance metric function

$$d_\infty(g, g') := \|g - g'\|_\infty = \sup_{x \in \mathcal{X}} |g(x) - g'(x)|$$

or

$$d_2(g, g') := \|g - g'\|_2 = \left(\mathbb{E}_{p(x)} \left[(g(x) - g'(x))^2 \right] \right)^{1/2},$$

and then we measure the distance between $\widehat{g}_{PN}$ and $g^\star$. In statistical learning, we usually wrap classifiers by a learning criterion such as $R(g)$ and then measure the performance gap $R(\widehat{g}_{PN}) - R(g^\star)$. The latter way (measuring the performance gap of classifiers) is preferred to the former way (measuring the distance between classifiers) in statistical learning for three reasons.

Firstly, there may be more than one global minimizer of a learning criterion due to the non-convexity of the learning criterion. The loss ℓ does not need to be convex: $\ell(\widehat{y}, y)$ can be non-convex in $\widehat{y}$, or if it is margin-based, $\ell(m)$ can be non-convex in m, such as the ramp loss and the sigmoid loss in section 2.1.3. Even if the loss ℓ is convex, a learning criterion can still be non-convex, and we will encounter such empirical risk estimators in the following chapters.

Secondly, a learning criterion may not be convex in the parameter vector of g, even if it is convex in g. Recall that we have to substitute the abstract form $g(x)$ with the explicit form $g(x; \alpha)$ for training g, which gives us optimization problems in α. Then, these optimization problems should not be convex in α if g is nonlinear in α. In recent years, the common practice is to use a neural network model, which basically gives us smooth non-convex optimizations.

Finally, our risk-minimizing definition of $g^\star$, namely $g_R^\star$, could be different from the original probabilistic definition of the Bayes optimal classifier:

$$g_I^\star(x) := \operatorname{argmax}_y p(y \mid x) = \operatorname{sign}(p(y = +1 \mid x) - p(y = -1 \mid x)).$$

They should be different as long as ℓ is not the logistic loss in section 2.1.3. Then, we cannot show that $\widehat{g}_{PN} \to g_I^\star$ even for the ideal case where $\widehat{g}_{PN} \to g_R^\star$ and $I(\widehat{g}_{PN}) \to I(g_R^\star) = I(g_I^\star)$.

Therefore, we will wrap classifiers by a learning criterion and then measure the performance gap, without caring about which classifiers lead to a specific performance gap. Along this way, there are a few critical concepts concerning different types of bounds: the estimation error, approximation error, excess risk, and uniform deviation.

The *estimation error* is defined as the gap between the classification risk of $\widehat{g}_{PN}$ and that of g^{*}:[2]

$$R(\widehat{g}_{PN}) - R(g^*).$$

This corresponds to the same concept in statistics written as $\|\widehat{g}_{PN} - g^*\|$, and it measures convergence from an empirical risk minimizer to a true risk minimizer. The source of the estimation error is the finite amount of training data, and this is unavoidable. The estimation error depends on two main factors: the amount of data and the complexity of the function class. It tends to be small given a large amount of data and to be large given a small amount of data. On the other hand, since g^* is optimal inside the function class $\mathscr{G}$, it tends to be small if $\mathscr{G}$ is small and to be large if $\mathscr{G}$ is large.

The *approximation error* is defined as the gap between the classification risk of g^* and that of $g^\star$:

$$R(g^*) - R(g^\star).$$

This corresponds to the same concept in statistics written as $\|g^* - g^\star\|$, and it measures convergence from a true risk minimizer to the Bayes optimal classifier. The source of the approximation error is the limited complexity of $\mathscr{G}$, but this is avoidable if $g^\star \in \mathscr{G}$, which can be achieved if we can know the form of $p(y\,|\,\boldsymbol{x})$ and design $\mathscr{G}$ accordingly. Note that $g^\star$ is fixed, and the approximation error depends only on the complexity of $\mathscr{G}$. It tends to be small if $\mathscr{G}$ is large and to be large if $\mathscr{G}$ is small. Hence, there is essentially a trade-off between the estimation error and approximation error.

The *excess risk* is defined as the gap between the classification error of $\widehat{g}_{PN}$ and that of $g^\star$:

$$I(\widehat{g}_{PN}) - I(g^\star).$$

This corresponds to the same concept in statistics written as $\|\widehat{g}_{PN} - g^\star\|$, and it measures the ultimate goal of machine learning—namely, convergence from an empirical risk minimizer to the Bayes optimal classifier. Because of the classification calibration of ℓ,

$$I(\widehat{g}_{PN}) - I^\star \leq \psi_\ell^{-1}(R(\widehat{g}_{PN}) - R^\star)$$

$$\leq \psi_\ell^{-1}\big((R(\widehat{g}_{PN}) - R(g^*)) + (R(g^*) - R^\star)\big),$$

we can see the excess risk comes from the estimation and approximation errors. It depends on the amount of data in a similar manner to the estimation error, and it depends on the complexity of $\mathscr{G}$ in a complex manner:

- If $\mathscr{G}$ is too small, there must be a small estimation error but a large approximation error, and the excess risk would be large due to underfitting.

- If $\mathscr{G}$ is too large, there must be a small approximation error but a large estimation error, and the excess risk would be large due to overfitting.

Thus, only if $\mathscr{G}$ has an appropriate complexity could the excess risk be small.

The *uniform deviation* is the maximum gap between the empirical risk of g and the true risk of g over $\mathcal{G}$:

$$\sup_{g \in \mathcal{G}} \left| \widehat{R}_{\mathrm{PN}}(g) - R(g) \right|.$$

As distinct from the previous performance gaps, this is a gap not between the same learning criterion of two classifiers but between two learning criteria of the same classifier. This concept is not from statistics; instead, it is original in machine learning, especially statistical learning, and it measures convergence of empirical risk estimators (Vapnik, 1998). Uniform deviation bounds can be thought of as uniform laws of large numbers:

- For a single g, the law of large numbers can ensure that $\widehat{R}_{\mathrm{PN}}(g)$ converges to $R(g)$.

- For a function class $\mathcal{G}$ that often contains as many as uncountably infinite g, uniform deviation bounds can ensure that $\widehat{R}_{\mathrm{PN}}(g)$ converges to $R(g)$ considering the worst-case g.

The uniform deviation depends (in the same manner as the estimation error does) on the amount of data and the complexity of $\mathcal{G}$. In fact, we will prove estimation error bounds through uniform deviation bounds in this book.

As estimation error bounds and uniform deviation bounds, we simply refer to upper bounds of the estimation error and uniform deviation, which hold with high probability. Given any reasonable $\mathcal{G}$, learning is said to be *consistent* (in a *weak* sense) if the estimation error approaches zero with high probability as the amount of data approaches infinity:

$$\lim_{n_{\mathrm{P}}, n_{\mathrm{N}} \to \infty} R(\widehat{g}_{\mathrm{PN}}) - R(g^*) = 0,$$

or more specifically, for any $\epsilon > 0$,

$$\lim_{n_{\mathrm{P}}, n_{\mathrm{N}} \to \infty} \mathrm{Pr}_{\mathcal{X}_{\mathrm{P}}, \mathcal{X}_{\mathrm{N}}} (R(\widehat{g}_{\mathrm{PN}}) - R(g^*) \le \epsilon) = 1,$$

where the probability is over the repeated sampling of data for obtaining $\widehat{g}_{\mathrm{PN}}$. On the other hand, given a series of function classes $\mathcal{G}_{n_{\mathrm{P}}, n_{\mathrm{N}}}$ that possess the increasing complexities with n_{P} and n_{N}, learning is said to be *consistent* (in a *strong* sense) if the excess risk approaches zero with high probability:

$$\lim_{n_{\mathrm{P}}, n_{\mathrm{N}} \to \infty} I(\widehat{g}_{\mathrm{PN}}) - I(g^*) = 0.$$

However, it is extremely difficult to preset appropriate complexities for $\mathcal{G}_{n_{\mathrm{P}}, n_{\mathrm{N}}}$ and to check whether g^* is included in $\mathcal{G}_{n_{\mathrm{P}}, n_{\mathrm{N}}}$ at last. Therefore, an estimation error bound approaching zero is usually the best we can achieve, and we refer to consistency as the weak one when we talk about it.

We have avoided saying a *generalization error bound* because it may refer to many different things. The *generalization error* is just the classification error, and then any upper bound of $I(g)$ can be called a generalization error bound, where it is unclear which g is considered or whether all $g \in \mathcal{G}$ are considered. Alternatively, the generalization error can be the difference between the classification error and the training error. Fortunately, generalization error bounds are basically equivalent given different definitions of the generalization error: just for the former/latter bound, the training error is in the right- or left-hand side.

In most cases, generalization error bounds are about

$$\sup_{g \in \mathcal{G}} I(g) - \widehat{R}_{\mathrm{PN}}(g),$$

where the loss inside $\widehat{R}_{\mathrm{PN}}(g)$ is the hinge loss in section 2.1.3, not necessarily the same as the one for training g. Indeed, those error bounds are from one-sided uniform deviation bounds, where training is not involved; the errors and risks are related by $\ell_{0\text{-}1}(m) \leq \ell(m)$ rather than by the classification calibration of ℓ, where ℓ is the hinge loss. Although training is not involved, those generalization error bounds represent a different philosophy: With high probability, any $g \in \mathcal{G}$ can generalize well; this is theoretically guaranteed before training and it does not matter which g is selected after training. When we say "generalize well," we imply that the classification error is only a bit worse than the empirical error/risk. In the end, the reference of $I(\widehat{g}_{\mathrm{PN}})$ or $R(\widehat{g}_{\mathrm{PN}})$ in generalization error bounds is an empirical performance of the same $\widehat{g}_{\mathrm{PN}}$, while the reference in estimation error or excess risk bounds is the same performance of g^* or $g^\star$. When we follow that line of thinking, the reference in generalization error bounds is computationally easy to evaluate, and the reference in estimation error or excess risk bounds is conceptually easy to understand and compare because g^* or $g^\star$ is the target of convergence.

We have finished our extensive discussion on the measures of convergence. Before going ahead, we would like to give a remark on the names: estimation error, approximation error, and excess risk. We can find that the first two errors are defined using the risk and the last risk is defined using the error. This is because those names are again statistical jargons, but they are assigned new meanings in machine learning.

3.1.2.3 Rademacher complexity

Based on the previous discussions, it is obvious that the complexity of $\mathcal{G}$ plays a role of vital importance. Thus, we want to select a measure of complexity that best fits the purpose. Among popular complexity measures, we adopt the *Rademacher complexity*, and we will explain why we made this decision after giving its definitions.

The Rademacher complexity has been written about by many (Bartlett and Mendelson, 2002; Mohri et al., 2012; Shalev-Shwartz and Ben-David, 2014) and has several definitions. A random variable σ is called a *Rademacher variable* if it follows the *Rademacher distribution*:

$$\Pr(\sigma = +1) = \Pr(\sigma = -1) = 1/2,$$

which is a symmetric version of the Bernoulli distribution in both the support and the probability. Rademacher variables serve as completely random noises for defining the complexity of a set and that of a function class. Given a set of n-dimensional vectors $S \subseteq \mathbb{R}^n$, the *Rademacher complexity* of S is defined as

$$\mathfrak{R}(S) := \mathbb{E}_{\sigma_1, \dots, \sigma_n} \left[\sup_{(s_1, \dots, s_n) \in S} \frac{1}{n} \sum_{i=1}^{n} \sigma_i s_i \right], \tag{3.3}$$

where $\sigma_1, \dots, \sigma_n$ are independent Rademacher variables. In (3.3), it does not matter whether the cardinality of S is finite, countably infinite, or uncountably infinite. To understand the implication of $\mathfrak{R}(S)$ in (3.3), let us denote by $s = (s_1, \dots, s_n)$ and $\sigma = (\sigma_1, \dots, \sigma_n)$, and $\mathfrak{R}(S)$ can be expressed as

$$\mathfrak{R}(S) = \mathbb{E}_\sigma \left[\sup_{s \in S} \frac{1}{n} \sigma \cdot s \right],$$

where $\sigma \cdot s$ is the dot product between σ and s. Since $\sigma \cdot s$ can be regarded as a similarity or fitness between σ and s, the Rademacher complexity of S can be interpreted as the expected fitness between randomly generated noises and the most fitted member of S for those noises.

Subsequently, given a set of n data $S_n = \{x_i\}_{i=1}^n$, the *empirical Rademacher complexity* of $\mathscr{G}$ given S_n is defined as

$$\mathfrak{R}_{S_n}(\mathscr{G}) := \mathfrak{R}(\mathscr{G} \circ S_n) = \mathbb{E}_{\sigma_1, \dots, \sigma_n} \left[\sup_{g \in \mathscr{G}} \frac{1}{n} \sum_{i=1}^{n} \sigma_i g(x_i) \right],$$

where $\mathscr{G} \circ S_n$ denotes the composition of $\mathscr{G}$ and S_n by evaluating the members of $\mathscr{G}$ at the members of S_n, i.e.,

$$\mathscr{G} \circ S_n := \{ (g(x_1), \dots, g(x_n)) \mid g \in \mathscr{G} \}.$$

Let $q(x)$ be an arbitrary density of x—in the case of PN classification, $q(x)$ is either $p_\mathrm{P}(x)$ or $p_\mathrm{N}(x)$. Then, the *Rademacher complexity* of $\mathscr{G}$ for the sample size n with respect to $q(x)$ is defined as

$$\mathfrak{R}_{n,q}(\mathscr{G}) := \mathbb{E}_{S_n \sim q^n} [\mathfrak{R}_{S_n}(\mathscr{G})]$$

$$= \mathbb{E}_{x_1, \dots, x_n} \mathbb{E}_{\sigma_1, \dots, \sigma_n} \left[\sup_{g \in \mathscr{G}} \frac{1}{n} \sum_{i=1}^{n} \sigma_i g(x_i) \right],$$

where $S_n \sim q^n$ denotes that S_n consists of n data independently sampled from $q(x)$, i.e.,

$$S_n = \{x_i\}_{i=1}^n \overset{\text{i.i.d.}}{\sim} q(x).$$

The Rademacher complexity of $\mathcal{G}$ can similarly be interpreted as the expected fitness between randomly generated noises and the most fitted member of $\mathcal{G}$ after the members of $\mathcal{G}$ have been evaluated at n independent data generated according to $q(\boldsymbol{x})$. For fixed n and $q(\boldsymbol{x})$, the smaller/larger $\mathfrak{R}_{n,q}(\mathcal{G})$ is, the smaller/larger complexity $\mathcal{G}$ has.

There is an alternative set of definitions of Rademacher complexities that have an additional absolute value of the average before taking the supremum:

$$\mathfrak{R}'(S) := \mathbb{E}_{\sigma_1,\ldots,\sigma_n}\left[\sup_{(s_1,\ldots,s_n)\in S}\left|\frac{1}{n}\sum_{i=1}^{n}\sigma_i s_i\right|\right],$$

$$\mathfrak{R}'_{S_n}(\mathcal{G}) := \mathfrak{R}'(\mathcal{G}\circ S_n) = \mathbb{E}_{\sigma_1,\ldots,\sigma_n}\left[\sup_{g\in\mathcal{G}}\left|\frac{1}{n}\sum_{i=1}^{n}\sigma_i g(\boldsymbol{x}_i)\right|\right],$$

$$\mathfrak{R}'_{n,q}(\mathcal{G}) := \mathbb{E}_{S_n\sim q^n}[\mathfrak{R}'_{S_n}(\mathcal{G})]$$

$$= \mathbb{E}_{\boldsymbol{x}_1,\ldots,\boldsymbol{x}_n}\,\mathbb{E}_{\sigma_1,\ldots,\sigma_n}\left[\sup_{g\in\mathcal{G}}\left|\frac{1}{n}\sum_{i=1}^{n}\sigma_i g(\boldsymbol{x}_i)\right|\right].$$

This alternative version comes from Koltchinskii (2001) and Bartlett and Mendelson (2002), whose authors are the pioneers of deriving error bounds based on the Rademacher complexity and upper-bounding the Rademacher complexity for various function classes.

The default and alternative Rademacher complexities can be related as follows. Without any composition, for arbitrary S or $\mathcal{G}$, the alternative one is no less than the default one:

$$\mathfrak{R}'(S) \geq \mathfrak{R}(S), \quad \mathfrak{R}'_{S_n}(\mathcal{G}) \geq \mathfrak{R}_{S_n}(\mathcal{G}), \quad \mathfrak{R}'_{n,q}(\mathcal{G}) \geq \mathfrak{R}_{n,q}(\mathcal{G});$$

if S or $\mathcal{G}$ is symmetric or, more precisely, closed under negation, namely, $s \in S$ if and only if $-s \in S$, or $g \in \mathcal{G}$ if and only if $-g \in \mathcal{G}$, then the alternative one is equal to the default one:

$$\mathfrak{R}'(S) = \mathfrak{R}(S), \quad \mathfrak{R}'_{S_n}(\mathcal{G}) = \mathfrak{R}_{S_n}(\mathcal{G}), \quad \mathfrak{R}'_{n,q}(\mathcal{G}) = \mathfrak{R}_{n,q}(\mathcal{G}).$$

However, with a composition

$$\ell\circ\mathcal{G} := \{\ell\circ g \mid g\in\mathcal{G}\} = \{(\boldsymbol{x},y)\mapsto \ell(g(\boldsymbol{x}),y) \mid g\in\mathcal{G}\},$$

where the loss function ℓ is non-negative, the Rademacher complexities of the composite function class $\ell\circ\mathcal{G}$ would generally satisfy

$$\mathfrak{R}'_{S_n}(\ell\circ\mathcal{G}) \neq \mathfrak{R}_{S_n}(\ell\circ\mathcal{G}), \quad \mathfrak{R}'_{n,q}(\ell\circ\mathcal{G}) \neq \mathfrak{R}_{n,q}(\ell\circ\mathcal{G}),$$

since $\ell\circ\mathcal{G}$ is generally not closed under negation.

A vital disagreement between the default and alternative Rademacher complexities arises when considering the famous *contraction principle/property*. Let $f : \mathbb{R} \to \mathbb{R}$ be a Lipschitz continuous function with a Lipschitz constant L_f and assume that f satisfies $f(0) = 0$. Then, according to *Talagrand's contraction lemma* (Ledoux and Talagrand, 1991, theorem 4.12), we only have

$$\mathfrak{R}'_{S_n}(f \circ \mathscr{G}) \leq 2L_f \mathfrak{R}'_{S_n}(\mathscr{G}), \quad \mathfrak{R}'_{n,q}(f \circ \mathscr{G}) \leq 2L_f \mathfrak{R}'_{n,q}(\mathscr{G}).$$

However, according to lemma 4.2 in Mohri et al. (2012) or lemma 26.9 in Shalev-Shwartz and Ben-David (2014), which are powerful extensions of the original contraction lemma, we further have

$$\mathfrak{R}_{S_n}(f \circ \mathscr{G}) \leq L_f \mathfrak{R}_{S_n}(\mathscr{G}), \quad \mathfrak{R}_{n,q}(f \circ \mathscr{G}) \leq L_f \mathfrak{R}_{n,q}(\mathscr{G}),$$

and the assumption $f(0) = 0$ can be safely dropped. We can see that the latter contraction principle/property is tighter than the former one and $\ell(0, y) = 0$ is not required, and thus the default set of definitions is easier to use for deriving error bounds.

Why adopt the Rademacher complexity? Let us explain. Intuitively, there are two steps or components of PN classification: first estimation of the true risk and then minimization of the empirical risk. Similarly, in subsequent chapters, we will meet the weak supervisions where the training and test data are sampled from two different distributions, and we will handle the challenge of those weak supervisions by focusing on the estimation of the true risk and designing empirical risk estimators accordingly. After we figure out the estimation of the true risk under a weak supervision, the minimization of the empirical risk will not change too much. This methodology means that learning criteria and models are decoupled in algorithm design, and bounding the estimation error with a model-dependent complexity measure and bounding this complexity measure for different models should also be decoupled in algorithm analysis. Hence, the Rademacher complexity best fits the purpose.

3.1.2.4   Rademacher complexity bounds

We have discussed that bounding the Rademacher complexity does not depend on the learning criterion, and thus it does not affect bounding the estimation error. From here on, we upper-bound the Rademacher complexities for the models in section 2.1.2.

Consider the easiest linear-in-input model $g(x) = \alpha^\top x$. We follow the common practice to assume that x and α are bounded. In practice, all x should be preprocessed by some outlier detection and normalization method, and all α should be regularized to mitigate overfitting. As a result, it is quite natural to assume that x and α are bounded.

Theorem 3.1 (Complexity for linear-in-input model)   *Assume that there exists a constant $C_x > 0$ such that $\|x\|_2 \leq C_x$ for all x. Given a constant $C_\alpha > 0$, define the linear-in-input function class as*

$$\mathscr{G}_{\mathrm{LII}} := \{x \mapsto \alpha^\top x \mid \|\alpha\|_2 \leq C_\alpha\}, \tag{3.4}$$

where LII is short for linear-in-input. Then, the Rademacher complexity of $\mathscr{G}_{\mathrm{LII}}$ can be bounded by

$$\mathfrak{R}_{n,q}(\mathscr{G}_{\mathrm{LII}}) \leq \frac{C_x C_\alpha}{\sqrt{n}}. \tag{3.5}$$

Proof Let $S_n \sim q^n$ and then

$$\mathfrak{R}_{S_n}(\mathscr{G}_{\mathrm{LII}}) = \mathbb{E}_{\sigma_1,\ldots,\sigma_n}\left[\sup_{\alpha:\|\alpha\|_2 \leq C_\alpha} \frac{1}{n}\sum_{i=1}^{n}\sigma_i\alpha^\top x_i\right]$$

$$= \frac{1}{n}\mathbb{E}_{\sigma_1,\ldots,\sigma_n}\left[\sup_{\alpha:\|\alpha\|_2 \leq C_\alpha} \alpha^\top\left(\sum_{i=1}^{n}\sigma_i x_i\right)\right]$$

$$= \frac{C_\alpha}{n}\mathbb{E}_{\sigma_1,\ldots,\sigma_n}\left[\left\|\sum_{i=1}^{n}\sigma_i x_i\right\|_2\right] \qquad \text{(solving the supremum)}$$

$$\leq \frac{C_\alpha}{n}\left(\mathbb{E}_{\sigma_1,\ldots,\sigma_n}\left[\left\|\sum_{i=1}^{n}\sigma_i x_i\right\|_2^2\right]\right)^{1/2} \qquad \text{(Jensen's inequality)}$$

$$= \frac{C_\alpha}{n}\left(\mathbb{E}_{\sigma_1,\ldots,\sigma_n}\left[\sum_{i,j=1}^{n}\sigma_i\sigma_j x_i^\top x_j\right]\right)^{1/2}$$

$$= \frac{C_\alpha}{n}\left(\sum_{i,j=1}^{n}\mathbb{E}_{\sigma_i,\sigma_j}[\sigma_i\sigma_j]x_i^\top x_j\right)^{1/2} \qquad \text{(linearity of expectation)}$$

$$= \frac{C_\alpha}{n}\left(\sum_{i=1}^{n}\|x_i\|_2^2\right)^{1/2} \qquad \text{(definition of } \sigma_i, \sigma_j)$$

$$\leq \frac{C_\alpha}{n}\left(nC_x^2\right)^{1/2} \qquad \text{(assumption on } x)$$

$$= C_x C_\alpha/\sqrt{n},$$

where in the third step we solved the supremum as a constrained maximization problem and obtained its solution

$$\alpha^* = \frac{C_\alpha \sum_{i=1}^{n}\sigma_i x_i}{\|\sum_{i=1}^{n}\sigma_i x_i\|_2},$$

and in the third from last step we relied on the fact that σ_i and σ_j are independent Rademacher variables and

$$\mathbb{E}_{\sigma_i,\sigma_j}[\sigma_i\sigma_j] = \begin{cases} \mathbb{E}_{\sigma_i}[1] = 1 & (i=j), \\ \mathbb{E}_{\sigma_i}[\sigma_i]\,\mathbb{E}_{\sigma_j}[\sigma_j] = 0 & (i \neq j). \end{cases}$$

Taking the expectation of $\mathfrak{R}_{S_n}(\mathscr{G}_{\mathrm{LII}})$ proves (3.5). $\qquad\square$

Next, consider the linear-in-parameter model $g(x) = \alpha^\top \phi(x)$ and the kernel model $g(x) = \sum_{i=1}^n \alpha_i K(x, x_i)$. Denote by

$$\phi_j(x; x_j) := K(x, x_j),$$

$$\phi(x; S_n) := (\phi_1(x; x_1), \ldots, \phi_n(x; x_n))^\top,$$

and then the kernel model can also be written as $g(x) = \alpha^\top \phi(x; S_n)$. We may imagine that the linear-in-parameter model can be obtained from the linear-in-input model by replacing $x \in \mathbb{R}^d$ with $\phi(x) \in \mathbb{R}^b$, and so can the kernel model with $\phi(x; S_n) \in \mathbb{R}^n$. Alternatively, let $\mathscr{F}$ be a (not necessarily unique) *feature space* corresponding to K, and let $\varphi \colon \mathscr{X} \to \mathscr{F}$ be a *kernel induced feature map*, such that $K(x, x') = \langle \varphi(x), \varphi(x') \rangle_{\mathscr{F}}$ where $\langle \cdot, \cdot \rangle_{\mathscr{F}}$ denotes the inner product in $\mathscr{F}$. Then, plugging $w = \sum_{i=1}^n \alpha_i \varphi(x_i)$ into $g(x) = \sum_{i=1}^n \alpha_i K(x, x_i)$, the kernel model can also be parameterized as

$$g(x) = \langle w, \varphi(x) \rangle_{\mathscr{F}},$$

which appears independent of S_n and is more similar to the linear-in-parameter model. In fact, according to kernel-based learning jargons, $g(x) = \langle w, \varphi(x) \rangle_{\mathscr{F}}$ is the *primal form*, but $g(x) = \sum_{i=1}^n \alpha_i K(x, x_i)$ is its *dual form* (Schölkopf and Smola, 2002). Regardless of which Hilbert space $\mathscr{F}$ is selected, g belongs to a fixed *reproducing kernel Hilbert space* $\mathscr{H}$ of functions $\mathscr{X} \to \mathbb{R}$.

Note that the input dimensionality d and the parameter dimensionality b are fixed, whereas the sample size n varies with S_n. However, n is not necessarily bigger than b: For the Gaussian kernel, the dimensionality of $\mathscr{F}$ is infinite and hence b should be bigger than any finite n. In order to bound the Rademacher complexities, we can modify our assumptions into that $\phi(x)$, $\phi(x; S_n)$, and α are bounded, or $\varphi(x)$ and w are bounded. While it might be slightly strong to assume that $\phi(x; S_n)$ is bounded independently of n, we can naturally assume that w is bounded even if w is high-dimensional. It is because either w should be regularized via $\|w\|_{\mathscr{F}}$ or g should be regularized via $\|g\|_{\mathscr{H}}$, and these two regularizers are exactly the same since

$$\|w\|_{\mathscr{F}}^2 = \left\| \sum_{i=1}^n \alpha_i \varphi(x_i) \right\|_{\mathscr{F}}^2$$

$$= \left\langle \sum_{i=1}^n \alpha_i \varphi(x_i), \sum_{j=1}^n \alpha_j \varphi(x_j) \right\rangle_{\mathscr{F}}$$

$$= \sum_{i,j=1}^n \alpha_i \alpha_j \langle \varphi(x_i), \varphi(x_j) \rangle_{\mathscr{F}}$$

$$= \sum_{i,j=1}^n \alpha_i \alpha_j K(x_i, x_j),$$

$$\|g\|_{\mathcal{H}}^2 = \left\| \sum_{i=1}^{n} \alpha_i K(\cdot, \mathbf{x}_i) \right\|_{\mathcal{H}}^2$$

$$= \left\langle \sum_{i=1}^{n} \alpha_i K(\mathbf{x}_i, \cdot), \sum_{j=1}^{n} \alpha_j K(\cdot, \mathbf{x}_j) \right\rangle_{\mathcal{H}}$$

$$= \sum_{i,j=1}^{n} \alpha_i \alpha_j \langle K(\mathbf{x}_i, \cdot), K(\cdot, \mathbf{x}_j) \rangle_{\mathcal{H}}$$

$$= \sum_{i,j=1}^{n} \alpha_i \alpha_j K(\mathbf{x}_i, \mathbf{x}_j).$$

Theorem 3.2 (Complexity for linear-in-parameter model)   *Assume that there exists $C_\phi > 0$ such that $\|\boldsymbol{\phi}(\mathbf{x})\|_2 \leq C_\phi$ for all $\mathbf{x}$. Given $C_\alpha > 0$, define the linear-in-parameter function class as*

$$\mathscr{G}_{\mathrm{LIP}} := \{\mathbf{x} \mapsto \boldsymbol{\alpha}^\top \boldsymbol{\phi}(\mathbf{x}) \mid \|\boldsymbol{\alpha}\|_2 \leq C_\alpha\}, \tag{3.6}$$

where LIP is short for linear-in-parameter. Then, the Rademacher complexity of $\mathscr{G}_{\mathrm{LIP}}$ can be bounded by

$$\mathfrak{R}_{n,q}(\mathscr{G}_{\mathrm{LIP}}) \leq \frac{C_\phi C_\alpha}{\sqrt{n}}. \tag{3.7}$$

Proof   The theorem follows by replacing $\mathbf{x}$ with $\boldsymbol{\phi}(\mathbf{x})$ in theorem 3.1. $\square$

Theorem 3.3 (Complexity for kernel model)   *Assume that there exists $C_\phi > 0$ independent of n such that $\|\boldsymbol{\phi}(\mathbf{x}; S_n)\|_2 \leq C_\phi$ for all S_n and $\mathbf{x}$. Given $C_\alpha > 0$, define the kernel function class in dual form (KD) as*

$$\mathscr{G}_{\mathrm{KD}} := \{\mathbf{x} \mapsto \boldsymbol{\alpha}^\top \boldsymbol{\phi}(\mathbf{x}; S_n) \mid S_n \sim q^n, \|\boldsymbol{\alpha}\|_2 \leq C_\alpha\}. \tag{3.8}$$

The Rademacher complexity of $\mathscr{G}_{\mathrm{KD}}$ can be bounded by

$$\mathfrak{R}_{n,q}(\mathscr{G}_{\mathrm{KD}}) \leq \frac{C_\phi C_\alpha}{\sqrt{n}}. \tag{3.9}$$

Alternatively, assume that there exists $C_K > 0$ such that $K(\mathbf{x}, \mathbf{x}) \leq C_K^2$ for all $\mathbf{x}$. Given $C_w > 0$, define the kernel function class in primal form (KP) as

$$\mathscr{G}_{\mathrm{KP}} := \{\mathbf{x} \mapsto \langle \mathbf{w}, \varphi(\mathbf{x}) \rangle_{\mathscr{F}} \mid \|\mathbf{w}\|_{\mathscr{F}} \leq C_w\}. \tag{3.10}$$

The Rademacher complexity of $\mathscr{G}_{\mathrm{KP}}$ can be bounded by

$$\mathfrak{R}_{n,q}(\mathscr{G}_{\mathrm{KP}}) \leq \frac{C_K C_w}{\sqrt{n}}. \tag{3.11}$$

Proof   The theorem follows from that $\|\varphi(\mathbf{x})\|_{\mathscr{F}} = \sqrt{K(\mathbf{x}, \mathbf{x})} \leq C_K$. $\square$

The proofs of theorems 3.1 to 3.3 can be found in many books and papers (e.g., Mohri et al., 2012; Shalev-Shwartz and Ben-David, 2014; Bartlett and Mendelson, 2002; Kakade et al., 2009), and they are included here for illustrating the ideas about how to upper-bound the Rademacher complexities. The complexity bounds in three theorems are *norm-based*, i.e., they are evaluated based on the norm bounds C_x, C_ϕ, C_K, C_α, and C_w. Thus, the complexity bounds are also *size-independent*, i.e., they are completely independent of the dimensionality of the model parameters α and w. This nice characteristic is our desideratum for decoupling learning criteria and models in algorithm design as much as possible.

Finally, consider the neural network model

$$g(x) = \alpha^{(t)} \phi(\alpha^{(t-1)} \phi(\cdots \alpha^{(2)} \phi(\alpha^{(1)} x))),$$

where $\alpha^{(1)} \in \mathbb{R}^{b_2 \times d}$, $\alpha^{(2)} \in \mathbb{R}^{b_3 \times b_2}, \ldots, \alpha^{(t-1)} \in \mathbb{R}^{b_t \times b_{t-1}}$, $\alpha^{(t)} \in \mathbb{R}^{1 \times b_t}$ are the parameter matrices (also known as the weight matrices), the activation $\phi(z)$ of a vector z is applied element-wise into a vector of the activation $\phi(z)$ of every element z in z, and the biases $\alpha^{(1)}_{:,0}, \ldots, \alpha^{(t)}_{:,0}$ are omitted. We omit the biases in our theoretical analysis since they are not regularized in learning algorithms, and we cannot define bounded function classes without excluding them from the neural network model. Concerning Rademacher complexity bounds, there are two cases depending on the choice of the activation function ϕ.

The first case is that ϕ is a bounded function, which is easy to handle. Let us define a basis function vector to be the activation of the penultimate layer:

$$\boldsymbol{\phi}(x; \alpha^{(1)}, \ldots, \alpha^{(t-1)}) := \phi(\alpha^{(t-1)} \phi(\cdots \alpha^{(2)} \phi(\alpha^{(1)} x))) \in \mathbb{R}^{b_t},$$

where $\boldsymbol{\phi}$ denotes the basis in the left-hand side and ϕ denotes the activation in the right-hand side. Since ϕ is bounded and the dimensionality b_t is fixed, we can know that $\boldsymbol{\phi}$ must also be bounded and then we can apply theorem 3.2.

The second case is that ϕ is an unbounded function. Among the unbounded ϕ, to the best of our knowledge, the only possibility is *positive-homogeneous* ϕ, i.e., $\phi(az) = a\phi(z)$ for all $a \geq 0$ and $z \in \mathbb{R}$ (or any ϕ that can be lower- and upper-bounded by two positive-homogeneous functions). The set of positive-homogeneous functions includes all ReLU-like activation functions and hence enables various deep ReLU networks. For simplicity, we assume that ϕ has a Lipschitz constant 1, as 1 is the Lipschitz constant for all ReLU-like activation functions. Although there exist size-independent complexity bounds for deep ReLU networks, the complexity bounds and their assumptions are very complicated, and then they do not look similar to the size-independent complexity bounds for previous models. Here, we present a width-independent but a little depth-dependent complexity bound from Golowich et al. (2018), which is most similar to the previous complexity bounds. Its proof is sophisticated and thus omitted; see Golowich et al. (2018) for the proof and more discussions.

Theorem 3.4 (Complexity for neural network model)   *Let ϕ be a fixed 1-Lipschitz positive-homogeneous activation function. Assume that there exists $C_x > 0$ such that*

$\|x\|_2 \le C_x$ *for all* x. *Given* t *constants* $C_\alpha^{(1)} > 0, \ldots, C_\alpha^{(t)} > 0$, *define the neural network function class as*

$$\mathscr{G}_{\mathrm{NN}} := \{x \mapsto \alpha^{(t)} \phi(\ldots \phi(\alpha^{(1)} x)) \mid \|\alpha^{(1)}\|_F \le C_\alpha^{(1)}, \ldots, \|\alpha^{(t)}\|_F \le C_\alpha^{(t)}\}, \tag{3.12}$$

where $\|\cdot\|_F$ *denotes the Frobenius norm. Then, the Rademacher complexity of* $\mathscr{G}_{\mathrm{NN}}$ *can be bounded by*

$$\mathfrak{R}_{n,q}(\mathscr{G}_{\mathrm{NN}}) \le \frac{C_x(\sqrt{2\ln 2}\sqrt{t} + 1) \prod_{j=1}^{t} C_\alpha^{(j)}}{\sqrt{n}}. \tag{3.13}$$

Combining theorems 3.1 to 3.4, we could obtain the following meaningful corollary to be used in the subsequent analysis:

Corollary 3.5 (Summary of complexity bounds)   *Let* $\mathscr{G}$ *be a function class chosen from* $\mathscr{G}_{\mathrm{LII}}, \mathscr{G}_{\mathrm{LIP}}, \mathscr{G}_{\mathrm{KD}}, \mathscr{G}_{\mathrm{KP}},$ *and* $\mathscr{G}_{\mathrm{NN}}$. *Then, there exists* $C_{\mathscr{G}} > 0$ *such that*

$$\mathfrak{R}_{n,q}(\mathscr{G}) \le C_{\mathscr{G}}/\sqrt{n}. \tag{3.14}$$

Proof   According to the definitions in (3.4), (3.6), (3.8), (3.10), and (3.12), we have

$$C_{\mathscr{G}} = \begin{cases} C_x C_\alpha & (\mathscr{G} = \mathscr{G}_{\mathrm{LII}}), \\ C_\phi C_\alpha & (\mathscr{G} = \mathscr{G}_{\mathrm{LIP}}), \\ C_\phi C_\alpha & (\mathscr{G} = \mathscr{G}_{\mathrm{KD}}), \\ C_K C_w & (\mathscr{G} = \mathscr{G}_{\mathrm{KP}}), \\ C_x(\sqrt{2\ln 2}\sqrt{t} + 1) \prod_{j=1}^{t} C_\alpha^{(j)} & (\mathscr{G} = \mathscr{G}_{\mathrm{NN}}), \end{cases}$$

based on (3.5), (3.7), (3.9), (3.11), and (3.13). $\qquad\qquad\square$

3.1.2.5   Estimation error bounds

We have derived the complexity bounds for the models in section 2.1.2 and also summarized the bounds as corollary 3.5. From here on, we upper-bound the estimation error of PN classification.

As a common practice, we assume that within the data domain and function class under consideration, the surrogate loss has bounded values and bounded derivatives in order to apply the *method of bounded differences* (McDiarmid, 1989) and the contraction lemma of the Rademacher complexity.

Theorem 3.6 (Estimation error of PN classification)   *Given a function class* $\mathscr{G}$, *we assume that the surrogate loss* ℓ *has bounded values, i.e.,*

- *either there exists* $C_\ell > 0$ *such that* $\ell(\widehat{y}, y) \le C_\ell$ *for all* $\widehat{y} \in \mathbb{R}$ *and* $y \in \mathscr{Y}$, *or*

- *there exist* $C_g > 0$ *and* $C_\ell > 0$ *such that* $\|g\|_\infty = \sup_{x \in \mathscr{X}} |g(x)| \le C_g$ *for all* $g \in \mathscr{G}$ *and* $\ell(\widehat{y}, y) \le C_\ell$ *for all* $|\widehat{y}| \le C_g$ *and* $y \in \mathscr{Y}$.[3]

We also assume that ℓ has bounded derivatives, i.e., $\ell(\widehat{y}, y)$ is Lipschitz continuous in $\widehat{y}$ for all $\widehat{y} \in \mathbb{R}$ or $|\widehat{y}| \leq C_g$ and $y \in \mathcal{Y}$ with a Lipschitz constant L_ℓ. Then, for any $\delta > 0$, with probability at least $1 - \delta$,[4]

$$R(\widehat{g}_{\mathrm{PN}}) - R(g^*) \leq 4L_\ell \pi_{\mathrm{P}} \mathfrak{R}_{n_{\mathrm{P}}, p_{\mathrm{P}}}(\mathcal{G}) + 4L_\ell \pi_{\mathrm{N}} \mathfrak{R}_{n_{\mathrm{N}}, p_{\mathrm{N}}}(\mathcal{G})$$

$$+ \sqrt{2 \ln \frac{2}{\delta}} C_\ell \left(\frac{\pi_{\mathrm{P}}}{\sqrt{n_{\mathrm{P}}}} + \frac{\pi_{\mathrm{N}}}{\sqrt{n_{\mathrm{N}}}} \right). \tag{3.15}$$

Our estimation error bound relies on the uniform deviation bound below.

Lemma 3.7 (Uniform deviation of PN classification)   *Under the same assumptions as theorem 3.6, for any $\delta > 0$, with probability at least $1 - \delta$,[5]*

$$\sup_{g \in \mathcal{G}} |\widehat{R}_{\mathrm{PN}}(g) - R(g)| \leq 2L_\ell \pi_{\mathrm{P}} \mathfrak{R}_{n_{\mathrm{P}}, p_{\mathrm{P}}}(\mathcal{G}) + 2L_\ell \pi_{\mathrm{N}} \mathfrak{R}_{n_{\mathrm{N}}, p_{\mathrm{N}}}(\mathcal{G})$$

$$+ \sqrt{\frac{1}{2} \ln \frac{2}{\delta}} C_\ell \left(\frac{\pi_{\mathrm{P}}}{\sqrt{n_{\mathrm{P}}}} + \frac{\pi_{\mathrm{N}}}{\sqrt{n_{\mathrm{N}}}} \right). \tag{3.16}$$

Proof of lemma 3.7   We prove this upper bound for $\sup_{g \in \mathcal{G}} \widehat{R}_{\mathrm{PN}}(g) - R(g)$ and $\sup_{g \in \mathcal{G}} R(g) - \widehat{R}_{\mathrm{PN}}(g)$ with probability at least $1 - \delta/2$, separately.

Consider how much $\sup_{g \in \mathcal{G}} \widehat{R}_{\mathrm{PN}}(g) - R(g)$ changes with a data replacement operation (i.e., a data point for evaluating $\widehat{R}_{\mathrm{PN}}(g)$ gets replaced with another data point). Actually, $\sup_{g \in \mathcal{G}} \widehat{R}_{\mathrm{PN}}(g) - R(g)$ shares the same upper bounds of changes with $\widehat{R}_{\mathrm{PN}}(g)$. The change of $\widehat{R}_{\mathrm{PN}}(g)$ will be no more than $C_\ell \pi_{\mathrm{P}}/n_{\mathrm{P}}$ if some x_i^{P} is replaced, or no more than $C_\ell \pi_{\mathrm{N}}/n_{\mathrm{N}}$ if some x_i^{N} is replaced. Therefore, according to *McDiarmid's inequality* (McDiarmid, 1989), for any $\epsilon > 0$,

$$\Pr\{\sup_{g \in \mathcal{G}} \widehat{R}_{\mathrm{PN}}(g) - R(g) - \mathbb{E}[\sup_{g \in \mathcal{G}} \widehat{R}_{\mathrm{PN}}(g) - R(g)] \geq \epsilon\}$$

$$\leq \exp\left(-\frac{2\epsilon^2}{n_{\mathrm{P}}(C_\ell \pi_{\mathrm{P}}/n_{\mathrm{P}})^2 + n_{\mathrm{N}}(C_\ell \pi_{\mathrm{N}}/n_{\mathrm{N}})^2} \right)$$

$$= \exp(-2\epsilon^2/(C_\ell^2 \pi_{\mathrm{P}}^2/n_{\mathrm{P}} + C_\ell^2 \pi_{\mathrm{N}}^2/n_{\mathrm{N}})),$$

where uniform deviation and McDiarmid's inequality are both unidirectional. Let $\delta/2 = \exp(-2\epsilon^2/(C_\ell^2 \pi_{\mathrm{P}}^2/n_{\mathrm{P}} + C_\ell^2 \pi_{\mathrm{N}}^2/n_{\mathrm{N}}))$, then ϵ can be derived by solving this equation as

$$\epsilon = \sqrt{(1/2) \ln(2/\delta) C_\ell^2 (\pi_{\mathrm{P}}^2/n_{\mathrm{P}} + \pi_{\mathrm{N}}^2/n_{\mathrm{N}})}$$

$$\leq \sqrt{(1/2) \ln(2/\delta)} C_\ell (\pi_{\mathrm{P}}/\sqrt{n_{\mathrm{P}}} + \pi_{\mathrm{N}}/\sqrt{n_{\mathrm{N}}}),$$

due to the subadditivity of the square root. Hence, for any $\delta > 0$, with probability at least $1 - \delta/2$,

$$\sup_{g \in \mathcal{G}} \widehat{R}_{\mathrm{PN}}(g) - R(g) \leq \mathbb{E}[\sup_{g \in \mathcal{G}} \widehat{R}_{\mathrm{PN}}(g) - R(g)]$$

$$+ \sqrt{(1/2) \ln(2/\delta)} C_\ell (\pi_{\mathrm{P}}/\sqrt{n_{\mathrm{P}}} + \pi_{\mathrm{N}}/\sqrt{n_{\mathrm{N}}}),$$

which finishes our application of the method of bounded differences.

It remains to show that

$$\mathbb{E}[\sup_{g \in \mathcal{G}} \widehat{R}_{\mathrm{PN}}(g) - R(g)] \leq 2L_\ell \pi_{\mathrm{P}} \mathfrak{R}_{n_{\mathrm{P}}, p_{\mathrm{P}}}(\mathcal{G}) + 2L_\ell \pi_{\mathrm{N}} \mathfrak{R}_{n_{\mathrm{N}}, p_{\mathrm{N}}}(\mathcal{G}).$$

To this end, we resort to *symmetrization* (Vapnik, 1998), which is a technique in statistical learning theory. Notice that in $\mathbb{E}[\sup_{g \in \mathcal{G}} \widehat{R}_{\mathrm{PN}}(g) - R(g)]$, $\mathbb{E}$ is over the repeated sampling of $(\mathcal{X}_{\mathrm{P}}, \mathcal{X}_{\mathrm{N}})$ for evaluating $\widehat{R}_{\mathrm{PN}}(g)$; in other words, $\mathbb{E}$ and $\widehat{R}_{\mathrm{PN}}(g)$ can be explicitly written as $\mathbb{E}_{(\mathcal{X}_{\mathrm{P}}, \mathcal{X}_{\mathrm{N}})}$ and $\widehat{R}_{\mathrm{PN}}(g; \mathcal{X}_{\mathrm{P}}, \mathcal{X}_{\mathrm{N}})$. Then, suppose that $(\mathcal{X}'_{\mathrm{P}}, \mathcal{X}'_{\mathrm{N}})$ is a *ghost sample*, and

$$\mathbb{E}_{(\mathcal{X}_{\mathrm{P}}, \mathcal{X}_{\mathrm{N}})}[\sup_{g \in \mathcal{G}} \widehat{R}_{\mathrm{PN}}(g) - R(g)]$$

$$= \mathbb{E}_{(\mathcal{X}_{\mathrm{P}}, \mathcal{X}_{\mathrm{N}})} \left[\sup_{g \in \mathcal{G}} \widehat{R}_{\mathrm{PN}}(g) - \mathbb{E}_{(\mathcal{X}'_{\mathrm{P}}, \mathcal{X}'_{\mathrm{N}})} \widehat{R}_{\mathrm{PN}}(g) \right]$$

$$\leq \mathbb{E}_{(\mathcal{X}_{\mathrm{P}}, \mathcal{X}_{\mathrm{N}}),(\mathcal{X}'_{\mathrm{P}}, \mathcal{X}'_{\mathrm{N}})}[\sup_{g \in \mathcal{G}} \widehat{R}_{\mathrm{PN}}(g; \mathcal{X}_{\mathrm{P}}, \mathcal{X}_{\mathrm{N}}) - \widehat{R}_{\mathrm{PN}}(g; \mathcal{X}'_{\mathrm{P}}, \mathcal{X}'_{\mathrm{N}})],$$

where we applied *Jensen's inequality* since the supremum is convex. Next, by decomposing the difference

$$\widehat{R}_{\mathrm{PN}}(g; \mathcal{X}_{\mathrm{P}}, \mathcal{X}_{\mathrm{N}}) - \widehat{R}_{\mathrm{PN}}(g; \mathcal{X}'_{\mathrm{P}}, \mathcal{X}'_{\mathrm{N}})$$

$$= \pi_{\mathrm{P}} \widehat{R}_{\mathrm{P}}(g; \mathcal{X}_{\mathrm{P}}) + \pi_{\mathrm{N}} \widehat{R}_{\mathrm{N}}(g; \mathcal{X}_{\mathrm{N}}) - \pi_{\mathrm{P}} \widehat{R}_{\mathrm{P}}(g; \mathcal{X}'_{\mathrm{P}}) - \pi_{\mathrm{N}} \widehat{R}_{\mathrm{N}}(g; \mathcal{X}'_{\mathrm{N}})$$

$$= \frac{\pi_{\mathrm{P}}}{n_{\mathrm{P}}} \left(\sum_{i=1}^{n_{\mathrm{P}}} \ell(g(x_i^{\mathrm{P}}), +1) - \ell(g(x_i^{\mathrm{P}'}), +1) \right)$$

$$+ \frac{\pi_{\mathrm{N}}}{n_{\mathrm{N}}} \left(\sum_{i=1}^{n_{\mathrm{N}}} \ell(g(x_i^{\mathrm{N}}), -1) - \ell(g(x_i^{\mathrm{N}'}), -1) \right),$$

we can know that

$$\mathbb{E}_{(\mathcal{X}_{\mathrm{P}}, \mathcal{X}_{\mathrm{N}}),(\mathcal{X}'_{\mathrm{P}}, \mathcal{X}'_{\mathrm{N}})}[\sup_{g \in \mathcal{G}} \widehat{R}_{\mathrm{PN}}(g; \mathcal{X}_{\mathrm{P}}, \mathcal{X}_{\mathrm{N}}) - \widehat{R}_{\mathrm{PN}}(g; \mathcal{X}'_{\mathrm{P}}, \mathcal{X}'_{\mathrm{N}})]$$

$$\leq \frac{\pi_{\mathrm{P}}}{n_{\mathrm{P}}} \mathbb{E}_{\mathcal{X}_{\mathrm{P}}, \mathcal{X}'_{\mathrm{P}}} \left[\sup_{g \in \mathcal{G}} \sum_{i=1}^{n_{\mathrm{P}}} \ell(g(x_i^{\mathrm{P}}), +1) - \ell(g(x_i^{\mathrm{P}'}), +1) \right]$$

$$+ \frac{\pi_{\mathrm{N}}}{n_{\mathrm{N}}} \mathbb{E}_{\mathcal{X}_{\mathrm{N}}, \mathcal{X}'_{\mathrm{N}}} \left[\sup_{g \in \mathcal{G}} \sum_{i=1}^{n_{\mathrm{N}}} \ell(g(x_i^{\mathrm{N}}), -1) - \ell(g(x_i^{\mathrm{N}'}), -1) \right],$$

since the supremum is subadditive.

Let us take a look at the distribution of $\ell(g(x_i^{\mathrm{P}}),+1) - \ell(g(x_i^{\mathrm{P}\prime}),+1)$. Since $x_i^{\mathrm{P}\prime}$ is a ghost data point of x_i^{P}, the difference $\ell(g(x_i^{\mathrm{P}}),+1) - \ell(g(x_i^{\mathrm{P}\prime}),+1)$ and the difference $\ell(g(x_i^{\mathrm{P}\prime}),+1) - \ell(g(x_i^{\mathrm{P}}),+1)$ share the distribution. Indeed, the same distribution is followed by $\sigma_i(\ell(g(x_i^{\mathrm{P}\prime}),+1) - \ell(g(x_i^{\mathrm{P}}),+1))$, where σ_i is an additional Rademacher variable, due to the symmetry of x_i^{P} and $x_i^{\mathrm{P}\prime}$ and the symmetry of σ_i itself, i.e.,

$$p(x_i^{\mathrm{P}}) = p(x_i^{\mathrm{P}\prime}) = p_{\mathrm{P}}(x), \quad \Pr(\sigma_i = +1) = \Pr(\sigma_i = -1) = 1/2.$$

Consequently, denoting by $L_{2,n_{\mathrm{P}}} = \sum_{i=2}^{n_{\mathrm{P}}} \ell(g(x_i^{\mathrm{P}}),+1) - \ell(g(x_i^{\mathrm{P}\prime}),+1)$,

$$\mathbb{E}_{\mathscr{X}_{\mathrm{P}},\mathscr{X}_{\mathrm{P}}'}\left[\sup_{g\in\mathscr{G}} \sum_{i=1}^{n_{\mathrm{P}}} \ell(g(x_i^{\mathrm{P}}),+1) - \ell(g(x_i^{\mathrm{P}\prime}),+1)\right]$$

$$= \mathbb{E}_{\mathscr{X}_{\mathrm{P}},\mathscr{X}_{\mathrm{P}}'}[\sup_{g\in\mathscr{G}} \ell(g(x_1^{\mathrm{P}}),+1) - \ell(g(x_1^{\mathrm{P}\prime}),+1) + L_{2,n_{\mathrm{P}}}]$$

$$= \frac{1}{2}\,\mathbb{E}_{\mathscr{X}_{\mathrm{P}},\mathscr{X}_{\mathrm{P}}'}[\sup_{g\in\mathscr{G}} \ell(g(x_1^{\mathrm{P}}),+1) - \ell(g(x_1^{\mathrm{P}\prime}),+1) + L_{2,n_{\mathrm{P}}}]$$

$$+ \frac{1}{2}\,\mathbb{E}_{\mathscr{X}_{\mathrm{P}},\mathscr{X}_{\mathrm{P}}'}[\sup_{g\in\mathscr{G}} \ell(g(x_1^{\mathrm{P}\prime}),+1) - \ell(g(x_1^{\mathrm{P}}),+1) + L_{2,n_{\mathrm{P}}}]$$

$$= \mathbb{E}_{\sigma_1,\mathscr{X}_{\mathrm{P}},\mathscr{X}_{\mathrm{P}}'}[\sup_{g\in\mathscr{G}} \sigma_1(\ell(g(x_1^{\mathrm{P}}),+1) - \ell(g(x_1^{\mathrm{P}\prime}),+1)) + L_{2,n_{\mathrm{P}}}].$$

By repeating this step n_{P} times, we obtain

$$\mathbb{E}_{\mathscr{X}_{\mathrm{P}},\mathscr{X}_{\mathrm{P}}'}\left[\sup_{g\in\mathscr{G}} \sum_{i=1}^{n_{\mathrm{P}}} \ell(g(x_i^{\mathrm{P}}),+1) - \ell(g(x_i^{\mathrm{P}\prime}),+1)\right]$$

$$= \mathbb{E}_{\sigma,\mathscr{X}_{\mathrm{P}},\mathscr{X}_{\mathrm{P}}'}\left[\sup_{g\in\mathscr{G}} \sum_{i=1}^{n_{\mathrm{P}}} \sigma_i(\ell(g(x_i^{\mathrm{P}}),+1) - \ell(g(x_i^{\mathrm{P}\prime}),+1))\right],$$

where $\sigma = (\sigma_1,\ldots,\sigma_{n_{\mathrm{P}}})$. Subsequently,

$$\mathbb{E}_{\sigma,\mathscr{X}_{\mathrm{P}},\mathscr{X}_{\mathrm{P}}'}\left[\sup_{g\in\mathscr{G}} \sum_{i=1}^{n_{\mathrm{P}}} \sigma_i(\ell(g(x_i^{\mathrm{P}}),+1) - \ell(g(x_i^{\mathrm{P}\prime}),+1))\right]$$

$$\leq \mathbb{E}_{\sigma,\mathscr{X}_{\mathrm{P}}}\left[\sup_{g\in\mathscr{G}} \sum_{i=1}^{n_{\mathrm{P}}} \sigma_i\ell(g(x_i^{\mathrm{P}}),+1)\right] + \mathbb{E}_{\sigma,\mathscr{X}_{\mathrm{P}}'}\left[\sup_{g\in\mathscr{G}} \sum_{i=1}^{n_{\mathrm{P}}} -\sigma_i\ell(g(x_i^{\mathrm{P}\prime}),+1)\right]$$

$$= 2\,\mathbb{E}_{\sigma,\mathscr{X}_{\mathrm{P}}}\left[\sup_{g\in\mathscr{G}} \sum_{i=1}^{n_{\mathrm{P}}} \sigma_i\ell(g(x_i^{\mathrm{P}}),+1)\right],$$

where the inequality is due to the supremum being subadditive, and the equality is due to that $\mathscr{X}_{\mathrm{P}}'$ shares the distribution with $\mathscr{X}_{\mathrm{P}}$ and $-\sigma_i$ shares the distribution with σ_i.

To sum up, we have proved that

$$\frac{\pi_{\mathrm{P}}}{n_{\mathrm{P}}}\,\mathbb{E}_{\mathcal{X}_{\mathrm{P}},\mathcal{X}_{\mathrm{P}}'}\left[\sup_{g\in\mathcal{G}}\sum_{i=1}^{n_{\mathrm{P}}}\ell(g(x_i^{\mathrm{P}}),+1)-\ell(g(x_i^{\mathrm{P}'}),+1)\right]$$

$$\leq 2\pi_{\mathrm{P}}\,\mathbb{E}_{\sigma,\mathcal{X}_{\mathrm{P}}}\left[\sup_{g\in\mathcal{G}}\frac{1}{n_{\mathrm{P}}}\sum_{i=1}^{n_{\mathrm{P}}}\sigma_i\ell(g(x_i^{\mathrm{P}}),+1)\right]$$

$$= 2\pi_{\mathrm{P}}\mathfrak{R}_{n_{\mathrm{P}},p_{\mathrm{P}}}(\ell(\cdot,+1)\circ\mathcal{G}).$$

According to the contraction lemma of the Rademacher complexity,

$$2\pi_{\mathrm{P}}\mathfrak{R}_{n_{\mathrm{P}},p_{\mathrm{P}}}(\ell(\cdot,+1)\circ\mathcal{G})\leq 2\pi_{\mathrm{P}}L_{\ell}\mathfrak{R}_{n_{\mathrm{P}},p_{\mathrm{P}}}(\mathcal{G}).$$

In the same way, we can prove that

$$\frac{\pi_{\mathrm{N}}}{n_{\mathrm{N}}}\,\mathbb{E}_{\mathcal{X}_{\mathrm{N}},\mathcal{X}_{\mathrm{N}}'}\left[\sup_{g\in\mathcal{G}}\sum_{i=1}^{n_{\mathrm{N}}}\ell(g(x_i^{\mathrm{N}}),-1)-\ell(g(x_i^{\mathrm{N}'}),-1)\right]\leq 2\pi_{\mathrm{N}}L_{\ell}\mathfrak{R}_{n_{\mathrm{N}},p_{\mathrm{N}}}(\mathcal{G}),$$

which means

$$\mathbb{E}[\sup_{g\in\mathcal{G}}\widehat{R}_{\mathrm{PN}}(g)-R(g)]\leq 2L_{\ell}\pi_{\mathrm{P}}\mathfrak{R}_{n_{\mathrm{P}},p_{\mathrm{P}}}(\mathcal{G})+2L_{\ell}\pi_{\mathrm{N}}\mathfrak{R}_{n_{\mathrm{N}},p_{\mathrm{N}}}(\mathcal{G}),$$

and with probability at least $1-\delta/2$,

$$\sup_{g\in\mathcal{G}}\widehat{R}_{\mathrm{PN}}(g)-R(g)\leq 2L_{\ell}\pi_{\mathrm{P}}\mathfrak{R}_{n_{\mathrm{P}},p_{\mathrm{P}}}(\mathcal{G})+2L_{\ell}\pi_{\mathrm{N}}\mathfrak{R}_{n_{\mathrm{N}},p_{\mathrm{N}}}(\mathcal{G})$$

$$+\sqrt{(1/2)\ln(2/\delta)}C_{\ell}(\pi_{\mathrm{P}}/\sqrt{n_{\mathrm{P}}}+\pi_{\mathrm{N}}/\sqrt{n_{\mathrm{N}}}).$$

Similarly, when considering the other direction, namely the upper bound for $\sup_{g\in\mathcal{G}}R(g)-\widehat{R}_{\mathrm{PN}}(g)$, it would be that with probability at least $1-\delta/2$,

$$\sup_{g\in\mathcal{G}}R(g)-\widehat{R}_{\mathrm{PN}}(g)\leq \mathbb{E}[\sup_{g\in\mathcal{G}}R(g)-\widehat{R}_{\mathrm{PN}}(g)]$$

$$+\sqrt{(1/2)\ln(2/\delta)}C_{\ell}(\pi_{\mathrm{P}}/\sqrt{n_{\mathrm{P}}}+\pi_{\mathrm{N}}/\sqrt{n_{\mathrm{N}}})$$

$$\leq 2L_{\ell}\pi_{\mathrm{P}}\mathfrak{R}_{n_{\mathrm{P}},p_{\mathrm{P}}}(\mathcal{G})+2L_{\ell}\pi_{\mathrm{N}}\mathfrak{R}_{n_{\mathrm{N}},p_{\mathrm{N}}}(\mathcal{G})$$

$$+\sqrt{(1/2)\ln(2/\delta)}C_{\ell}(\pi_{\mathrm{P}}/\sqrt{n_{\mathrm{P}}}+\pi_{\mathrm{N}}/\sqrt{n_{\mathrm{N}}}).$$

The two directions hold simultaneously with probability at least $1-\delta$, which completes the proof of the upper bound for $\sup_{g\in\mathcal{G}}|\widehat{R}_{\mathrm{PN}}(g)-R(g)|$. $\square$

Proof of theorem 3.6   Based on lemma 3.7, the estimation error bound in (3.15) is simply proven through

$$R(\widehat{g}_{\mathrm{PN}}) - R(g^*)$$

$$= \left(R(\widehat{g}_{\mathrm{PN}}) - \widehat{R}_{\mathrm{PN}}(\widehat{g}_{\mathrm{PN}})\right) + \left(\widehat{R}_{\mathrm{PN}}(\widehat{g}_{\mathrm{PN}}) - \widehat{R}_{\mathrm{PN}}(g^*)\right) + \left(\widehat{R}_{\mathrm{PN}}(g^*) - R(g^*)\right)$$

$$\leq \sup_{g \in \mathcal{G}} |R(g) - \widehat{R}_{\mathrm{PN}}(g)| + 0 + \sup_{g \in \mathcal{G}} |\widehat{R}_{\mathrm{PN}}(g) - R(g)|$$

$$\leq 4 L_\ell \pi_{\mathrm{P}} \mathfrak{R}_{n_{\mathrm{P}}, p_{\mathrm{P}}}(\mathcal{G}) + 4 L_\ell \pi_{\mathrm{N}} \mathfrak{R}_{n_{\mathrm{N}}, p_{\mathrm{N}}}(\mathcal{G})$$

$$+ \sqrt{2 \ln(2/\delta)} C_\ell (\pi_{\mathrm{P}}/\sqrt{n_{\mathrm{P}}} + \pi_{\mathrm{N}}/\sqrt{n_{\mathrm{N}}}),$$

where $\widehat{R}_{\mathrm{PN}}(\widehat{g}_{\mathrm{PN}}) \leq \widehat{R}_{\mathrm{PN}}(g^*)$ by the definition of $\widehat{g}_{\mathrm{PN}}$ as the empirical risk minimizer, and the last inequality holds with probability at least $1 - \delta$. $\qquad\qquad\square$

By plugging corollary 3.5 into theorem 3.6, we can simplify theorem 3.6 as follows. Let $\mathcal{G}$ be a function class (corresponding to a model in section 2.1.2) with $C_{\mathcal{G}} > 0$ such that $\mathfrak{R}_{n_{\mathrm{P}}, p_{\mathrm{P}}}(\mathcal{G}) \leq C_{\mathcal{G}}/\sqrt{n_{\mathrm{P}}}$ and $\mathfrak{R}_{n_{\mathrm{N}}, p_{\mathrm{N}}}(\mathcal{G}) \leq C_{\mathcal{G}}/\sqrt{n_{\mathrm{N}}}$. For any $\delta > 0$, define

$$C_\delta := 4 L_\ell C_{\mathcal{G}} + \sqrt{2 \ln(2/\delta)} C_\ell,$$

and then with probability at least $1 - \delta$,

$$R(\widehat{g}_{\mathrm{PN}}) - R(g^*) \leq C_\delta (\pi_{\mathrm{P}}/\sqrt{n_{\mathrm{P}}} + \pi_{\mathrm{N}}/\sqrt{n_{\mathrm{N}}}). \tag{3.17}$$

Equation (3.17) establishes the consistency of PN classification, i.e., the weak consistency of $\widehat{g}_{\mathrm{PN}}$. The convergence rate (from $R(\widehat{g}_{\mathrm{PN}})$ to $R(g^*)$) is $\mathcal{O}_p(1/\sqrt{n_{\mathrm{P}}} + 1/\sqrt{n_{\mathrm{N}}})$, where $\mathcal{O}_p$ denotes the *order in probability*.

Note that the above estimation error bound has been proved based on a uniform deviation bound. Therefore, even with small regularization, the same estimation error bound still holds (Mohri et al., 2012; Shalev-Shwartz and Ben-David, 2014).

3.2   Multi-Class Classification

Next, we extend the PN classification method to c-class scenarios for $c > 2$.

3.2.1   Formulation

Similarly, there are two different cases of the formulation, depending on how the data of different classes are drawn. The first case uses a set of data drawn from all classes, and the second case uses c sets of data, each drawn from one class. We will focus on the second case because of its generality.

Specifically, we make use of training data from each class:

$$\mathcal{X}_1 := \{x_i^{(1)}\}_{i=1}^{n_1} \overset{\text{i.i.d.}}{\sim} p_1(x) := p(x \mid y = 1),$$

$$\vdots$$

$$\mathcal{X}_c := \{x_i^{(c)}\}_{i=1}^{n_c} \overset{\text{i.i.d.}}{\sim} p_c(x) := p(x \mid y = c).$$

We can see that $\mathscr{X}_1, \ldots, \mathscr{X}_c$ are sampled separately as c independent sets and their sample sizes $n_1, \ldots, n_c$ are c independent variables, and thus $\mathscr{X}_1, \ldots, \mathscr{X}_c$ can again be collected in different ways from different sources.

The goal is to obtain a classifier that minimizes the classification risk (here, everything is c-class, including the classifier, loss, and risk):

$$R(\boldsymbol{g}) := \mathbb{E}_{p(\boldsymbol{x},y)}[\mathscr{L}(\boldsymbol{g}(\boldsymbol{x}),y)],$$

where $\mathscr{L} : \mathbb{R}^c \times \mathscr{Y} \to \mathbb{R}$ is a properly chosen loss function (see section 2.2.2). In the c-sample case, $R(\boldsymbol{g})$ can be decomposed into

$$R(\boldsymbol{g}) := \sum_{y=1}^{c} \pi_y R_y(\boldsymbol{g}),$$

where, for $y = 1, \ldots, c$,

$$\pi_y := p(y)$$

is the class-prior probability for class y and

$$R_y(\boldsymbol{g}) := \mathbb{E}_{p_y(\boldsymbol{x})}[\mathscr{L}(\boldsymbol{g}(\boldsymbol{x}),y)]$$

is the (partial) risk for class y.

Subsequently, for $y = 1, \ldots, c$, $R_y(\boldsymbol{g})$ can be approximated using $\mathscr{X}_y$ as

$$\widehat{R}_y(\boldsymbol{g}) := \frac{1}{n_y} \sum_{i=1}^{n_y} \mathscr{L}(\boldsymbol{g}(\boldsymbol{x}_i^{(y)}),y),$$

and $R(\boldsymbol{g})$ can be empirically approximated as

$$\widehat{R}(\boldsymbol{g}) := \sum_{y=1}^{c} \pi_y \widehat{R}_y(\boldsymbol{g}). \tag{3.18}$$

This $\widehat{R}(\boldsymbol{g})$ is the empirical risk estimator for multi-class classification. Finally, a minimizer of $\widehat{R}(\boldsymbol{g})$ over a properly chosen function class $\mathscr{G}^c$ is obtained as

$$\widehat{\boldsymbol{g}} := \underset{\boldsymbol{g} \in \mathscr{G}^c}{\operatorname{argmin}} \, \widehat{R}(\boldsymbol{g}),$$

where any of the regularizers is omitted for simplicity. Note that the function class $\mathscr{G}^c$ means the function classes of $g_1, \ldots, g_c$ are the same $\mathscr{G}$. This $\widehat{\boldsymbol{g}}$ is the empirical risk minimizer for multi-class classification.

Because we rely on the c-sample empirical risk estimator/minimizer where $\mathscr{X}_1, \ldots, \mathscr{X}_c$ are independent sets and $n_1, \ldots, n_c$ are independent variables, we cannot estimate $\pi_1, \ldots, \pi_c$ from $\mathscr{X}_1, \ldots, \mathscr{X}_c$. As in PN classification, it is not difficult to know them by asking domain experts or by estimating them from a set of labeled/unlabeled validation data.

3.2.2   Theoretical Analysis

This subsection establishes the theoretical guarantee of multi-class classification. Since the contents here are mathematically highly advanced, those who are interested only in conceptual and algorithmic issues may skip section 3.2.2.

3.2.2.1   Estimation error bounds

In order to establish the theoretical guarantee of multi-class classification, let us observe $\widehat{R}_{\mathrm{PN}}(g)$ in (3.2) and $\widehat{R}(g)$ in (3.18). There are two terms in $\widehat{R}_{\mathrm{PN}}(g)$, where $\widehat{R}_{\mathrm{P}}(g)$ and $\widehat{R}_{\mathrm{N}}(g)$ converge to $R_{\mathrm{P}}(g)$ and $R_{\mathrm{N}}(g)$. Then, there are c terms in $\widehat{R}(\boldsymbol{g})$, and we hope that each $\widehat{R}_y(\boldsymbol{g})$ converges to the corresponding $R_y(\boldsymbol{g})$, respectively. We prove the convergence properties of the empirical risk minimizer $\widehat{\boldsymbol{g}}$ in the rest of this chapter.

First of all, we have the following theorem as a straightforward extension of theorem 3.6:

Theorem 3.8 (Estimation error of multi-class classification)   *Under the same assumptions as theorem 3.6, suppose that there exists $C_{\mathscr{L}} > 0$ such that $\mathscr{L}(\boldsymbol{g}(\boldsymbol{x}),y) \leq C_{\mathscr{L}}$, and there exists $L_{\mathscr{L}} > 0$ such that $\mathfrak{R}_{n_y,p_y}(\mathscr{L}(\cdot,y) \circ \mathscr{G}^c) \leq L_{\mathscr{L}} \mathfrak{R}_{n_y,p_y}(\mathscr{G})$.[6] Then, for any $\delta > 0$, with probability at least $1 - \delta$,*

$$R(\widehat{\boldsymbol{g}}) - R(\boldsymbol{g}^*) \leq \sum_{y=1}^{c} 4L_{\mathscr{L}}\pi_y \mathfrak{R}_{n_y,p_y}(\mathscr{G}) + \sqrt{2\ln\frac{2}{\delta}}C_{\mathscr{L}}\left(\sum_{y=1}^{c}\frac{\pi_y}{\sqrt{n_y}}\right), \tag{3.19}$$

where $\boldsymbol{g}^* := \mathrm{argmin}_{\boldsymbol{g}\in\mathscr{G}^c} R(\boldsymbol{g})$ is the true risk minimizer.

In the above theorem, the constants $C_{\mathscr{L}}$ and $L_{\mathscr{L}}$ appear. Here, let us express them explicitly to the three surrogate losses introduced in section 2.2.2—that is, the *one-versus-the-rest* (OVR) loss, the *pairwise comparison* (PC) loss, and the *softmax cross-entropy* (CE) loss. Since the OVR and PC losses are defined based on ℓ, we can upper-bound them and their derivatives under the assumptions of theorem 3.6. Indeed, we can do the same for the CE loss under those assumptions.

Let C_ℓ be the norm bound of ℓ and C_{g} be the norm bound of g_y. Then we can immediately express $C_{\mathscr{L}}$ as

$$C_{\mathscr{L}} = \begin{cases} 2C_\ell & \text{(for the OVR loss)}, \\ (c-1)C_\ell & \text{(for the PC loss)}, \\ 2C_{\mathrm{g}} + \ln c & \text{(for the CE loss)}. \end{cases}$$

On the other hand, $L_{\mathscr{L}}$ is more involved. Since L_ℓ was defined as a Lipschitz constant of ℓ, $L_{\mathscr{L}}$ in theorem 3.8 may look like a Lipschitz constant of $\mathscr{L}$. However, it is actually the Lipschitz constant for $g_1,\ldots,g_c$ uniformly rather than $\boldsymbol{g}$. When we talk about the Lipschitz continuity of ℓ, we mean $\ell(g(\boldsymbol{x}),y)$ is Lipschitz continuous in $g(\boldsymbol{x})$. However, the Lipschitz continuity of $\mathscr{L}(\boldsymbol{g}(\boldsymbol{x}),y)$ in $\boldsymbol{g}(\boldsymbol{x})$ is not especially useful, as $\mathscr{L}(\boldsymbol{g}(\boldsymbol{x}),y)$ is expressed in

$g_y(\boldsymbol{x})$ and $g_{y'}(\boldsymbol{x})$ instead of $\boldsymbol{g}(\boldsymbol{x})$ as a whole. Furthermore, $\boldsymbol{g}(\boldsymbol{x})$ is vector-valued and $\mathfrak{R}(\mathcal{G}^c)$ is not well defined. As a consequence, we will upper-bound in the following theorem the composed complexity $\mathfrak{R}(\mathcal{L}(\cdot,y)\circ\mathcal{G}^c)$ through the composed complexity $\mathfrak{R}(\ell\circ\mathcal{G})$ without using the Lipschitz continuity of $\mathcal{L}$.

Theorem 3.9 (Composed complexities) *Fix $y\in\{1,\ldots,c\}$. Let*

$$L_{\mathcal{L}} = \begin{cases} 2L_\ell & \textit{(for the OVR loss),} \\ 2(c-1)L_\ell & \textit{(for the PC loss),} \\ c & \textit{(for the CE loss),} \end{cases}$$

where L_ℓ is the Lipschitz constant of ℓ. Then, we have

$$\mathfrak{R}_{n_y,p_y}(\mathcal{L}(\cdot,y)\circ\mathcal{G}^c) \le L_{\mathcal{L}}\mathfrak{R}_{n_y,p_y}(\mathcal{G}).$$

Proof First, for the OVR loss, we have

$$\mathfrak{R}_{n_y,p_y}(\mathcal{L}(\cdot,y)\circ\mathcal{G}^c)$$

$$= \mathbb{E}_{\boldsymbol{\sigma},\mathscr{X}_y}\left[\sup_{\boldsymbol{g}\in\mathcal{G}^c}\frac{1}{n_y}\sum_{i=1}^{n_y}\sigma_i\mathcal{L}(\boldsymbol{g}(\boldsymbol{x}_i^{(y)}),y)\right]$$

$$= \mathbb{E}_{\boldsymbol{\sigma},\mathscr{X}_y}\left[\sup_{\boldsymbol{g}\in\mathcal{G}^c}\frac{1}{n_y}\sum_{i=1}^{n_y}\sigma_i\left(\ell(g_y(\boldsymbol{x}_i^{(y)}),+1)+\frac{1}{c-1}\sum_{y'\ne y}\ell(g_{y'}(\boldsymbol{x}_i^{(y)}),-1)\right)\right]$$

$$\le \mathfrak{R}_{n_y,p_y}(\ell(\cdot,+1)\circ\mathcal{G})+\frac{1}{c-1}\sum_{y'\ne y}\mathfrak{R}_{n_y,p_y}(\ell(\cdot,-1)\circ\mathcal{G})$$

$$\le 2L_\ell\mathfrak{R}_{n_y,p_y}(\mathcal{G}),$$

where the first inequality is due to the subadditivity of the supremum, and the second inequality is due to the Lipschitz continuity of ℓ.

Second, for the PC loss, we have

$$\mathfrak{R}_{n_y,p_y}(\mathcal{L}(\cdot,y)\circ\mathcal{G}^c)$$

$$= \mathbb{E}_{\boldsymbol{\sigma},\mathscr{X}_y}\left[\sup_{\boldsymbol{g}\in\mathcal{G}^c}\frac{1}{n_y}\sum_{i=1}^{n_y}\sigma_i\sum_{y'\ne y}\ell\left(g_y(\boldsymbol{x}_i^{(y)})-g_{y'}(\boldsymbol{x}_i^{(y)})\right)\right]$$

$$\le \sum_{y'\ne y}L_\ell\,\mathbb{E}_{\boldsymbol{\sigma},\mathscr{X}_y}\left[\sup_{g_y,g_{y'}\in\mathcal{G}}\frac{1}{n_y}\sum_{i=1}^{n_y}\sigma_i\left(g_y(\boldsymbol{x}_i^{(y)})-g_{y'}(\boldsymbol{x}_i^{(y)})\right)\right]$$

$$\leq \sum_{y' \neq y} L_\ell (\mathfrak{R}_{n_y, p_y}(\mathcal{G}) + \mathfrak{R}_{n_y, p_y}(\mathcal{G}))$$

$$= 2(c-1) L_\ell \mathfrak{R}_{n_y, p_y}(\mathcal{G}),$$

where the second inequality is because $\sigma_i(-g_{y'}(x_i^{(y)}))$ and $\sigma_i g_{y'}(x_i^{(y)})$ have the same distribution and g_y and $g_{y'}$ also have the same function class.

Finally, for the CE loss, note that

$$\frac{\partial \mathcal{L}(g(x), y)}{\partial g_{y'}(x)} = \begin{cases} -1 + \dfrac{\exp(g_y(x))}{\sum_{z=1}^{c} \exp(g_z(x))} & (y' = y), \\[4mm] \dfrac{\exp(g_{y'}(x))}{\sum_{z=1}^{c} \exp(g_z(x))} & (y' \neq y). \end{cases}$$

Both of them can be bounded between -1 and $+1$, which implies that $\mathcal{L}(g(x), y)$ is 1-Lipschitz in $g_y(x)$ and $g_{y'}(x)$. As a result,

$$\mathfrak{R}_{n_y, p_y}(\mathcal{L}(\cdot, y) \circ \mathcal{G}^c)$$

$$= \mathbb{E}_{\sigma, \mathcal{X}_y} \left[\sup_{g \in \mathcal{G}^c} \frac{1}{n_y} \sum_{i=1}^{n_y} \sigma_i \mathcal{L}(g(x_i^{(y)}), y) \right]$$

$$\leq \mathbb{E}_{\sigma, \mathcal{X}_y} \left[\sup_{g \in \mathcal{G}^c} \frac{1}{n_y} \sum_{i=1}^{n_y} \sigma_i \left(\sum_{y'=1}^{c} g_{y'}(x_i^{(y)}) \right) \right]$$

$$\leq c \mathfrak{R}_{n_y, p_y}(\mathcal{G}),$$

where the first inequality is due to the contraction lemma of the Rademacher complexity (see the proof of lemma 4.2 in Mohri et al. (2012) for details). □

In theorem 3.9, it is critical to fix y; otherwise, the corresponding $L_{\mathcal{L}}$ may become larger. For example, in the OVR loss, the class y has weight 1, and the classes y' have weight $1/(c-1)$. If we work on $\mathfrak{R}_{n, p(x,y)}(\mathcal{L} \circ \mathcal{G}^c)$, its upper bound would be $c L_\ell \mathfrak{R}_{n, p(x)}(\mathcal{G})$ but not $2 L_\ell \mathfrak{R}_{n, p(x)}(\mathcal{G})$, as every class may be the one with the larger weight 1. This is another reason why we focus on the c-sample empirical risk estimator/ minimizer.

By plugging corollary 3.5 into theorem 3.8, we can simplify theorem 3.8 as follows. Once more, let $\mathcal{G}$ be a function class (corresponding to a model in section 2.1.2) with $C_{\mathcal{G}} > 0$, satisfying for $y = 1, \ldots, c$, $\mathfrak{R}_{n_y, p_y}(\mathcal{G}) \leq C_{\mathcal{G}} / \sqrt{n_y}$. For any $\delta > 0$, define

$$C_\delta = 4 L_{\mathcal{L}} C_{\mathcal{G}} + \sqrt{2 \ln(2/\delta)} C_{\mathcal{L}},$$

and then with probability at least $1 - \delta$,

$$R(\widehat{\boldsymbol{g}}) - R(\boldsymbol{g}^*) \leq C_\delta \sum_{y=1}^{c} \frac{\pi_y}{\sqrt{n_y}}. \tag{3.20}$$

Equation (3.20) establishes the consistency of multi-class classification, with convergence rate $\mathcal{O}_p(\sum_{y=1}^{c} 1/\sqrt{n_y})$.

At first glance, the convergence rate above seems slower than $\mathcal{O}_p(1/\sqrt{n})$, as $\mathcal{O}_p(\sum_{y=1}^{c} 1/\sqrt{n_y}) = \mathcal{O}_p(1/\sqrt{\min_y n_y})$. Nevertheless, remember that $n_1, \ldots, n_c$ are independent variables, and we can explicitly control their balance by more effectively allocating the resources for data collection to different classes. In fact, $\mathcal{O}_p(1/\sqrt{n_y})$ is exactly the convergence rate from $R_y(\widehat{\boldsymbol{g}})$ to $R_y(\boldsymbol{g}^*)$, and if we expect $R_1(\boldsymbol{g}^*), \ldots, R_c(\boldsymbol{g}^*)$ to have similar values (e.g., tiny numbers close to zero), $R_1(\widehat{\boldsymbol{g}}), \ldots, R_c(\widehat{\boldsymbol{g}})$ should ideally also have similar values. Thus, if we find that some $R_y(\widehat{\boldsymbol{g}})$ is remarkably larger than the average, we should allocate more resources (if we still have some now or will have some later) to class y.

Our analysis can cover the one-sample empirical risk estimator/minimizer, where the data collection is basically labeling according to $p(y \mid \boldsymbol{x})$ instead of sampling according to $p(\boldsymbol{x} \mid y)$, so that $n_1, \ldots, n_c$ become proportional to each other. Let n be the total sample size in the one-sample case: for $y = 1, \ldots, c$, it holds that $n_y = n\pi_y$ in expectation. Hence, (3.20) can be reduced to

$$R(\widehat{\boldsymbol{g}}) - R(\boldsymbol{g}^*) \leq C_\delta \left(\sum_{y=1}^{c} \sqrt{\pi_y} \right) \frac{1}{\sqrt{n}},$$

whose convergence rate is $\mathcal{O}_p(1/\sqrt{n})$. This is the same convergence rate as the one we obtain if we directly work on the one-sample empirical risk estimator/minimizer. However, C_δ can be smaller due to a smaller $L_{\mathscr{L}}$, since y is fixed in theorem 3.9.

3.2.2.2   Classification calibration

Last but not least, we would like to discuss the classification calibration of $\mathscr{L}$. Recall that, in binary classification, popular surrogate losses are calibrated. Unfortunately, in multi-class classification, popular surrogate losses might not be calibrated. Even though ℓ is calibrated to $\ell_{0\text{-}1}$, there is no guarantee that $\mathscr{L}$ is calibrated to $\mathscr{L}_{0\text{-}1}$, since we cannot obtain $\mathscr{L}_{0\text{-}1}$ from $\mathscr{L}$ by replacing ℓ with $\ell_{0\text{-}1}$ in $\mathscr{L}$ when considering the OVR, PC, or CE loss.

In multi-class classification, the difference between the estimated scores of true and competitive labels,

$$m = g_y(\boldsymbol{x}) - \max_{y' \neq y} g_{y'}(\boldsymbol{x}),$$

is called the margin of g for (x, y); given a binary margin-based loss ℓ, plugging m into ℓ,

$$\mathscr{L}(g(x), y) = \ell\left(g_y(x) - \max_{y' \neq y} g_{y'}(x)\right) = \max_{y' \neq y} \ell(g_y(x) - g_{y'}(x)),$$

is called a multi-class margin-based loss. For margin-based losses, we indeed obtain $\mathscr{L}_{0\text{-}1}$ from $\mathscr{L}$ by replacing ℓ with $\ell_{0\text{-}1}$ in $\mathscr{L}$. Surprisingly, a multi-class margin-based loss $\mathscr{L}$ is not calibrated when the binary margin-based loss ℓ is the hinge loss or is differentiable and convex (Tewari and Bartlett, 2007).

Subsequently, the OVR and PC losses could be calibrated when ℓ satisfies a few nice conditions, but they are not calibrated for any non-convex ℓ (Tewari and Bartlett, 2007; Zhang, 2004), including the upper-bounded ramp and sigmoid losses which are useful and thus popular in weakly supervised classification. If we replace $\max_{y' \neq y} \ell(\cdot)$ with $\sum_{y' \neq y} \ell(\cdot)$, we can obtain the PC loss from any margin-based loss, and we may expect that the PC loss is not calibrated. The PC loss is calibrated if ℓ is non-negative, convex, calibrated, and differentiable (almost surely differentiable is not enough, so that ℓ cannot be the hinge loss). The OVR loss is calibrated if ℓ further satisfies that $\ell(m) < \ell(-m)$ for $m > 0$, where the OVR loss needs to be defined using a margin-based ℓ. For instance, the squared-hinge loss satisfies all those conditions and enables the classification calibration of the OVR and PC losses, but it increases so quickly and may be very sensitive to large negative margins. When ℓ enables the classification calibration of $\mathscr{L}$, we can guarantee the strong consistency of $\widehat{g}$; otherwise, we can only guarantee the weak consistency of $\widehat{g}$.

The CE loss is always classification-calibrated, and it guarantees the strong consistency of $\widehat{g}$. Additionally, it is also *confidence-calibrated*, meaning that we not only obtain the minimizer of $I(g)$ but also recover $p(y \mid x)$ by minimizing $R(g)$, as shown in theorem 3.10 below. This advocates the use of the CE loss as the dominant surrogate loss in multi-class classification.

Theorem 3.10 (Classification and confidence calibration of CE loss) *Let $g^\star :=$ $\operatorname{argmin}_g R(g)$ be the Bayes optimal classifier under the CE loss, where the minimization is over the set of all vector-valued measurable functions. Then, $g^\star$ is also a Bayes optimal classifier under the multi-class zero-one loss, i.e., $g^\star = \operatorname{argmin}_g I(g)$. Moreover, $g^\star$ recovers the class-posterior probability $p(y \mid x)$, i.e.,*

$$\frac{\exp(g_y^\star(x))}{\sum_{y'=1}^{c} \exp(g_{y'}^\star(x))} = p(y \mid x). \tag{3.21}$$

Proof   To begin with, unlike the logistic loss $\ell(g(x), y)$ which is strictly convex in g, the CE loss $\mathscr{L}(g(x), y)$ is convex but not strictly convex in g. It means that every stationary point of $R(g)$ must be a global minimizer of $R(g)$, but there may be multiple global minimizers. Thus, we are going to prove that every minimizer of $R(g)$ is a minimizer of $I(g)$ and recovers the class-posterior probability.

Since $\boldsymbol{g}^\star = \mathrm{argmin}_{\boldsymbol{g}}\, \mathbb{E}_{p(\boldsymbol{x},y)}[\mathcal{L}(\boldsymbol{g}(\boldsymbol{x}),y)]$ is a vector-valued measurable function, this minimization can be decomposed into pointwise minimization problems as

$$\boldsymbol{g}^\star(\boldsymbol{x}) = \mathrm{argmin}_{\boldsymbol{g}(\boldsymbol{x})}\, \mathbb{E}_{p(y|\boldsymbol{x})}[\mathcal{L}(\boldsymbol{g}(\boldsymbol{x}),y)]$$

$$= \mathrm{argmin}_{\boldsymbol{g}(\boldsymbol{x})} \sum_{y=1}^{c} p(y\,|\,\boldsymbol{x})\mathcal{L}(\boldsymbol{g}(\boldsymbol{x}),y)$$

$$= \mathrm{argmin}_{\boldsymbol{g}(\boldsymbol{x})} \ln\left(\sum_{y=1}^{c} \exp(g_y(\boldsymbol{x}))\right) - \sum_{y=1}^{c} p(y\,|\,\boldsymbol{x})g_y(\boldsymbol{x}).$$

The partial derivative of $\mathbb{E}_{p(y|\boldsymbol{x})}[\mathcal{L}(\boldsymbol{g}(\boldsymbol{x}),y)]$ with respect to $g_y(\boldsymbol{x})$ is

$$\frac{\partial\, \mathbb{E}_{p(y|\boldsymbol{x})}[\mathcal{L}(\boldsymbol{g}(\boldsymbol{x}),y)]}{\partial g_y(\boldsymbol{x})} = \frac{\exp(g_y(\boldsymbol{x}))}{\sum_{y'=1}^{c} \exp(g_{y'}(\boldsymbol{x}))} - p(y\,|\,\boldsymbol{x}).$$

In order for $\boldsymbol{g}^\star(\boldsymbol{x})$ to be a stationary point of $\mathbb{E}_{p(y|\boldsymbol{x})}[\mathcal{L}(\boldsymbol{g}(\boldsymbol{x}),y)]$, it must satisfy that for $y=1,\ldots,c$,

$$\frac{\partial\, \mathbb{E}_{p(y|\boldsymbol{x})}[\mathcal{L}(\boldsymbol{g}^\star(\boldsymbol{x}),y)]}{\partial g_y^\star(\boldsymbol{x})} = 0,$$

which proves equation (3.21).

Let us solve (3.21) as a system of nonlinear equations. The system has c equations in c variables, while its rank is $c-1$ due to the additional constraint $\sum_{y=1}^{c} p(y\,|\,\boldsymbol{x}) = 1$, which means that the system has infinitely many solutions. It is easy to see that $\exp(g_y^\star(\boldsymbol{x})) = \alpha p(y\,|\,\boldsymbol{x})$ for $\alpha > 0$ or equivalently $g_y^\star(\boldsymbol{x}) = \ln(p(y\,|\,\boldsymbol{x})) + \beta$ for $\beta \in \mathbb{R}$. For any $\beta \in \mathbb{R}$, $\mathrm{argmax}_y\, g_y^\star(\boldsymbol{x}) = \mathrm{argmax}_y\, p(y\,|\,\boldsymbol{x})$, which verifies that $\boldsymbol{g}^\star$ is also a minimizer of $I(\boldsymbol{g})$. $\square$

II WEAKLY SUPERVISED LEARNING FOR BINARY CLASSIFICATION

4 Positive-Unlabeled (PU) Classification

In this chapter, we discuss the problem of positive-unlabeled (PU) classification, which is a natural first step toward weakly supervised classification from positive-negative (PN) classification.

4.1 Introduction

Suppose that collecting negative (N) data is expensive, while positive (P) and unlabeled (U) data can be gathered abundantly. The objective of PU classification is to learn a PN classifier from only PU data.

Such PU classification scenarios are conceivable in various real-world classification tasks. A typical example is click prediction on the web—we want to classify whether a user likes the link or not. In this scenario, positive ("like") samples can be easily obtained from the click log of a user. On the other hand, it is also possible to easily collect links that were displayed on the screen but were not clicked. Naively, such unclicked links look like negative ("dislike") samples. However, it is not clear whether the user did not click the links because the user did not like the links or the user liked the links but the user was too busy to click the links. For this reason, such unclicked links displayed on the screen should be treated as unlabeled samples.

In this chapter, we introduce PU classification methods using *unbiased risk estimation* and discuss their theoretical properties. The mathematical derivations of these methods epitomize our empirical risk minimization approach to machine learning from weak supervision. The same derivation technique will be frequently applied in subsequent chapters. Therefore, this chapter serves more as a foundation for the following chapters than as the introduction of PU classification methods using unbiased risk estimation itself. For readers who are interested in practical applications of PU classification, see chapter 11 for more advanced and powerful methods.

4.2   Formulation

First of all, we formulate the problem of PU classification. There are again two famous cases of the formulation depending on how positive data are obtained:

- In the one-sample case, a set of unlabeled data is drawn according to $p(x)$ and split into two sets: All the unlabeled positive data join the set of (labeled) positive data with a constant probability, i.e., positive data are *selected completely at random*; the remaining unlabeled positive data and all negative data join the set of unlabeled data.

- In the two-sample case, positive data and unlabeled data are directly drawn from two different distributions of x separately as two independent sets.

We will focus on the two-sample case of PU classification in this chapter, because of its simplicity and generality.

PU classification uses positive and unlabeled data:

$$\mathscr{X}_{\mathrm{P}} := \{x_i^{\mathrm{P}}\}_{i=1}^{n_{\mathrm{P}}} \overset{\text{i.i.d.}}{\sim} p_{\mathrm{P}}(x) := p(x \mid y = +1),$$

$$\mathscr{X}_{\mathrm{U}} := \{x_i^{\mathrm{U}}\}_{i=1}^{n_{\mathrm{U}}} \overset{\text{i.i.d.}}{\sim} p_{\mathrm{U}}(x) := \pi_{\mathrm{P}} p_{\mathrm{P}}(x) + \pi_{\mathrm{N}} p_{\mathrm{N}}(x),$$

where $p_{\mathrm{N}}(x) := p(x \mid y = -1)$ is the negative class-conditional density, i.e., the probability density function for negative data (which are expensive to collect and not available in the current setup), and where

$$\pi_{\mathrm{P}} := p(y = +1),$$

$$\pi_{\mathrm{N}} := p(y = -1),$$

are the class-prior probabilities for the positive and negative classes, such that

$$\pi_{\mathrm{P}} + \pi_{\mathrm{N}} = 1.$$

Note that $p_{\mathrm{U}}(x)$ agrees with $p(x)$. The goal of PU classification is to obtain a PN classifier that minimizes the original classification risk

$$R(g) := \mathbb{E}_{p(x,y)}[\ell(g(x), y)],$$

where $\ell \colon \mathbb{R} \times \mathcal{Y} \to \mathbb{R}$ is a loss (see section 2.1.3) and $g \colon \mathcal{X} \to \mathbb{R}$ is a binary classifier (see section 2.1.2).

Let us take a closer look at $\mathscr{X}_{\mathrm{P}}$ and $\mathscr{X}_{\mathrm{U}}$ in the two-sample case. Positive and unlabeled data must follow certain distributions, and positive data must be independent and unlabeled data must be independent. However, it is never required that positive and unlabeled data (i.e., $\mathscr{X}_{\mathrm{P}} \cup \mathscr{X}_{\mathrm{U}}$) are independent. This means that, in practice, $\mathscr{X}_{\mathrm{U}}$ may contain a part of or the whole $\mathscr{X}_{\mathrm{P}}$. Furthermore, it is easy to convert training data in the one-sample case to training data in the two-sample case by adding the set of positive data back to the set of unlabeled data. Conversely, converting training data in a two-sample case to the one-sample case is hard in general when $\mathscr{X}_P \cap \mathscr{X}_U = \emptyset$. To ensure that positive and unlabeled data as a union of two sets follow $p(x)$, some positive data have to be removed from $\mathscr{X}_U$ and they should be selected completely at random among all the positive data in $\mathscr{X}_U$. If so, we could already identify all the positive data in $\mathscr{X}_U$ perfectly and could already reduce PU classification to PN classification. As a consequence, any method solving the two-sample case is also able to solve the one-sample case, but a method solving the one-sample case may not be able to solve the two-sample case without largely modifying its algorithm design. The discussion above demonstrates the generality of the two-sample case of PU classification.

4.3 Unbiased Risk Estimation from PU Data

As explained in chapter 3, in the PN classification scenario where we can use both positive and negative data for training a classifier g, the classification risk $R(g)$ was decomposed into

$$R(g) = \pi_P R_P(g) + \pi_N R_N(g),$$

where

$$R_P(g) := \mathbb{E}_P[\ell(g(x), +1)],$$

$$R_N(g) := \mathbb{E}_N[\ell(g(x), -1)],$$

are the risks for positive and negative data. In order to train g, the risks $R_P(g)$ and $R_N(g)$ were empirically estimated based on the positive and negative data and the empirical risk (with a regularizer) was minimized.

However, in the PU classification scenario where we have no negative data, we cannot empirically estimate the negative risk $R_N(g)$ straightforwardly. We should empirically estimate it from only positive and unlabeled data, and then we can minimize the empirical risk as in PN classification. In this section, we show how to empirically estimate the entire risk $R(g)$ from only PU data.

4.3.1 General Approach

The entire risk $R(g)$ was expressed by the expectations over $p_P(x)$ and $p_N(x)$. To empirically estimate it from only PU data, we need $R(g)$ to be expressed by the expectations over $p_P(x)$ and $p_U(x)$. The process of rewriting the risk into an expression that matches the problem setting under consideration is called "risk rewriting," and it is the core technology of the empirical risk minimization approach to machine learning from weak supervision.

To this end, a simple trick to overcome the issue of having no negative data is to use the fact that the unlabeled data density is a mixture of the positive and negative data densities:

$$p_U(x) = \pi_P p_P(x) + \pi_N p_N(x),$$

which yields

$$\pi_{\mathrm{N}} p_{\mathrm{N}}(\boldsymbol{x}) = p_{\mathrm{U}}(\boldsymbol{x}) - \pi_{\mathrm{P}} p_{\mathrm{P}}(\boldsymbol{x}). \tag{4.1}$$

More specifically, based on (4.1), the negative risk can be expressed as

$$\pi_{\mathrm{N}} R_{\mathrm{N}}(g) = \pi_{\mathrm{N}} \, \mathbb{E}_{\mathrm{N}}[\ell(g(\boldsymbol{x}), -1)]$$

$$= \pi_{\mathrm{N}} \int \ell(g(\boldsymbol{x}), -1) p_{\mathrm{N}}(\boldsymbol{x}) \mathrm{d}\boldsymbol{x}$$

$$= \int \ell(g(\boldsymbol{x}), -1) p_{\mathrm{U}}(\boldsymbol{x}) \mathrm{d}\boldsymbol{x} - \pi_{\mathrm{P}} \int \ell(g(\boldsymbol{x}), -1) p_{\mathrm{P}}(\boldsymbol{x}) \mathrm{d}\boldsymbol{x}$$

$$= \mathbb{E}_{\mathrm{U}}[\ell(g(\boldsymbol{x}), -1)] - \pi_{\mathrm{P}} \, \mathbb{E}_{\mathrm{P}}[\ell(g(\boldsymbol{x}), -1)].$$

Then, we can eliminate the expectation over the negative data density $p_{\mathrm{N}}(\boldsymbol{x})$ from the risk $R(g)$, and express it only with the expectations over the positive and unlabeled data densities $p_{\mathrm{P}}(\boldsymbol{x})$ and $p_{\mathrm{U}}(\boldsymbol{x})$ as

$$R(g) = \pi_{\mathrm{P}} \, \mathbb{E}_{\mathrm{P}}[\widetilde{\ell}(g(\boldsymbol{x}), +1)] + \mathbb{E}_{\mathrm{U}}[\ell(g(\boldsymbol{x}), -1)], \tag{4.2}$$

where

$$\widetilde{\ell}(g(\boldsymbol{x}), y) := \ell(g(\boldsymbol{x}), y) - \ell(g(\boldsymbol{x}), -y)$$

is called a *composite loss* (du Plessis et al., 2015). As a result, from only a set of positive samples $\mathscr{X}_{\mathrm{P}}$ and a set of unlabeled samples $\mathscr{X}_{\mathrm{U}}$, the risk $R(g)$ can be empirically approximated as

$$\widehat{R}_{\mathrm{PU}}(g) := \frac{\pi_{\mathrm{P}}}{n_{\mathrm{P}}} \sum_{i=1}^{n_{\mathrm{P}}} \widetilde{\ell}(g(\boldsymbol{x}_i^{\mathrm{P}}), +1) + \frac{1}{n_{\mathrm{U}}} \sum_{i=1}^{n_{\mathrm{U}}} \ell(g(\boldsymbol{x}_i^{\mathrm{U}}), -1). \tag{4.3}$$

The estimator $\widehat{R}_{\mathrm{PU}}(g)$ of the entire risk $R(g)$ is the most general empirical risk estimator in PU classification. Equation (4.3) means that, in PU classification, while the unlabeled samples are classified as negative samples by the original loss ℓ, the positive samples are now classified by the composite loss $\widetilde{\ell}$ (Natarajan et al., 2013; du Plessis et al., 2015). In terms of the original loss ℓ, we regard the positive samples as positive and also subtract the empirical risk regarding the positive samples as negative, which exactly cancels the bias resulted from regarding the unlabeled samples as negative. Thus, $\widehat{R}_{\mathrm{PU}}(g)$ in equation (4.3) stands for unbiased risk estimation from PU data, in a sense that

$$\mathbb{E}_{\mathscr{X}_{\mathrm{P}}, \mathscr{X}_{\mathrm{U}}}[\widehat{R}_{\mathrm{PU}}(g)] = \mathbb{E}_{\mathscr{X}_{\mathrm{P}}} \left[\frac{\pi_{\mathrm{P}}}{n_{\mathrm{P}}} \sum_{i=1}^{n_{\mathrm{P}}} \widetilde{\ell}(g(\boldsymbol{x}_i^{\mathrm{P}}), +1) \right] + \mathbb{E}_{\mathscr{X}_{\mathrm{U}}} \left[\frac{1}{n_{\mathrm{U}}} \sum_{i=1}^{n_{\mathrm{U}}} \ell(g(\boldsymbol{x}_i^{\mathrm{U}}), -1) \right]$$

$$= \frac{\pi_{\mathrm{P}}}{n_{\mathrm{P}}} \sum_{i=1}^{n_{\mathrm{P}}} \mathbb{E}_{\mathrm{P}}[\widetilde{\ell}(g(\boldsymbol{x}_i^{\mathrm{P}}), +1)] + \frac{1}{n_{\mathrm{U}}} \sum_{i=1}^{n_{\mathrm{U}}} \mathbb{E}_{\mathrm{U}}[\ell(g(\boldsymbol{x}_i^{\mathrm{U}}), -1)]$$

$$= R(g).$$

Here, we have assumed that π_P is known for estimating $R(g)$ and training g. In practice, π_P should also be estimated from PU data (Elkan and Noto, 2008; Blanchard et al., 2010; du Plessis and Sugiyama, 2014a; Liu and Tao, 2015; Ramaswamy et al., 2016; Jain et al., 2016; du Plessis et al., 2017; Bekker and Davis, 2018; Ivanov, 2019; Zeiberg et al., 2020; Yao et al., 2022), which will be discussed in detail in chapter 12.

4.3.2   Cost-Sensitive Approach

If the original loss ℓ satisfies a symmetric condition,

$$\ell(g(\boldsymbol{x}), +1) + \ell(g(\boldsymbol{x}), -1) = 1, \tag{4.4}$$

we have

$$\mathbb{E}_P[\widetilde{\ell}(g(\boldsymbol{x}), +1)] = \mathbb{E}_P[\ell(g(\boldsymbol{x}), +1) - (1 - \ell(g(\boldsymbol{x}), +1))]$$

$$= 2\,\mathbb{E}_P[\ell(g(\boldsymbol{x}), +1)] - 1,$$

due to the linearity of expectations. Then, equation (4.2) is reduced to the following form (du Plessis et al., 2014; Ghosh et al., 2015):

$$R(g) = 2\pi_P\,\mathbb{E}_P[\ell(g(\boldsymbol{x}), +1)] + \mathbb{E}_U[\ell(g(\boldsymbol{x}), -1)] - \pi_P,$$

and equation (4.3) is reduced to

$$\widehat{R}_{\mathrm{PU}}(g) = \frac{2\pi_P}{n_P} \sum_{i=1}^{n_P} \ell(g(\boldsymbol{x}_i^P), +1) + \frac{1}{n_U} \sum_{i=1}^{n_U} \ell(g(\boldsymbol{x}_i^U), -1) - \pi_P.$$

We can see that, given (4.4), the same loss ℓ is used for both positive and unlabeled data with different weights $2\pi_P$ and 1 (i.e., the loss for positive data remains the same while the weight increases from π_P to $2\pi_P$), which is called *cost-sensitive classification* (Elkan, 2001).

The symmetric condition (4.4) is satisfied by, for example, the following margin-based losses $\ell(g(\boldsymbol{x}), y) = \ell(m)$ with $m = yg(\boldsymbol{x})$:[1]

$$\text{Zero-one loss: } \ell(m) = \begin{cases} 0 & (m > 0), \\ 1 & (m \leq 0), \end{cases}$$

$$\text{Ramp loss: } \ell(m) = \frac{1}{2} \min\{2, \max\{0, 1 - m\}\},$$

$$\text{Sigmoid loss: } \ell(m) = \frac{1}{1 + \exp(m)},$$

$$\text{Linear loss: } \ell(m) = -m + \frac{1}{2},$$

where the linear loss is shifted by a constant to sum to one. The zero-one loss, the ramp loss, and the sigmoid loss are all *non-convex*.

4.3.3　Convex Approach

Given $\ell(\widehat{y}, y)$ that is convex in $\widehat{y}$, the composite loss $\widetilde{\ell}(\widehat{y}, y)$ is often non-convex in $\widehat{y}$ due to the concavity of $-\ell(\widehat{y}, -y)$. Fortunately, it is known how to easily check the convexity of the composite loss.

Theorem 4.1 (Convex composite loss)　*For any $\ell(\widehat{y}, y)$ that is convex in $\widehat{y}$, $\widetilde{\ell}(\widehat{y}, y)$ is also convex in $\widehat{y}$ if and only if ℓ is linear in $\widehat{y}$.*

Proof　The same theoretical result has been proved for margin-based losses in du Plessis et al. (2015), and this theorem simply extends it to general losses.

The "if" direction is straightforward: If ℓ is linear in $\widehat{y}$, it must be convex in $\widehat{y}$. The "only if" direction is because any convex odd function must be linear. Specifically, the composite loss $\widetilde{\ell}$ is an odd function:

$$\widetilde{\ell}(\widehat{y}, -y) = \ell(\widehat{y}, -y) - \ell(\widehat{y}, y) = -\widetilde{\ell}(\widehat{y}, y).$$

Since $\widetilde{\ell}$ is convex in $\widehat{y}$ regardless of y, we have

$$\frac{\partial^2 \widetilde{\ell}(\widehat{y}, y)}{\partial \widehat{y}^2} \geq 0,$$

$$\frac{\partial^2 \widetilde{\ell}(\widehat{y}, -y)}{\partial \widehat{y}^2} \geq 0.$$

As a result, it also holds that

$$\frac{\partial^2 \widetilde{\ell}(\widehat{y}, y)}{\partial \widehat{y}^2} = \frac{\partial^2 (-\widetilde{\ell}(\widehat{y}, -y))}{\partial \widehat{y}^2} = -\frac{\partial^2 \widetilde{\ell}(\widehat{y}, -y)}{\partial \widehat{y}^2} \leq 0.$$

The only possibility is that $\partial^2 \widetilde{\ell}(\widehat{y}, y)/\partial \widehat{y}^2 = 0$ and $\partial^2 \widetilde{\ell}(\widehat{y}, -y)/\partial \widehat{y}^2 = 0$ without leading to contradictions, which implies that $\widetilde{\ell}$ is actually linear in $\widehat{y}$.　$\square$

Then, suppose that the composite loss is given by

$$\widetilde{\ell}(\widehat{y}, y) = \ell(\widehat{y}, y) - \ell(\widehat{y}, -y) = -\widehat{y}, \tag{4.5}$$

so that if ℓ is convex in $\widehat{y}$, the entire risk $R(g)$ is convex with respect to g:

$$R(g) = -\pi_{\mathrm{P}} \, \mathbb{E}_{\mathrm{P}}[g(\boldsymbol{x})] + \mathbb{E}_{\mathrm{U}}[\ell(g(\boldsymbol{x}), -1)],$$

and so is the corresponding empirical risk:

$$\widehat{R}_{\mathrm{PU}}(g) = -\frac{\pi_{\mathrm{P}}}{n_{\mathrm{P}}} \sum_{i=1}^{n_{\mathrm{P}}} g(\boldsymbol{x}_i^{\mathrm{P}}) + \frac{1}{n_{\mathrm{U}}} \sum_{i=1}^{n_{\mathrm{U}}} \ell(g(\boldsymbol{x}_i^{\mathrm{U}}), -1).$$

This means that when g is a linear-in-input model, linear-in-parameter model, or kernel model (see section 2.1.2), the risk and empirical risk become convex with respect to model parameters α, and thus the global minimizer (which is unique if ℓ is strictly convex) can be obtained easily.

The linearity condition (4.5) is satisfied by, for example, the following margin-based losses $\ell(g(x), y) = \ell(m)$ with $m = yg(x)$:[2]

$$\text{Squared loss: } \ell(m) = \tfrac{1}{4}(1 - m)^2,$$

$$\text{Logistic loss: } \ell(m) = \ln(1 + \exp(-m)),$$

$$\text{Double-hinge loss: } \ell(m) = \max\left\{-m, \tfrac{1}{2}\max\{0, 1 - m\}\right\},$$

$$\text{Huber-hinge loss: } \ell(m) = \begin{cases} \tfrac{1}{4}\max\{0, 1 - m\}^2 & (m \geq -1), \\ -m & (m < -1), \end{cases}$$

$$\text{Perceptron loss: } \ell(m) = \max\{0, -m\},$$

where the squared, double-hinge, and Huber-hinge losses are scaled to have the unit linear coefficient (cf. section 2.1.3).

For the squared loss $\ell(m) = \tfrac{1}{4}(1 - m)^2$, the risk is expressed as

$$R(g) = -\pi_{\mathrm{P}} \, \mathbb{E}_{\mathrm{P}}[g(x)] + \frac{1}{4} \, \mathbb{E}_{\mathrm{U}}[(1 + g(x))^2]$$

$$= -\int g(x)\pi_{\mathrm{P}} p_{\mathrm{P}}(x)\mathrm{d}x + \frac{1}{4}\int \left(g(x)^2 + 2g(x)\right) p_{\mathrm{U}}(x)\mathrm{d}x + \frac{1}{4}$$

$$= \frac{1}{4}\int g(x)^2 p_{\mathrm{U}}(x)\mathrm{d}x - \frac{1}{2}\int g(x)r(x)p_{\mathrm{U}}(x)\mathrm{d}x + \frac{1}{4}$$

$$= \frac{1}{4}\int \left(g(x) - r(x)\right)^2 p_{\mathrm{U}}(x)\mathrm{d}x + C, \tag{4.6}$$

where

$$r(x) := 2\pi_{\mathrm{P}}\frac{p_{\mathrm{P}}(x)}{p_{\mathrm{U}}(x)} - 1,$$

and C is a constant independent of g. Note that $r(x)$ can be expressed as

$$r(x) = \frac{2\pi_{\mathrm{P}} p_{\mathrm{P}}(x) - p_{\mathrm{U}}(x)}{p_{\mathrm{U}}(x)}$$

$$= \frac{2p(y = +1)p(x \mid y = +1)}{p(x)}$$

$$- \frac{p(y = +1)p(x \mid y = +1) + p(y = -1)p(x \mid y = -1)}{p(x)}$$

$$= \frac{p(x, y = +1)}{p(x)} - \frac{p(x, y = -1)}{p(x)}$$

$$= p(y = +1 \mid x) - p(y = -1 \mid x),$$

which is the difference of the class-posterior probabilities. As a result, the sign of $r(x)$ corresponds to the original probabilistic Bayes optimal classifier:

$$\mathrm{sign}(p(y = +1 \mid x) - p(y = -1 \mid x)) = \mathrm{argmax}_y \, p(y \mid x),$$

so that $r(x)$ is one of the Bayes optimal classifiers that minimize the classification error:

$$r = \mathrm{argmin}_g \, \mathbb{E}_{p(x,y)}[\ell_{0\text{-}1}(yg(x))],$$

where $\ell_{0\text{-}1}(m)$ denotes the zero-one loss:

$$\ell_{0\text{-}1}(m) = \begin{cases} 0 & (m > 0), \\ 1 & (m \le 0). \end{cases}$$

Hence, equation (4.6) gives a natural interpretation that PU classification with the squared loss fits model g to the difference of the class-posterior probabilities, which is an optimal strategy for minimizing the classification error.

Now, let us consider the linear-in-parameter model g that was introduced in section 2.1.2.2:

$$g(x) = \sum_{j=1}^{b} \alpha_j \phi_j(x) = \boldsymbol{\alpha}^\top \boldsymbol{\phi}(x),$$

where b denotes the number of parameters,

$$\boldsymbol{\alpha} := (\alpha_1, \ldots, \alpha_b)^\top \in \mathbb{R}^b$$

is a parameter vector, and

$$\boldsymbol{\phi}(x) := (\phi_1(x), \ldots, \phi_b(x))^\top \in \mathbb{R}^b$$

is a vector of basis functions. For this model g and the squared loss, the risk is expressed as

$$R(\boldsymbol{\alpha}) = -\pi_{\mathrm{P}} \, \mathbb{E}_{\mathrm{P}}[\boldsymbol{\alpha}^\top \boldsymbol{\phi}(x)] + \frac{1}{4} \, \mathbb{E}_{\mathrm{U}}[(1 + (\boldsymbol{\alpha}^\top \boldsymbol{\phi}(x)))^2].$$

Thus, its empirical approximation with weighted ℓ_2-regularization (see section 2.1.5) is a quadratic function of $\boldsymbol{\alpha}$ given as

$$\widehat{R}_{\mathrm{PU}}(\boldsymbol{\alpha}) = -\frac{\pi_{\mathrm{P}}}{n_{\mathrm{P}}} \sum_{i=1}^{n_{\mathrm{P}}} \boldsymbol{\alpha}^\top \boldsymbol{\phi}\left(x_i^{\mathrm{P}}\right) + \frac{1}{4n_{\mathrm{U}}} \sum_{i=1}^{n_{\mathrm{U}}} \left(1 + \left(\boldsymbol{\alpha}^\top \boldsymbol{\phi}(x_i^{\mathrm{U}})\right)\right)^2 + \frac{\lambda}{2} \boldsymbol{\alpha}^\top R \boldsymbol{\alpha}$$

$$= \frac{1}{4n_{\mathrm{U}}} \boldsymbol{\alpha}^\top \boldsymbol{\Phi}_{\mathrm{U}}^\top \boldsymbol{\Phi}_{\mathrm{U}} \boldsymbol{\alpha} + \frac{1}{2n_{\mathrm{U}}} \mathbf{1}_{n_{\mathrm{U}}}^\top \boldsymbol{\Phi}_{\mathrm{U}} \boldsymbol{\alpha} - \frac{\pi_{\mathrm{P}}}{n_{\mathrm{P}}} \mathbf{1}_{n_{\mathrm{P}}}^\top \boldsymbol{\Phi}_{\mathrm{P}} \boldsymbol{\alpha} + \frac{\lambda}{2} \boldsymbol{\alpha}^\top R \boldsymbol{\alpha} + \frac{1}{4},$$

where $\lambda \geq 0$ is the regularization parameter, $\boldsymbol{R}$ is the regularization matrix, $\mathbf{1}_n$ is the n-dimensional vector with all ones, and $\boldsymbol{\Phi}_U$ and $\boldsymbol{\Phi}_P$ are the $n_U \times b$ and $n_P \times b$ matrices whose (i,j)-entries are given respectively by

$$[\boldsymbol{\Phi}_U]_{i,j} := \phi_j(x_i^U),$$

$$[\boldsymbol{\Phi}_P]_{i,j} := \phi_j(x_i^P).$$

Assuming λ and $\boldsymbol{R}$ are carefully chosen, such that $\lambda\boldsymbol{R}$ is positive definite and $\widehat{R}_{PU}(\boldsymbol{\alpha})$ is a strongly convex quadratic function with respect to $\boldsymbol{\alpha}$, its unique global minimizer $\widehat{\boldsymbol{\alpha}}$ can be obtained *analytically* as

$$\widehat{\boldsymbol{\alpha}} := \underset{\boldsymbol{\alpha} \in \mathbb{R}^b}{\arg\min}\, \widehat{R}_{PU}(\boldsymbol{\alpha}) = \left(\frac{1}{2n_U} \boldsymbol{\Phi}_U^\top \boldsymbol{\Phi}_U + \lambda\boldsymbol{R} \right)^{-1} \left(\frac{\pi_P}{n_P} \boldsymbol{\Phi}_P^\top \mathbf{1}_{n_P} - \frac{1}{2n_U} \boldsymbol{\Phi}_U^\top \mathbf{1}_{n_U} \right).$$

4.4   Theoretical Analysis

In this section, we derive an estimation error bound for PU classification (in the same way as section 3.1.2). Then, we drive a similar bound for *NU classification* (which stands for negative-unlabeled classification and is the counterpart of PU classification) and compare PU/NU classification with PN classification based on the bounds.

4.4.1   PU Classification

Recall that in the general approach, the empirical risk estimator is given by

$$\widehat{R}_{PU}(g) := \frac{\pi_P}{n_P} \sum_{i=1}^{n_P} \widetilde{\ell}\left(g(x_i^P), +1\right) + \frac{1}{n_U} \sum_{i=1}^{n_U} \ell\left(g(x_i^U), -1\right).$$

For a properly chosen function class $\mathscr{G}$, the corresponding empirical risk minimizer is defined as

$$\widehat{g}_{PU} := \underset{g \in \mathscr{G}}{\arg\min}\, \widehat{R}_{PU}(g).$$

According to theorem 4.1, $\widehat{R}_{PU}$ is not convex if $\widetilde{\ell}$ is not linear, and thus let $\widehat{g}_{PU}$ be an arbitrary global minimizer of $\widehat{R}_{PU}$ when it is not unique. The setting for our theoretical analysis here follows section 3.1.2. Specifically the target of convergence is the true risk minimizer:

$$g^* := \underset{g \in \mathscr{G}}{\arg\min}\, R(g),$$

which is the optimal binary classifier in $\mathscr{G}$ (where the optimality is in terms of minimizing the classification risk but not the classification error); the measure of convergence is the estimation error:

$$R(\widehat{g}_{PU}) - R(g^*),$$

because $\widehat{g}_{\mathrm{PU}}$ may not be unique and thus we should not directly study $\|\widehat{g}_{\mathrm{PU}} - g^*\|$; and the measure of complexity is the Rademacher complexity of $\mathscr{G}$ for the sample size n with respect to $q(\boldsymbol{x})$:

$$\mathfrak{R}_{n,q}(\mathscr{G}) := \mathbb{E}_{\boldsymbol{x}_1,\dots,\boldsymbol{x}_n}\, \mathbb{E}_{\sigma_1,\dots,\sigma_n}\left[\sup_{g\in\mathscr{G}} \frac{1}{n}\sum_{i=1}^{n} \sigma_i g(\boldsymbol{x}_i)\right],$$

where $\sigma_1,\dots,\sigma_n$ are independent Rademacher variables, i.e., $\Pr(\sigma_i=+1)=\Pr(\sigma_i=-1)=1/2$.

Theorem 4.2 (Estimation error of PU classification)   *Given a function class $\mathscr{G}$, we assume that the surrogate loss ℓ has bounded values, i.e.,*

- *either there exists $C_\ell > 0$ such that $\ell(\widehat{y}, y) \le C_\ell$ for all $\widehat{y}\in\mathbb{R}$ and $y \in \mathscr{Y}$, or*

- *there exist $C_\mathrm{g} > 0$ and $C_\ell > 0$ such that $\|g\|_\infty = \sup_{\boldsymbol{x}\in\mathscr{X}} |g(\boldsymbol{x})| \le C_\mathrm{g}$ for all $g\in\mathscr{G}$ and $\ell(\widehat{y}, y) \le C_\ell$ for all $|\widehat{y}| \le C_\mathrm{g}$ and $y \in \mathscr{Y}$.*

We also assume that ℓ has bounded derivatives, i.e., $\ell(\widehat{y}, y)$ is Lipschitz continuous in $\widehat{y}$ for all $\widehat{y}\in\mathbb{R}$ or $|\widehat{y}| \le C_\mathrm{g}$ and $y \in \mathscr{Y}$ with a Lipschitz constant L_ℓ. Then, for any $\delta > 0$, with probability at least $1 - \delta$,

$$R(\widehat{g}_{\mathrm{PU}}) - R(g^*) \le 8L_\ell\pi_\mathrm{P}\mathfrak{R}_{n_\mathrm{P},p_\mathrm{P}}(\mathscr{G}) + 4L_\ell\mathfrak{R}_{n_\mathrm{U},p_\mathrm{U}}(\mathscr{G})$$

$$+ \sqrt{2\ln\frac{2}{\delta}}\, C_\ell\left(\frac{2\pi_\mathrm{P}}{\sqrt{n_\mathrm{P}}} + \frac{1}{\sqrt{n_\mathrm{U}}}\right). \tag{4.7}$$

Proof   A similar theoretical result has been proved under the symmetric condition (4.4) in Niu et al. (2016), and this theorem simply extends it to general losses.

    The proof of this theorem shares the same structure as that of theorem 3.6. The change of $\widehat{R}_{\mathrm{PU}}(g)$ will be no more than $2C_\ell\pi_\mathrm{P}/n_\mathrm{P}$ if some $\boldsymbol{x}_i^\mathrm{P}$ is replaced, or no more than C_ℓ/n_U if some $\boldsymbol{x}_i^\mathrm{U}$ is replaced. Hence, for any $\delta > 0$, with probability at least $1 - \delta/2$,

$$\sup_{g\in\mathscr{G}} \widehat{R}_{\mathrm{PU}}(g) - R(g) \le \mathbb{E}[\sup_{g\in\mathscr{G}} \widehat{R}_{\mathrm{PU}}(g) - R(g)]$$

$$+ \sqrt{(1/2)\ln(2/\delta)}\, C_\ell(2\pi_\mathrm{P}/\sqrt{n_\mathrm{P}} + 1/\sqrt{n_\mathrm{U}}).$$

By the same proof technique as that in theorem 3.6 (i.e., symmetrization), we can obtain that

$$\mathbb{E}[\sup_{g\in\mathscr{G}} \widehat{R}_{\mathrm{PU}}(g) - R(g)]$$

$$\le 2\pi_\mathrm{P}\mathfrak{R}_{n_\mathrm{P},p_\mathrm{P}}(\widetilde{\ell}(\cdot, +1)\circ\mathscr{G}) + 2\mathfrak{R}_{n_\mathrm{U},p_\mathrm{U}}(\ell(\cdot, -1)\circ\mathscr{G})$$

$$\le 4L_\ell\pi_\mathrm{P}\mathfrak{R}_{n_\mathrm{P},p_\mathrm{P}}(\mathscr{G}) + 2L_\ell\mathfrak{R}_{n_\mathrm{U},p_\mathrm{U}}(\mathscr{G}),$$

since $\widetilde{\ell}$ must be Lipschitz continuous with a Lipschitz constant $2L_\ell$. The other direction, $\sup_{g \in \mathcal{G}} R(g) - \widehat{R}_{\mathrm{PU}}(g)$, has exactly the same upper bound with probability at least $1 - \delta/2$. As a result, we have two unidirectional uniform deviation bounds to create a necessary bidirectional uniform deviation bound, i.e., for any $\delta > 0$, with probability at least $1 - \delta$,

$$\sup_{g \in \mathcal{G}} |\widehat{R}_{\mathrm{PU}}(g) - R(g)| \leq 4L_\ell \pi_{\mathrm{P}} \mathfrak{R}_{n_{\mathrm{P}}, p_{\mathrm{P}}}(\mathcal{G}) + 2L_\ell \mathfrak{R}_{n_{\mathrm{U}}, p_{\mathrm{U}}}(\mathcal{G})$$

$$+ \sqrt{(1/2)\ln(2/\delta)}C_\ell(2\pi_{\mathrm{P}}/\sqrt{n_{\mathrm{P}}} + 1/\sqrt{n_{\mathrm{U}}}).$$

This proves (4.7) since $R(\widehat{g}_{\mathrm{PU}}) - R(g^*) \leq 2\sup_{g \in \mathcal{G}} |\widehat{R}_{\mathrm{PU}}(g) - R(g)|$. $\square$

By plugging corollary 3.5 into theorem 4.2, we can simplify theorem 4.2 as follows. Let $\mathcal{G}$ be a function class (corresponding to a model in section 2.1.2) with $C_{\mathcal{G}} > 0$ such that $\mathfrak{R}_{n_{\mathrm{P}}, p_{\mathrm{P}}}(\mathcal{G}) \leq C_{\mathcal{G}}/\sqrt{n_{\mathrm{P}}}$ and $\mathfrak{R}_{n_{\mathrm{U}}, p_{\mathrm{U}}}(\mathcal{G}) \leq C_{\mathcal{G}}/\sqrt{n_{\mathrm{U}}}$. Recall that we have defined

$$C_\delta := 4L_\ell C_{\mathcal{G}} + \sqrt{2\ln(2/\delta)}C_\ell,$$

and then with probability at least $1 - \delta$,

$$R(\widehat{g}_{\mathrm{PU}}) - R(g^*) \leq C_\delta(2\pi_{\mathrm{P}}/\sqrt{n_{\mathrm{P}}} + 1/\sqrt{n_{\mathrm{U}}}). \tag{4.8}$$

Equation (4.8) establishes the consistency of PU classification, i.e., the weak consistency of $\widehat{g}_{\mathrm{PU}}$. The convergence rate (from $R(\widehat{g}_{\mathrm{PU}})$ to $R(g^*)$) is $\mathcal{O}_p(1/\sqrt{n_{\mathrm{P}}} + 1/\sqrt{n_{\mathrm{U}}})$, where $\mathcal{O}_p$ denotes the order in probability.

4.4.2   NU Classification

Consider the counterpart of PU classification called *negative-unlabeled* (NU) classification, where only NU data is available for classifier training:

$$\mathcal{X}_{\mathrm{N}} := \{\boldsymbol{x}_i^{\mathrm{N}}\}_{i=1}^{n_{\mathrm{N}}} \overset{\mathrm{i.i.d.}}{\sim} p_{\mathrm{N}}(\boldsymbol{x}) := p(\boldsymbol{x} \mid y = -1),$$

$$\mathcal{X}_{\mathrm{U}} := \{\boldsymbol{x}_i^{\mathrm{U}}\}_{i=1}^{n_{\mathrm{U}}} \overset{\mathrm{i.i.d.}}{\sim} p_{\mathrm{U}}(\boldsymbol{x}) := \pi_{\mathrm{P}} p_{\mathrm{P}}(\boldsymbol{x}) + \pi_{\mathrm{N}} p_{\mathrm{N}}(\boldsymbol{x}).$$

Similarly to PU classification, the positive risk can be expressed as

$$\pi_{\mathrm{P}} R_{\mathrm{P}}(g) = \pi_{\mathrm{P}} \, \mathbb{E}_{\mathrm{P}}[\ell(g(\boldsymbol{x}), +1)]$$

$$= \pi_{\mathrm{P}} \int \ell(g(\boldsymbol{x}), +1) p_{\mathrm{P}}(\boldsymbol{x}) d\boldsymbol{x}$$

$$= \int \ell(g(\boldsymbol{x}), +1) p_{\mathrm{U}}(\boldsymbol{x}) d\boldsymbol{x} - \pi_{\mathrm{N}} \int \ell(g(\boldsymbol{x}), +1) p_{\mathrm{N}}(\boldsymbol{x}) d\boldsymbol{x}$$

$$= \mathbb{E}_{\mathrm{U}}[\ell(g(\boldsymbol{x}), +1)] - \pi_{\mathrm{N}} \, \mathbb{E}_{\mathrm{N}}[\ell(g(\boldsymbol{x}), +1)].$$

Then, we can eliminate the expectation over the positive data density $p_{\mathrm{P}}(\boldsymbol{x})$, and express $R(g)$ only with the expectations over the negative and unlabeled data densities $p_{\mathrm{N}}(\boldsymbol{x})$ and $p_{\mathrm{U}}(\boldsymbol{x})$ as

$$R(g) = \pi_{\mathrm{N}}\,\mathbb{E}_{\mathrm{N}}[\widetilde{\ell}(g(\boldsymbol{x}), -1)] + \mathbb{E}_{\mathrm{U}}[\ell(g(\boldsymbol{x}), +1)].$$

Subsequently, the empirical risk estimator is

$$\widehat{R}_{\mathrm{NU}}(g) := \frac{\pi_{\mathrm{N}}}{n_{\mathrm{N}}}\sum_{i=1}^{n_{\mathrm{N}}}\widetilde{\ell}(g(\boldsymbol{x}_i^{\mathrm{N}}), -1) + \frac{1}{n_{\mathrm{U}}}\sum_{i=1}^{n_{\mathrm{U}}}\ell(g(\boldsymbol{x}_i^{\mathrm{U}}), +1), \tag{4.9}$$

and the corresponding empirical risk minimizer is

$$\widehat{g}_{\mathrm{NU}} := \operatorname*{argmin}_{g \in \mathcal{G}} \widehat{R}_{\mathrm{NU}}(g),$$

where $\widehat{g}_{\mathrm{NU}}$ can be an arbitrary global minimizer of $\widehat{R}_{\mathrm{NU}}$ when it is not unique. $\widehat{R}_{\mathrm{NU}}(g)$ in (4.9) stands for unbiased risk estimation from NU data, in the sense that $\mathbb{E}_{\mathscr{X}_{\mathrm{N}}, \mathscr{X}_{\mathrm{U}}}[\widehat{R}_{\mathrm{NU}}(g)] = R(g)$, which is important for deriving the estimation error bound for NU classification.

Theorem 4.3 (Estimation error of NU classification)   *Under the same assumptions as theorem 4.2, for any $\delta > 0$, with probability at least $1 - \delta$,*

$$R(\widehat{g}_{\mathrm{NU}}) - R(g^*) \leq 8L_\ell \pi_{\mathrm{N}} \mathfrak{R}_{n_{\mathrm{N}}, p_{\mathrm{N}}}(\mathcal{G}) + 4L_\ell \mathfrak{R}_{n_{\mathrm{U}}, p_{\mathrm{U}}}(\mathcal{G})$$

$$+ \sqrt{2\ln\frac{2}{\delta}}\,C_\ell\left(\frac{2\pi_{\mathrm{N}}}{\sqrt{n_{\mathrm{N}}}} + \frac{1}{\sqrt{n_{\mathrm{U}}}}\right). \tag{4.10}$$

Proof   For NU classification, we have that, for any $\delta > 0$, with probability at least $1 - \delta/2$,

$$\sup_{g \in \mathcal{G}} \widehat{R}_{\mathrm{NU}}(g) - R(g) \leq \mathbb{E}[\sup_{g \in \mathcal{G}} \widehat{R}_{\mathrm{NU}}(g) - R(g)]$$

$$+ \sqrt{(1/2)\ln(2/\delta)}\,C_\ell(2\pi_{\mathrm{N}}/\sqrt{n_{\mathrm{N}}} + 1/\sqrt{n_{\mathrm{U}}}),$$

as well as that

$$\mathbb{E}[\sup_{g \in \mathcal{G}} \widehat{R}_{\mathrm{NU}}(g) - R(g)]$$

$$\leq 2\pi_{\mathrm{N}} \mathfrak{R}_{n_{\mathrm{N}}, p_{\mathrm{N}}}(\widetilde{\ell}(\cdot, -1) \circ \mathcal{G}) + 2\mathfrak{R}_{n_{\mathrm{U}}, p_{\mathrm{U}}}(\ell(\cdot, +1) \circ \mathcal{G})$$

$$\leq 4L_\ell \pi_{\mathrm{N}} \mathfrak{R}_{n_{\mathrm{N}}, p_{\mathrm{N}}}(\mathcal{G}) + 2L_\ell \mathfrak{R}_{n_{\mathrm{U}}, p_{\mathrm{U}}}(\mathcal{G}).$$

The proof of (4.10) is then almost the same as that of (4.7). $\square$

By plugging corollary 3.5 into theorem 4.3, we can simplify theorem 4.3 as well. With probability at least $1 - \delta$,

$$R(\widehat{g}_{\mathrm{NU}}) - R(g^*) \leq C_\delta(2\pi_{\mathrm{N}}/\sqrt{n_{\mathrm{N}}} + 1/\sqrt{n_{\mathrm{U}}}). \tag{4.11}$$

Equation (4.11) establishes the consistency of NU classification, i.e., the weak consistency of $\widehat{g}_{\mathrm{NU}}$. The convergence rate is $\mathcal{O}_p(1/\sqrt{n_{\mathrm{N}}} + 1/\sqrt{n_{\mathrm{U}}})$.

4.4.3 Comparisons with PN Classification

Note that theorems 4.2 and 4.3 rely on exactly the same assumptions as theorem 3.6 (we repeated those assumptions in theorem 4.2 just to make it self-contained). The theorems have been simplified in the same way and let us put the simplified estimation error bounds (3.17), (4.8), and (4.11) together. With probability at least $1 - \delta$, the following inequalities hold independently:

$$R(\widehat{g}_{\mathrm{PN}}) - R(g^*) \leq C_\delta(\pi_{\mathrm{P}}/\sqrt{n_{\mathrm{P}}} + \pi_{\mathrm{N}}/\sqrt{n_{\mathrm{N}}}),$$

$$R(\widehat{g}_{\mathrm{PU}}) - R(g^*) \leq C_\delta(2\pi_{\mathrm{P}}/\sqrt{n_{\mathrm{P}}} + 1/\sqrt{n_{\mathrm{U}}}),$$

$$R(\widehat{g}_{\mathrm{NU}}) - R(g^*) \leq C_\delta(2\pi_{\mathrm{N}}/\sqrt{n_{\mathrm{N}}} + 1/\sqrt{n_{\mathrm{U}}}),$$

where $C_\delta = 4L_\ell C_{\mathcal{G}} + \sqrt{2\ln(2/\delta)}C_\ell$.

4.4.3.1 Finite-sample comparisons
By comparing these simplified estimation error bounds, we can find that the PU bound is smaller than the PN bound if

$$\pi_{\mathrm{P}}/\sqrt{n_{\mathrm{P}}} + 1/\sqrt{n_{\mathrm{U}}} < \pi_{\mathrm{N}}/\sqrt{n_{\mathrm{N}}},$$

while the NU bound is smaller than the PN bound if

$$\pi_{\mathrm{N}}/\sqrt{n_{\mathrm{N}}} + 1/\sqrt{n_{\mathrm{U}}} < \pi_{\mathrm{P}}/\sqrt{n_{\mathrm{P}}}.$$

This is a surprising result, since PU/NU classification seems hitherto to be an inferior alternative to PN classification when negative/positive data is unavailable. Nevertheless, the above observation indicates that PU/NU classification can actually be better than PN classification under some conditions.

To further analyze the conditions when PU/NU classification can be better, define two indicators as

$$\alpha_{\mathrm{PU,PN}} := \frac{\pi_{\mathrm{P}}/\sqrt{n_{\mathrm{P}}} + 1/\sqrt{n_{\mathrm{U}}}}{\pi_{\mathrm{N}}/\sqrt{n_{\mathrm{N}}}}, \tag{4.12}$$

$$\alpha_{\mathrm{NU,PN}} := \frac{\pi_{\mathrm{N}}/\sqrt{n_{\mathrm{N}}} + 1/\sqrt{n_{\mathrm{U}}}}{\pi_{\mathrm{P}}/\sqrt{n_{\mathrm{P}}}}. \tag{4.13}$$

Intuitively, they measure the relative quality of the PU and NU bounds against the PN bound. We can see that

$$\alpha_{\mathrm{PU,PN}}\alpha_{\mathrm{NU,PN}} = \left(1 + \frac{\sqrt{n_{\mathrm{P}}}}{\pi_{\mathrm{P}}\sqrt{n_{\mathrm{U}}}}\right)\left(1 + \frac{\sqrt{n_{\mathrm{N}}}}{\pi_{\mathrm{N}}\sqrt{n_{\mathrm{U}}}}\right) > 1,$$

Table 4.1
Properties of the indicators $\alpha_{\mathrm{PU,PN}}$ and $\alpha_{\mathrm{NU,PN}}$.

	No specification	
	Monotonic increasing	Monotonic decreasing
$\alpha_{\mathrm{PU,PN}}$	$\pi_{\mathrm{P}}, n_{\mathrm{N}}$	$n_{\mathrm{P}}, n_{\mathrm{U}}$
$\alpha_{\mathrm{NU,PN}}$	n_{P}	$\pi_{\mathrm{P}}, n_{\mathrm{N}}, n_{\mathrm{U}}$
	Sample sizes are proportional	
	Monotonic increasing	Monotonic decreasing
$\alpha_{\mathrm{PU,PN}}$	$\pi_{\mathrm{P}}, \rho_{\mathrm{PU}}$	ρ_{PN}
$\alpha_{\mathrm{NU,PN}}$	$\rho_{\mathrm{PN}}, \rho_{\mathrm{NU}}$	π_{P}
	Further enforce $\rho_{\mathrm{PN}} = \pi_{\mathrm{P}}/\pi_{\mathrm{N}}$	
	Monotonic increasing	Minimum
$\alpha_{\mathrm{PU,PN}}$	ρ_{PU}	$2\sqrt{\rho_{\mathrm{PU}}} + \sqrt{\rho_{\mathrm{PU}}}$
$\alpha_{\mathrm{NU,PN}}$	ρ_{NU}	$2\sqrt{\rho_{\mathrm{NU}}} + \sqrt{\rho_{\mathrm{NU}}}$

and consequently it is impossible to have $\alpha_{\mathrm{PU,PN}} < 1$ and $\alpha_{\mathrm{NU,PN}} < 1$ simultaneously. Therefore, the two indicators play their roles as follows:

- If $\alpha_{\mathrm{PU,PN}} < 1$, then $\alpha_{\mathrm{NU,PN}} > 1$, so that the PU bound is smallest, the PN bound is in the middle, and the NU bound is largest.

- If $\alpha_{\mathrm{NU,PN}} < 1$, then $\alpha_{\mathrm{PU,PN}} > 1$, so that the NU bound is smallest, the PN bound is in the middle, and the PU bound is largest.

- If $\alpha_{\mathrm{PU,PN}} > 1$ and $\alpha_{\mathrm{NU,PN}} > 1$, then the PN bound is smallest.

In order to better understand the theoretical comparisons of PU/NU classification with PN classification, let us further investigate $\alpha_{\mathrm{PU,PN}}$ and $\alpha_{\mathrm{NU,PN}}$. The indicator $\alpha_{\mathrm{PU,PN}}$ has the following properties:

- It depends only on π_{P}, n_{P}, n_{N}, and n_{U} (since $\pi_{\mathrm{N}} = 1 - \pi_{\mathrm{P}}$).

- It is independent of ℓ, $\mathcal{G}$, and $p(\boldsymbol{x}, y)$—and thus independent of $p_{\mathrm{P}}(\boldsymbol{x})$, $p_{\mathrm{N}}(\boldsymbol{x})$, and $p_{\mathrm{U}}(\boldsymbol{x})$ derived from $p(\boldsymbol{x}, y)$—as long as $\mathfrak{R}_{n,q}(\mathcal{G}) \leq C_{\mathcal{G}}/\sqrt{n}$ can be satisfied (cf. corollary 3.5).

- It is a monotonic function of π_{P}, n_{P}, n_{N}, and n_{U}.

- It is not an upper-bounded function, no matter if π_{P} is fixed or not.

The indicator $\alpha_{\mathrm{NU,PN}}$ also has similar properties, as summarized in table 4.1.

Implications of the monotonicity of $\alpha_{\mathrm{PU,PN}}$ are given as follows. Intuitively, when π_{P} and n_{P} are fixed, larger n_{U} improves the PU bound and larger n_{N} improves the PN bound, respectively. The phenomenon can be straightforwardly seen from the simplified estimation error bounds.

However, it is complicated as to why $\alpha_{\mathrm{PU,PN}}$ is monotonically increasing with π_{P} and decreasing with n_{P}. In PU classification, $\mathscr{X}_{\mathrm{P}}$ also joins the estimation of $\pi_{\mathrm{N}} R_{\mathrm{N}}(g)$, which makes it more impactful than it is in PN classification. In the sense of risk estimation, ℓ has a range from 0 to C_ℓ and a Lipschitz constant L_ℓ, whereas $\widetilde{\ell}$ has a doubled range, from $-C_\ell$ to C_ℓ, and a doubled Lipschitz constant, $2L_\ell$. Then, the same $\mathscr{X}_{\mathrm{P}}$ can affect the estimation of $\mathbb{E}_{\mathrm{P}}[\widetilde{\ell}(g(x), +1)]$ more than the estimation of $\mathbb{E}_{\mathrm{P}}[\ell(g(x), +1)]$. As a consequence, larger n_{P} improves the PU bound more than the PN bound. Additionally, larger π_{P} means larger influence of $\mathbb{E}_{\mathrm{P}}[\widetilde{\ell}(g(x), +1)]$ or $\mathbb{E}_{\mathrm{P}}[\ell(g(x), +1)]$ in the entire risk $R(g)$ so that larger π_{P} impairs the PU bound more than the PN bound (by enlarging the error of estimating $\mathbb{E}_{\mathrm{P}}[\widetilde{\ell}(g(x), +1)]$ or $\mathbb{E}_{\mathrm{P}}[\ell(g(x), +1)]$).

It might be easier to see why $\alpha_{\mathrm{PU,PN}}$ is monotonically increasing with π_{P} if we decompose $\pi_{\mathrm{P}} \mathbb{E}_{\mathrm{P}}[\widetilde{\ell}(g(x), +1)]$ and focus on the estimation of $\pi_{\mathrm{N}} R_{\mathrm{N}}(g)$. In PN classification, we rely on

$$\pi_{\mathrm{N}} R_{\mathrm{N}}(g) = \pi_{\mathrm{N}} \, \mathbb{E}_{\mathrm{N}}[\ell(g(x), -1)],$$

and estimate the right-hand side from $\mathscr{X}_{\mathrm{N}}$ that converges in $\mathcal{O}_p(\pi_{\mathrm{N}}/\sqrt{n_{\mathrm{N}}})$. If π_{P} is larger, then $\pi_{\mathrm{N}} = 1 - \pi_{\mathrm{P}}$ is smaller and the convergence becomes faster. In PU classification, we rely on

$$\pi_{\mathrm{N}} R_{\mathrm{N}}(g) = \mathbb{E}_{\mathrm{U}}[\ell(g(x), -1)] - \pi_{\mathrm{P}} \, \mathbb{E}_{\mathrm{P}}[\ell(g(x), -1)],$$

and estimate it from $\mathscr{X}_{\mathrm{P}}$ and $\mathscr{X}_{\mathrm{U}}$ that converges in $\mathcal{O}_p(\pi_{\mathrm{P}}/\sqrt{n_{\mathrm{P}}} + 1/\sqrt{n_{\mathrm{U}}})$. If π_{P} is larger, the convergence inevitably becomes slower. Therefore, when the sample sizes n_{P}, n_{N}, and n_{U} are all fixed, PU classification becomes more/less favorable as π_{P} decreases/increases.

At first glance, this phenomenon (i.e., the larger π_{P} is, the less favorable PU classification is) may be caused by the empirical risk minimization approach. It is inevitable as long as risk rewriting is employed. That being said, the phenomenon is not an artifact of risk rewriting but inherited in PU classification. Even if we follow other machine learning principles than empirical risk minimization, we will still regard unlabeled data as negative data and extract the information about the negative class from unlabeled data. When the sample sizes n_{P}, n_{N}, and n_{U} are all fixed, less information about the negative class is contained in and can be extracted from unlabeled data as π_{P} increases. This is an intrinsic property of the problem of PU classification itself and is unavoidable for all PU classification methods/algorithms when solving the problem.

Next, a natural question arises: What would the monotonicity of $\alpha_{\mathrm{PU,PN}}$ be, if we enforce n_{P}, n_{N}, and n_{U} to be proportional? To answer the question, we assume that

$$n_{\mathrm{P}}/n_{\mathrm{N}} = \rho_{\mathrm{PN}}, \quad n_{\mathrm{P}}/n_{\mathrm{U}} = \rho_{\mathrm{PU}}, \quad \text{and} \quad n_{\mathrm{N}}/n_{\mathrm{U}} = \rho_{\mathrm{NU}},$$

where ρ_{PN}, ρ_{PU}, and ρ_{NU} are certain positive constants. Then, the two indicators in (4.12) and (4.13) are reduced to

$$\alpha_{\mathrm{PU,PN}} = \frac{\pi_{\mathrm{P}} + \sqrt{\rho_{\mathrm{PU}}}}{\pi_{\mathrm{N}}\sqrt{\rho_{\mathrm{PN}}}} \quad \text{and} \quad \alpha_{\mathrm{NU,PN}} = \frac{\pi_{\mathrm{N}} + \sqrt{\rho_{\mathrm{NU}}}}{\pi_{\mathrm{P}}/\sqrt{\rho_{\mathrm{PN}}}}.$$

It is obvious that $\alpha_{\mathrm{PU,PN}}$ is now increasing with ρ_{PU} and decreasing with ρ_{PN}. It is because, for instance, when ρ_{PN} is fixed and ρ_{PU} increases/decreases, n_{U} is meant to decrease/increase relatively to n_{P} and n_{N}. For $\alpha_{\mathrm{NU,PN}}$, the monotonicity is also shown in table 4.1.

Finally, the properties would dramatically change, if we further enforce

$$\rho_{\mathrm{PN}} = \pi_{\mathrm{P}}/\pi_{\mathrm{N}}$$

that approximately holds in the one-sample case of PN classification. Under this constraint, we have

$$\alpha_{\mathrm{PU,PN}} = \frac{\pi_{\mathrm{P}} + \sqrt{\rho_{\mathrm{PU}}}}{\sqrt{\pi_{\mathrm{P}}\pi_{\mathrm{N}}}} \geq 2\sqrt{\rho_{\mathrm{PU}} + \sqrt{\rho_{\mathrm{PU}}}},$$

$$\alpha_{\mathrm{NU,PN}} = \frac{\pi_{\mathrm{N}} + \sqrt{\rho_{\mathrm{NU}}}}{\sqrt{\pi_{\mathrm{P}}\pi_{\mathrm{N}}}} \geq 2\sqrt{\rho_{\mathrm{NU}} + \sqrt{\rho_{\mathrm{NU}}}},$$

where the equality is achieved at

$$\bar{\pi}_{\mathrm{P}} = \frac{\sqrt{\rho_{\mathrm{PU}}}}{2\sqrt{\rho_{\mathrm{PU}}} + 1} \quad \text{and} \quad \bar{\pi}_{\mathrm{N}} = \frac{\sqrt{\rho_{\mathrm{NU}}}}{2\sqrt{\rho_{\mathrm{NU}}} + 1},$$

respectively. Here, $\alpha_{\mathrm{PU,PN}}$ decreases with π_{P} if $\pi_{\mathrm{P}} < \bar{\pi}_{\mathrm{P}}$ and increases with π_{P} if $\pi_{\mathrm{P}} > \bar{\pi}_{\mathrm{P}}$, though it is not convex in π_{P} when ρ_{PU} is sufficiently small. When n_{U} is sufficiently larger than n_{P}, or equivalently ρ_{PU} is sufficiently small, it is still possible to have $\alpha_{\mathrm{PU,PN}} < 1$ if $\pi_{\mathrm{P}} < 1/2$.

4.4.3.2   Asymptotic comparisons

In practice, we may find that $\widehat{g}_{\mathrm{PU}}$ performs worse than $\widehat{g}_{\mathrm{PN}}$ given $\mathcal{X}_{\mathrm{P}}$, $\mathcal{X}_{\mathrm{N}}$, and $\mathcal{X}_{\mathrm{U}}$. This is probably the consequence of not having large enough n_{U} to make $\widehat{g}_{\mathrm{PU}}$ comparable or better. Should we try to collect more unlabeled data for improving PU classification, or should we give up? Moreover, if we are able to have as many unlabeled data as possible, is there any solution that (in terms of the estimation error bounds) would be provably better than PN classification?

The finite-sample analysis has revealed an intriguing phenomenon—namely, if the sample sizes are proportional and $\rho_{\mathrm{PN}} = \pi_{\mathrm{P}}/\pi_{\mathrm{N}}$, we can observe that

$$\alpha_{\mathrm{PU,PN}} \to \sqrt{\pi_{\mathrm{P}}/\pi_{\mathrm{N}}} \quad \text{as} \quad \rho_{\mathrm{PU}} \to 0,$$

$$\alpha_{\mathrm{NU,PN}} \to \sqrt{\pi_{\mathrm{N}}/\pi_{\mathrm{P}}} \quad \text{as} \quad \rho_{\mathrm{NU}} \to 0.$$

Under those conditions, compared with the PN bound, either the PU bound is smaller if $\pi_{\mathrm{P}} < \pi_{\mathrm{N}}$, or the NU bound is smaller if $\pi_{\mathrm{P}} > \pi_{\mathrm{N}}$. Indeed, a similar conclusion holds without those conditions, as shown by the asymptotic analysis below.

To prepare an asymptotic analysis, consider the following two limiting cases:

- n_{P} and n_{N} stay finite, and n_{U} goes to infinity.

- n_{P} and n_{N} go to infinity in the same order, and n_{U} goes to infinity faster in order than n_{P} and n_{N}.

These limiting cases are sensible when positive, negative, and unlabeled data are all available: Positive and negative data should be roughly equally expensive, while unlabeled data should be much cheaper than positive and negative data. Define two limit indicators (i.e., the limits of $\alpha_{\mathrm{PU,PN}}$ and $\alpha_{\mathrm{NU,PN}}$) as

$$\alpha^*_{\mathrm{PU,PN}} := \lim_{n_{\mathrm{P}},n_{\mathrm{N}},n_{\mathrm{U}}\to\infty} \alpha_{\mathrm{PU,PN}},$$

$$\alpha^*_{\mathrm{NU,PN}} := \lim_{n_{\mathrm{P}},n_{\mathrm{N}},n_{\mathrm{U}}\to\infty} \alpha_{\mathrm{NU,PN}},$$

where n_{P}, n_{N}, and n_{U} are independent variables and approach infinity independently. Intuitively, they measure the relative quality of the limits of the PU and NU bounds against the limit of the PN bound. Note that the existence of them can be guaranteed in both limiting cases.

Let us figure out the limiting values accordingly. For the first limiting case where n_{P} and n_{N} stay finite, it is clear that

$$\alpha^*_{\mathrm{PU,PN}} = \frac{\pi_{\mathrm{P}}\sqrt{n_{\mathrm{N}}}}{\pi_{\mathrm{N}}\sqrt{n_{\mathrm{P}}}} \quad \text{and} \quad \alpha^*_{\mathrm{NU,PN}} = \frac{\pi_{\mathrm{N}}\sqrt{n_{\mathrm{P}}}}{\pi_{\mathrm{P}}\sqrt{n_{\mathrm{N}}}}.$$

Thus, $\alpha^*_{\mathrm{PU,PN}}\alpha^*_{\mathrm{NU,PN}} = 1$, which implies that either $\alpha^*_{\mathrm{PU,PN}} < 1$ or $\alpha^*_{\mathrm{NU,PN}} < 1$ unless $n_{\mathrm{P}}/n_{\mathrm{N}} = \pi_{\mathrm{P}}^2/\pi_{\mathrm{N}}^2$. For the second limiting case, as n_{P} and n_{N} go to infinity in the same order, we need to define

$$\rho^*_{\mathrm{PN}} := \lim_{n_{\mathrm{P}},n_{\mathrm{N}}\to\infty} n_{\mathrm{P}}/n_{\mathrm{N}},$$

so that $0 < \rho^*_{\mathrm{PN}} < \infty$; we can then obtain that

$$\alpha^*_{\mathrm{PU,PN}} = \frac{\pi_{\mathrm{P}}}{\pi_{\mathrm{N}}\sqrt{\rho^*_{\mathrm{PN}}}} \quad \text{and} \quad \alpha^*_{\mathrm{NU,PN}} = \frac{\pi_{\mathrm{N}}\sqrt{\rho^*_{\mathrm{PN}}}}{\pi_{\mathrm{P}}}.$$

Again, either $\alpha^*_{\mathrm{PU,PN}} < 1$ or $\alpha^*_{\mathrm{NU,PN}} < 1$ unless $\rho^*_{\mathrm{PN}} = \pi_{\mathrm{P}}^2/\pi_{\mathrm{N}}^2$.

It is amazing that, for both of the limiting cases, in terms of the estimation error bounds, either PU or NU classification will improve on PN classification. The limit indicators $\alpha^*_{\mathrm{PU,PN}}$ and $\alpha^*_{\mathrm{NU,PN}}$ reflect the limit power and the potential of PU and NU classification, respectively. Therefore, we can answer the persist-or-give-up question two ways:

- If we have confirmed that $\alpha^*_{\mathrm{PU,PN}} < 1$, PU classification is promising and we should collect more unlabeled data to improve PU classification.[3]

- Otherwise, we should give up PU classification; NU classification is promising instead and we may collect more unlabeled data as well to improve NU classification.

The only exception is $n_\mathrm{P}/n_\mathrm{N} = \pi_\mathrm{P}^2/\pi_\mathrm{N}^2$ or $\rho_\mathrm{PN}^* = \pi_\mathrm{P}^2/\pi_\mathrm{N}^2$ in the first and second limiting cases; when it occurs, we have $\alpha_\mathrm{PU,PN}^* = \alpha_\mathrm{NU,PN}^* = 1$, which implies that PN, PU, and NU classification are asymptotically equivalent to each other. In principle, the exception should be exceptionally rare (at least for the first limiting case, since $n_\mathrm{P}/n_\mathrm{N}$ is a rational number whereas $\pi_\mathrm{P}^2/\pi_\mathrm{N}^2$ is a real number).

In summary, PU/NU classification as a whole can serve as a comparable or superior alternative to PN classification when $\mathscr{X}_\mathrm{P}$, $\mathscr{X}_\mathrm{N}$, and $\mathscr{X}_\mathrm{U}$ are available but n_U is sufficiently larger than n_P and n_N; in particular, asymptotically, one of them can improve on PN classification sooner or later as n_U increases. More practically, if $\mathscr{X}_\mathrm{P}$, $\mathscr{X}_\mathrm{N}$, and $\mathscr{X}_\mathrm{U}$ are available, it would be natural to use all of them for training a classifier. This topic, which is called *PNU classification*, will be explored in chapter 5.

The finite-sample and asymptotic comparisons have originally been presented in Niu et al. (2016) under the symmetric condition (4.4), and the discussions here extend the previous comparisons to general losses. Since then, risk rewriting has been a core technology and the empirical risk minimization approach has been a key methodology and a hot research topic in machine learning from weak supervision, together with *regularization* (e.g., Belkin et al., 2006; Grandvalet and Bengio, 2005; Miyato et al., 2019) and *pseudo-labeling* (e.g., McLachlan, 1975; Blum and Mitchell, 1998; Sohn et al., 2020). The success of risk rewriting in PU classification opens the door to a new world for researchers and engineers who are interested in weakly supervised classification. Indeed, a lot of risk-rewriting-based weakly supervised classification methods/algorithms have been designed and developed in recent years. Some representative ones are covered in chapters 5–10.

5 Positive-Negative-Unlabeled (PNU) Classification

A classification scenario where a large number of inexpensive unlabeled data is available in addition to a limited number of expensive labeled data is called *semi-supervised classification* (Chapelle et al., 2006). Semi-supervised classification is one of the central topics of classification from weak supervision and has been extensively studied in the last decade. In binary classification scenarios where patterns belong to either the positive class or the negative class, semi-supervised classification uses positive, negative, and unlabeled data. For this reason, we also refer to semi-supervised binary classification as *positive-negative-unlabeled (PNU) classification*. This chapter explores methods of PNU classification.

5.1 Introduction

Collecting a large amount of labeled data is a critical bottleneck in real-world machine learning applications because of laborious manual annotation, as explained in chapter 1. On the other hand, unlabeled data can often be collected automatically and abundantly, e.g., by a web crawler. This has led to the development of various semi-supervised classification algorithms over the past decades.

To leverage unlabeled data in classifier training, most of the existing semi-supervised classification methods rely on particular assumptions on the data distribution (Chapelle et al., 2006). For example, the *cluster assumption* requires that samples in the same cluster are likely to share the same label (Chapelle et al., 2003), and the *manifold assumption* supposes that samples are distributed on a low-dimensional manifold in the data space (Belkin et al., 2006). In the existing framework of semi-supervised classification, such a distributional assumption is typically encoded as a *regularizer*; by that, the learned classifier is *biased* to fit the distributional assumption. However, if the assumption contradicts the true data distribution, the bias behaves adversely. As a result, the obtained classifier performs even worse than naive supervised classification from limited labeled data (Cozman et al., 2003; Sokolovska et al., 2008; Li and Zhou, 2015; Krijthe and Loog, 2017). Therefore, how to theoretically guarantee the performance of semi-supervised classification is a key challenge.

In chapter 4, we showed that the risk estimators of PU classification utilize unlabeled data for *risk evaluation*, implying that label information is directly extracted from unlabeled data without restrictive distributional assumptions. Besides, the theoretical analysis in section 4.4.3 showed that PU classification, or its counterpart negative-unlabeled (NU) classification, is likely to outperform positive-negative (PN) classification, depending on the number of positive, negative, and unlabeled samples (Niu et al., 2016). It is thus naturally expected that combining PN, PU, and NU classification can be a promising approach to semi-supervised classification without restrictive distributional assumptions.

In this chapter, we first formulate the problem of PNU classification in section 5.2. We then introduce two semi-supervised classification methods based on the manifold assumption (Belkin et al., 2006) in section 5.3 and the *information-maximization principle* (Niu et al., 2013) in section 5.4, both of which rely on strong distributional assumptions. Then, in section 5.5, we introduce a semi-supervised classification method based on PU classification (Sakai et al., 2017). For this method, we show through estimation error analysis that unlabeled data really helps to improve the performance without any distributional assumption. We also present numerical results and several extensions.

5.2   Formulation

In this section, we formulate the problem of PNU classification.

PNU classification uses positive, negative, and unlabeled data:

$$\mathcal{X}_{\mathrm{P}} := \{x_i^{\mathrm{P}}\}_{i=1}^{n_{\mathrm{P}}} \overset{\text{i.i.d.}}{\sim} p_{\mathrm{P}}(x) := p(x \mid y = +1),$$

$$\mathcal{X}_{\mathrm{N}} := \{x_i^{\mathrm{N}}\}_{i=1}^{n_{\mathrm{N}}} \overset{\text{i.i.d.}}{\sim} p_{\mathrm{N}}(x) := p(x \mid y = -1),$$

$$\mathcal{X}_{\mathrm{U}} := \{x_i^{\mathrm{U}}\}_{i=1}^{n_{\mathrm{U}}} \overset{\text{i.i.d.}}{\sim} p_{\mathrm{U}}(x) := \pi_{\mathrm{P}} p_{\mathrm{P}}(x) + \pi_{\mathrm{N}} p_{\mathrm{N}}(x),$$

where

$$\pi_{\mathrm{P}} := p(y = +1),$$

$$\pi_{\mathrm{N}} := p(y = -1)$$

are the class-prior probabilities for the positive and negative classes such that

$$\pi_{\mathrm{P}} + \pi_{\mathrm{N}} = 1.$$

Mixing all the positive, negative, and unlabeled samples, we also use the following notation:

$$\mathcal{X} := \{\underbrace{x_1, \ldots, x_{n_{\mathrm{P}}}}_{\text{positive samples}}, \underbrace{x_{n_{\mathrm{P}}+1}, \ldots, x_{n_{\text{labeled}}}}_{\text{negative samples}}, \underbrace{x_{n_{\text{labeled}}+1}, \ldots, x_{n_{\text{all}}}}_{\text{unlabeled samples}}\},$$

$$y_i := \begin{cases} +1 & (i=1,\ldots,n_{\mathrm{P}}), \\ -1 & (i=n_{\mathrm{P}}+1,\ldots,n_{\mathrm{labeled}}), \\ 0 & (i=n_{\mathrm{labeled}}+1,\ldots,n_{\mathrm{all}}), \end{cases}$$

where

$$n_{\mathrm{labeled}} := n_{\mathrm{P}} + n_{\mathrm{N}},$$

$$n_{\mathrm{all}} := n_{\mathrm{P}} + n_{\mathrm{N}} + n_{\mathrm{U}}.$$

The goal of PNU classification is to obtain a PN classifier that minimizes the classification risk:

$$R(g) := \mathbb{E}_{p(\boldsymbol{x},y)}[\ell(g(\boldsymbol{x}),y)],$$

where $\ell \colon \mathbb{R} \to \mathbb{R}$ is a loss (see Section 2.1.3) and $g \colon \mathbb{R}^d \to \mathbb{R}$ is a binary classifier.

5.3 Manifold-Based Semi-Supervised Classification

In this section, a semi-supervised classification method based on *manifold regularization* (Chapelle et al., 2006; Belkin et al., 2006) is introduced.

5.3.1 Laplacian Regularization

Manifold-based semi-supervised classification is based on the assumption that class labels are the same on a manifold.[1] In the simplest and most intuitive case, input patterns are clustered, and samples in the same cluster are assumed to belong to the same class (figure 5.1).

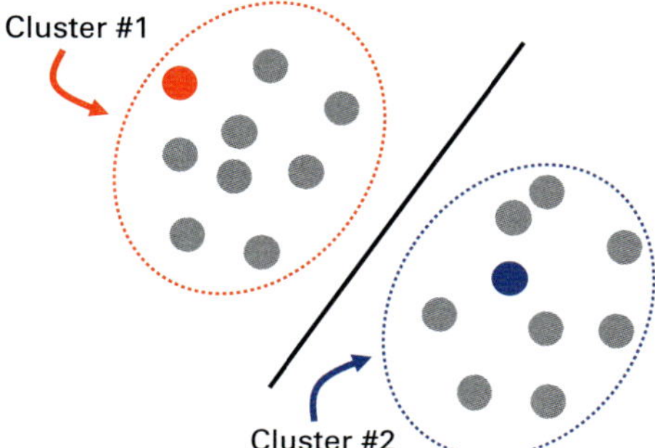

Figure 5.1
Cluster assumption in semi-supervised classification. There are two clusters. A cluster assumption requires that points in cluster #1 belong to the red class while those in cluster #2 belong to the blue class. A classifier based on the cluster assumption then draws a decision boundary between two clusters.

In ordinary supervised classification, a classifier g is trained to minimize the empirical classification risk (see section 3.1.1):

$$\frac{1}{n_{\text{labeled}}} \sum_{i=1}^{n_{\text{labeled}}} \ell(g(\boldsymbol{x}_i), y_i).$$

Manifold regularization is a way to incorporate a cluster assumption in classifier training through regularization. More specifically, we minimize

$$\frac{1}{n_{\text{labeled}}} \sum_{i=1}^{n_{\text{labeled}}} \ell(g(\boldsymbol{x}_i), y_i) + \gamma \sum_{i,i'=1}^{n_{\text{all}}} W_{i,i'} \left(g(\boldsymbol{x}_i) - g(\boldsymbol{x}_{i'}) \right)^2, \tag{5.1}$$

where $\gamma \geq 0$ is a smoothing parameter. We denote similarity between $\boldsymbol{x}_i$ and $\boldsymbol{x}_{i'}$ by $W_{i,i'} \in [0, 1]$, which takes a large/small value if $\boldsymbol{x}_i$ and $\boldsymbol{x}_{i'}$ are similar/dissimilar. Thus, minimizing (5.1) encourages $g(\boldsymbol{x}_i)$ and $g(\boldsymbol{x}_{i'})$ to take similar values if $\boldsymbol{x}_i$ and $\boldsymbol{x}_{i'}$ are similar.

Typically, $W_{i,i'}$ is defined as follows:

- Gaussian similarity:

$$W_{i,i'} := \exp\left(-\frac{\|\boldsymbol{x}_i - \boldsymbol{x}_{i'}\|^2}{2t^2} \right),$$

where $t > 0$ is a tuning parameter to control the Gaussian bandwidth.

- k-nearest-neighbor similarity:

$$W_{i,i'} := \begin{cases} 1 & (\boldsymbol{x}_i \in \mathscr{N}_k(\boldsymbol{x}_{i'}) \text{ or } \boldsymbol{x}_{i'} \in \mathscr{N}_k(\boldsymbol{x}_i)), \\ 0 & (\text{otherwise}), \end{cases}$$

where $\mathscr{N}_k(\boldsymbol{x})$ denotes the set of k nearest neighbor samples of $\boldsymbol{x}$ in $\{\boldsymbol{x}_i\}_{i=1}^{n_{\text{all}}}$ and $1 \leq k \leq n_{\text{all}}$ is a tuning parameter to control the locality. Note that this definition yields a *sparse* similarity matrix $\boldsymbol{W}$, which often contributes to high computational efficiency.

- Local scaling similarity (Zelnik-Manor and Perona, 2005):

$$W_{i,i'} := \exp\left(-\frac{\|\boldsymbol{x}_i - \boldsymbol{x}_{i'}\|^2}{2t_i t_{i'}} \right),$$

where t_i is the local scaling defined by

$$t_i := \|\boldsymbol{x}_i - \boldsymbol{x}_i^{(k)}\|$$

and $\boldsymbol{x}_i^{(k)}$ denotes the k-th nearest neighbor of $\boldsymbol{x}_i$ in $\{\boldsymbol{x}_i\}_{i=1}^{n_{\text{all}}}$. It is also possible to combine k-nearest-neighbor similarity with local scaling similarity.

Below, the similarity matrix $\boldsymbol{W}$ is assumed to be symmetric (i.e., $W_{i,i'} = W_{i',i}$).

5.3.2   Implementation

Let us consider a linear-in-parameter model (see section 2.1.2.2):

$$g(x) = \sum_{j=1}^{b} \alpha_j \phi_j(x) = \boldsymbol{\alpha}^\top \boldsymbol{\phi}(x),$$

where b denotes the number of parameters,

$$\boldsymbol{\alpha} := (\alpha_1, \ldots, \alpha_b)^\top \in \mathbb{R}^b$$

is a parameter vector, and

$$\boldsymbol{\phi}(x) := (\phi_1(x), \ldots, \phi_b(x))^\top \in \mathbb{R}^b$$

is a vector of basis functions. Here we show how to compute the minimizer of (5.1) for the squared loss (see section 2.1.3) with ℓ_2-regularization (see section 2.1.5), i.e.,

$$\frac{1}{2n_{\text{labeled}}} \sum_{i=1}^{n_{\text{labeled}}} \left(y_i - \boldsymbol{\phi}(x_i)^\top \boldsymbol{\alpha}\right)^2 + \frac{\lambda}{2} \|\boldsymbol{\alpha}\|^2$$

$$+ \frac{\gamma}{4} \sum_{i,i'=1}^{n_{\text{all}}} W_{i,i'} \left(\boldsymbol{\phi}(x_i)^\top \boldsymbol{\alpha} - \boldsymbol{\phi}(x_{i'})^\top \boldsymbol{\alpha}\right)^2, \tag{5.2}$$

where we used the fact that, for $y = \pm 1$,

$$(1 - y\widehat{y})^2 = (y - \widehat{y})^2.$$

Let $\boldsymbol{D}$ be the diagonal matrix whose diagonal elements are given by the row-sums of matrix $\boldsymbol{W}$, as shown by

$$\boldsymbol{D} := \operatorname{diag}\left(\sum_{i=1}^{n_{\text{all}}} W_{1,i}, \ldots, \sum_{i=1}^{n_{\text{all}}} W_{n_{\text{all}},i}\right),$$

and let

$$\boldsymbol{L} := \boldsymbol{D} - \boldsymbol{W}.$$

Then the third term in (5.2) can be rewritten as

$$\sum_{i,i'=1}^{n_{\text{all}}} W_{i,i'} \left(\boldsymbol{\phi}(x_i)^\top \boldsymbol{\alpha} - \boldsymbol{\phi}(x_{i'})^\top \boldsymbol{\alpha}\right)^2$$

$$= \sum_{i=1}^{n_{\text{all}}} D_{i,i}(\boldsymbol{\phi}(x_i)^\top \boldsymbol{\alpha})^2 - 2 \sum_{i,i'=1}^{n_{\text{all}}} W_{i,i'} \boldsymbol{\phi}(x_i)^\top \boldsymbol{\alpha}\boldsymbol{\phi}(x_{i'})^\top \boldsymbol{\alpha} + \sum_{i'=1}^{n_{\text{all}}} D_{i',i'}(\boldsymbol{\phi}(x_{i'})^\top \boldsymbol{\alpha})^2$$

$$= 2\boldsymbol{\alpha}^\top \left(\sum_{i,i'=1}^{n_{\text{all}}} L_{i,i'} \boldsymbol{\phi}(x_i)^\top \boldsymbol{\phi}(x_{i'})^\top\right) \boldsymbol{\alpha}.$$

Thus, (5.2) can be expressed as

$$\frac{1}{2n_{\text{labeled}}}\left\|\boldsymbol{\Phi}_{\text{labeled}}\boldsymbol{\alpha}-\boldsymbol{y}_{\text{labeled}}\right\|^2+\frac{\lambda}{2}\|\boldsymbol{\alpha}\|^2+\frac{\gamma}{2}\boldsymbol{\alpha}^\top\boldsymbol{\Phi}_{\text{all}}^\top\boldsymbol{L}\boldsymbol{\Phi}_{\text{all}}\boldsymbol{\alpha},\tag{5.3}$$

where $\boldsymbol{\Phi}_{\text{labeled}}$ is the $n_{\text{labeled}}\times b$ matrix and $\boldsymbol{\Phi}_{\text{all}}$ is the $n_{\text{all}}\times b$ matrix whose (i,j)-entries are respectively given by

$$[\boldsymbol{\Phi}_{\text{labeled}}]_{i,j}:=\phi_j(\boldsymbol{x}_i)\text{ for }i=1,\ldots,n_{\text{labeled}}\text{ and }j=1,\ldots,b,$$

$$[\boldsymbol{\Phi}_{\text{all}}]_{i,j}:=\phi_j(\boldsymbol{x}_i)\text{ for }i=1,\ldots,n_{\text{all}}\text{ and }j=1,\ldots,b,$$

and

$$\boldsymbol{y}_{\text{labeled}}=(y_1,\ldots,y_{n_{\text{labeled}}})^\top.$$

The minimizer of (5.3), denoted by $\widehat{\boldsymbol{\alpha}}$, can be obtained analytically as

$$\widehat{\boldsymbol{\alpha}}=(\boldsymbol{\Phi}_{\text{labeled}}\boldsymbol{\Phi}_{\text{labeled}}^\top+\lambda n_{\text{labeled}}\boldsymbol{I}+\gamma n_{\text{labeled}}\boldsymbol{\Phi}_{\text{all}}^\top\boldsymbol{L}\boldsymbol{\Phi}_{\text{all}})^{-1}\boldsymbol{\Phi}_{\text{labeled}}^\top\boldsymbol{y}_{\text{labeled}},$$

where $\boldsymbol{I}$ denotes the identity matrix.

The term $\boldsymbol{\alpha}^\top\boldsymbol{\Phi}_{\text{all}}^\top\boldsymbol{L}\boldsymbol{\Phi}_{\text{all}}\boldsymbol{\alpha}$ is called the *Laplacian regularizer*, following the terminology in *spectral graph theory*.[2] Note that the same Laplacian regularizer can be applied to multi-class classification problems. Thus, the above semi-supervised classification method can be naturally extended to multi-class classification scenarios just by changing the model and loss to multi-class ones (see section 2.2).

5.4   Information-Theoretic Semi-Supervised Classification

The above Laplacian-based semi-supervised classification method is based on the manifold assumption. However, the manifold setting is not the only way to go—the *information-maximization principle* is a useful alternative (see section 1.3.2). Information-maximization clustering (Agakov and Barber, 2006; Gomes et al., 2010; Sugiyama et al., 2014; Hu et al., 2017) is a typical example of the information-maximization principle, where a probabilistic classifier is trained in an unsupervised manner so that a given information measure between data and cluster assignments is maximized. In this section, a semi-supervised classification method based on the information-maximization principle is introduced (Niu et al., 2013).

5.4.1   Squared-Loss Mutual Information Regularization

The basic idea of semi-supervised classification based on the information-maximization principle is to use an information measure such as *mutual information* (Shannon, 1948) and its variants (Sugiyama et al., 2012). Among them, the *squared-loss mutual information* (SMI) is useful in practice because of its good computational properties and high robustness to outliers (Suzuki et al., 2009; Sugiyama, 2013). For pattern x and label y equipped with joint density $p(x, y)$, marginal density $p(x)$ and marginal probability $p(y)$, SMI between x and y is defined as

$$\text{SMI} := \frac{1}{2} \int \sum_y p(x)p(y) \left(\frac{p(x,y)}{p(x)p(y)} - 1 \right)^2 dx.$$

SMI is the *Pearson divergence* (Pearson, 1900) of $p(x, y)$ from $p(x)p(y)$, while ordinary *mutual information* (Shannon, 1948) is the *Kullback-Leibler divergence* (Kullback and Leibler, 1951) of $p(x, y)$ from $p(x)p(y)$:

$$\text{MI} := \int \sum_y p(x,y) \log \frac{p(x,y)}{p(x)p(y)} dx.$$

They both belong to the class of f-divergences (Ali and Silvey, 1966; Csiszár, 1967), and thus they share similar mathematical properties. For instance, both of them are nonnegative, and take zero if and only if x and y are statistically independent, i.e., $p(x, y) = p(x)p(y)$.

In Sugiyama et al. (2014), a computationally efficient unsupervised SMI approximator was proposed. Under the uniform distribution on the class-prior probability $p(y)$, SMI can be expressed as

$$\text{SMI} = \frac{c}{2} \int \sum_y (p(y \mid x))^2 p(x) dx - \frac{1}{2}, \tag{5.4}$$

where c is the number of classes (in the binary case, $c = 2$). Then, $p(y \mid x)$ is approximated by a kernel model (see section 2.1.2.3):

$$\sum_{i=1}^{n_{\text{all}}} \alpha_{y,i} K(x, x_i), \tag{5.5}$$

where $\alpha = \{\alpha_1, \ldots, \alpha_c\}$ and $\alpha_y = (\alpha_{y,1}, \ldots, \alpha_{y,n_{\text{all}}})^\top$ are model parameters and $K(x, x')$ is a kernel. After approximating the expectation with respect to $p(x)$ in (5.4) by the empirical average, an SMI approximator is derived as

$$\frac{c}{2n_{\text{all}}} \sum_y \alpha_y^\top K^2 \alpha_y - \frac{1}{2},$$

where $K_{i,i'} = K(x_i, x_{i'})$ is the kernel matrix.

When K is full-rank, a variation is to use

$$q(y \mid x; \alpha) := \alpha_y^\top E^{-1/2} K^{-1/2} k_{n_{\text{all}}}(x) \tag{5.6}$$

as a model of $p(y \mid x)$, where

$$k_{n_{\text{all}}}(x) = (K(x, x_1), \ldots, K(x, x_{n_{\text{all}}}))^\top$$

and E is the diagonal matrix with i-th diagonal element given by $\sum_{i'=1}^{n_{\text{all}}} K(x_i, x_{i'})$. Plugging (5.6) into (5.4) and approximating the expectation with respect to $p(x)$ by the empirical average gives the following SMI approximator:

$$\widehat{\text{SMI}} := \frac{c}{2n_{\text{all}}} \text{tr}\left(A^\top E^{-1/2} K E^{-1/2} A\right) - \frac{1}{2}, \tag{5.7}$$

where $A = (\alpha_1, \ldots, \alpha_c) \in \mathbb{R}^{n_{\text{all}} \times c}$.

Below, we employ (5.7) as a regularizer and introduce a method called *SMI regularization* (SMIR) (Niu et al., 2013). More specifically, there are three objectives in the SMIR method: (i) minimize a loss function $\Delta(p, q)$ to make the model q closer to $p(y \mid x)$; (ii) maximize $\widehat{\text{SMI}}$; (iii) regularize α. As a result, the optimization problem of SMIR is given as

$$\min_{\alpha_1,\ldots,\alpha_c \in \mathbb{R}^n} \Delta(p, q) - \gamma \widehat{\text{SMI}} + \lambda \sum_{y \in \mathcal{Y}} \frac{1}{2} \|\alpha_y\|_2^2, \tag{5.8}$$

where $\gamma, \lambda > 0$ are regularization parameters.

A remarkable characteristic of optimization (5.8) is its convexity, as long as the kernel function K is non-negative and $\lambda > \gamma c / n_{\text{all}}$.

Theorem 5.1 *(Niu et al., 2013) Assume that $K : \mathcal{X} \times \mathcal{X} \mapsto \mathbb{R}_+$, $\Delta(p, q)$ is convex with respect to q, and $\lambda > \gamma c / n_{\text{all}}$. Then optimization (5.8) is strictly convex, and there exists a unique globally optimal solution.[3]*

Proof   The loss function $\Delta(p, q)$ is convex with respect to $q(y \mid x; \alpha)$, and $q(y \mid x; \alpha)$ is linear with respect to α_y. Therefore, $\Delta(p, q)$ is convex with respect to α_y. The ℓ_2-norm of α_y is strictly convex with respect to α_y, i.e., it takes zero if and only if α_y is identically zero. Then, the rest is whether and when $\widehat{\text{SMI}} = \frac{c}{2n_{\text{all}}} \sum_y \alpha_y^\top E^{-1/2} K E^{-1/2} \alpha_y - 1/2$ is convex with respect to α_y. Recall that E is the diagonal matrix with i-th diagonal element given by $\sum_{i'=1}^{n_{\text{all}}} K(x_i, x_{i'})$ and the kernel function K is non-negative. Thus, $E^{-1/2}$ is positive definite, and because K is assumed to be full-rank, $E^{-1/2} K E^{-1/2}$ is positive definite. Therefore, $\widehat{\text{SMI}}$ is convex.

Finally, to make the objective function in (5.8) convex, $-\gamma c / (2n_{\text{all}}) + \lambda/2 > 0$ needs to be satisfied, which is $\lambda > \gamma c / n_{\text{all}}$. $\square$

Equation (5.6) is introduced for the following reasons: First, in principle, any kernel model, which is linear with respect to α_y, may be used to approximate $p(y \mid x)$, and maximizing $\widehat{\text{SMI}}$ alone must be non-convex. However, optimization (5.8) becomes convex if λ is large enough. Hence, only λ above a certain threshold is acceptable: The threshold of (5.6) is $\gamma c / n_{\text{all}}$. The threshold of (5.5) is $\|K\|_2^2 \cdot \gamma c / n_{\text{all}}$, where $\|K\|_2$ is the spectral norm of K. It depends on all the training data thoroughly and is usually much larger than $\gamma c / n_{\text{all}}$. Second, equation (5.6) experimentally outperformed (5.5) (cf. Niu et al., 2013).

5.4.2 Implementation

We choose the squared difference of probabilities p and q as the loss function (Sugiyama, 2010), since it yields an analytical solution and facilitates theoretical analysis:

$$\Delta_2(p, q) := \frac{1}{2} \int_{\mathcal{X}} \sum_{y \in \mathcal{Y}} (p(y \mid x) - q(y \mid x; \alpha))^2 p(x) dx.$$

Its empirical version is given as

$$\widehat{\Delta}_2 = \text{Const.} - \frac{1}{n_{\text{labeled}}} \sum_{i=1}^{n_{\text{labeled}}} q(y_i \mid x_i) + \frac{1}{2n_{\text{labeled}}} \sum_{i=1}^{n_{\text{labeled}}} \sum_{y=1}^{c} (q(y \mid x_i))^2. \tag{5.9}$$

Let $Y \in \mathbb{R}^{n_{\text{labeled}} \times c}$ be the class indicator matrix for n_{labeled} labeled data and $B = (I_{n_{\text{labeled}}}; 0_{n_{\text{U}} \times n_{\text{labeled}}}) \in \mathbb{R}^{n_{\text{all}} \times n_{\text{labeled}}}$. Subsequently, (5.9) can be expressed by

$$\widehat{\Delta}_2 = \text{Const.} - \frac{1}{n_{\text{labeled}}} \text{tr}(Y^\top B^\top K^{1/2} E^{-1/2} A)$$

$$+ \frac{1}{2n_{\text{labeled}}} \text{tr}(A^\top E^{-1/2} K^{1/2} BB^\top K^{1/2} E^{-1/2} A). \tag{5.10}$$

Substituting (5.10) into optimization (5.8), we will get the following objective function:

$$\mathcal{F}(A) = -\frac{1}{n_{\text{labeled}}} \text{tr}(Y^\top B^\top K^{1/2} E^{-1/2} A)$$

$$+ \frac{1}{2n_{\text{labeled}}} \text{tr}(A^\top E^{-1/2} K^{1/2} BB^\top K^{1/2} E^{-1/2} A)$$

$$- \frac{\gamma c}{2n_{\text{all}}} \text{tr}(A^\top E^{-1/2} K E^{-1/2} A) + \frac{\lambda}{2} \text{tr}(A^\top A).$$

At last, by equating $\nabla \mathcal{F}$ to the zero matrix, we obtain the analytical solution to unconstrained optimization problem (5.8):

$$A_{\mathcal{F}}^* = n_{\text{all}} \left(n_{\text{all}} E^{-1/2} K^{1/2} BB^\top K^{1/2} E^{-1/2} + \lambda n_{\text{all}} n_{\text{labeled}} I_{n_{\text{all}}} \right.$$

$$\left. - \gamma n_{\text{labeled}} c E^{-1/2} K E^{-1/2} \right)^{-1} E^{-1/2} K^{1/2} BY.$$

It is recommended to post-process the model parameters as

$$\bar{\alpha}_y = n_{\text{all}} \pi_y \cdot \frac{K^{-1/2} E^{-1/2} \alpha_y^*}{1_{n_{\text{all}}}^\top K^{1/2} E^{-1/2} \alpha_y^*},$$

where $\bar{\alpha}_y$ is a normalized version of α_y^* and π_y is an estimate of $p(y)$ based on labeled data. In addition, probability estimates should be non-negative, and thus the final solution may be obtained as follows (cf. Yamada et al., 2011):

$$\hat{p}(y\,|\,\boldsymbol{x}) = \frac{\max\left\{0, \bar{\boldsymbol{\alpha}}_y^\top \boldsymbol{k}_{n_{\mathrm{all}}}(\boldsymbol{x})\right\}}{\sum_{y'=1}^c \max\left\{0, \bar{\boldsymbol{\alpha}}_{y'}^\top \boldsymbol{k}_{n_{\mathrm{all}}}(\boldsymbol{x})\right\}}.$$

5.5   PU+PN Classification

Semi-supervised classification methods introduced above are based on specific assumptions on underlying data distributions, such as the cluster assumption in section 5.3 and the information-maximization principle in section 5.4. However, such a strong assumption could be violated in practice. In this section, we introduce a semi-supervised classification method based on PU classification (Sakai et al., 2017), which does not rely on such strong distributional assumptions.

5.5.1   PNU and PU+NU Risk Estimators

The key idea in Sakai et al. (2017) is to combine PU classification with PN classification, which we refer to as "PU+PN" classification (see figure 5.2). The empirical risk estimator in PU+PN classification is defined as

$$\widehat{R}^\gamma_{\mathrm{PU+PN}}(g) := \gamma \widehat{R}_{\mathrm{PU}}(g) + (1 - \gamma)\widehat{R}_{\mathrm{PN}}(g),$$

where $0 \leq \gamma \leq 1$ is a hyper-parameter to control the trade-off between the PU and PN risks. Since

$$\mathbb{E}[\widehat{R}_{\mathrm{PU}}(g)] = \mathbb{E}[\widehat{R}_{\mathrm{PN}}(g)] = R(g),$$

$\widehat{R}^\gamma_{\mathrm{PU+PN}}$ is also unbiased to the classification risk R, i.e.,

$$\mathbb{E}[\widehat{R}^\gamma_{\mathrm{PU+PN}}(g)] = R(g)$$

for any $0 \leq \gamma \leq 1$. In this PU+PN classification, P, N, and U data are all incorporated into risk estimation without any strong assumption on the underlying data distribution.

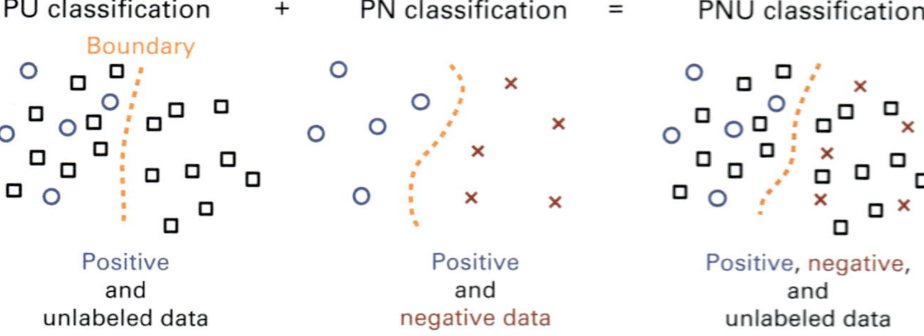

Figure 5.2
Illustration of PU+PN classification.

Although U data is naturally included in PU+PN classification, PU classification and PN classification are not necessarily the best combination. Indeed, as shown in section 4.4.3, which was originally provided in Niu et al. (2016), the NU classifier is superior to the PU classifier, e.g., if $\pi_P = \pi_N$, $n_N \gg n_P$, and n_U is sufficiently large. Motivated by this fact, Sakai et al. (2017) proposed to further combine NU+PN classification with PU+PN classification, which is called *PNU classification*. Let us define the empirical risk estimator for NU+PN classification as

$$\widehat{R}^{\gamma}_{\mathrm{NU+PN}}(g) := \gamma \widehat{R}_{\mathrm{NU}}(g) + (1 - \gamma)\widehat{R}_{\mathrm{PN}}(g),$$

where $0 \leq \gamma \leq 1$ is a hyper-parameter to control the trade-off between the NU and PN risks. Based on this, the empirical risk estimator for PNU classification is defined as

$$\widehat{R}^{\eta}_{\mathrm{PNU}}(g) := \begin{cases} \widehat{R}^{\eta}_{\mathrm{PU+PN}}(g) & (\eta \geq 0), \\ \widehat{R}^{-\eta}_{\mathrm{NU+PN}}(g) & (\eta < 0), \end{cases}$$

where $-1 \leq \eta \leq 1$ is a hyper-parameter to control the trade-off between PU+PN classification and NU+PN classification.

In addition to PU+PN classification and NU+PN classification, PU+NU classification can also be considered in the same way. Its empirical risk estimator is defined as

$$\widehat{R}^{\gamma}_{\mathrm{PU+NU}}(g) := (1 - \gamma)\widehat{R}_{\mathrm{PU}}(g) + \gamma \widehat{R}_{\mathrm{NU}}(g),$$

where $0 \leq \gamma \leq 1$ is a hyper-parameter to control the trade-off between the PU and NU risks. Although PU+NU classification incorporates P, N, and U data into empirical risk approximation in a symmetric and natural way, we will show below that this combination is actually not promising from both the theoretical and empirical viewpoints.

In chapter 11, we will see a *non-negative correction* technique to further improve the performance.

5.5.2 PNU vs. PU+NU Classification

Here, based on the estimation error bounds derived in section 4.4.3, we discuss which of PU+NU classification and PNU classification is more promising (Niu et al., 2016).

Recall that $\widehat{g}_{\mathrm{PN}}$, $\widehat{g}_{\mathrm{PU}}$, and $\widehat{g}_{\mathrm{NU}}$ are the minimizers of $\widehat{R}_{\mathrm{PN}}(g)$, $\widehat{R}_{\mathrm{PU}}(g)$, and $\widehat{R}_{\mathrm{NU}}(g)$, respectively. If

$$\alpha_{\mathrm{PU,PN}} = \frac{\pi_P/\sqrt{n_P} + 1/\sqrt{n_U}}{\pi_N/\sqrt{n_N}} > 1,$$

PN classification is more promising than PU classification in the following sense:

$$R(\widehat{g}_{\mathrm{PN}}) < R(\widehat{g}_{\mathrm{PU}}).$$

Similarly, if

$$\alpha_{\mathrm{NU,PN}} = \frac{\pi_N/\sqrt{n_N} + 1/\sqrt{n_U}}{\pi_P/\sqrt{n_P}} > 1,$$

PN classification is more promising than NU classification in the following sense:

$$R(\widehat{g}_{\mathrm{PN}}) < R(\widehat{g}_{\mathrm{NU}}).$$

If n_{U} is not so large compared with n_{P} and n_{N}, $\alpha_{\mathrm{PU,PN}} > 1$ and $\alpha_{\mathrm{NU,PN}} > 1$ tend to hold. Then we have

$$R(\widehat{g}_{\mathrm{PN}}) < R(\widehat{g}_{\mathrm{PU}}) < R(\widehat{g}_{\mathrm{NU}})$$

or

$$R(\widehat{g}_{\mathrm{PN}}) < R(\widehat{g}_{\mathrm{NU}}) < R(\widehat{g}_{\mathrm{PU}}).$$

This means that PN classification is the most promising, and either PU classification or NU classification is the second best.

On the other hand, if n_{U} is sufficiently larger than n_{P} and n_{N} in the sense that $n_{\mathrm{U}} \to \infty$ faster than $n_{\mathrm{P}}, n_{\mathrm{N}} \to \infty$, we have

$$\alpha^*_{\mathrm{PU,PN}} \cdot \alpha^*_{\mathrm{NU,PN}} = 1, \tag{5.11}$$

where

$$\alpha^*_{\mathrm{PU,PN}} = \lim_{n_{\mathrm{P}},n_{\mathrm{N}},n_{\mathrm{U}} \to \infty} \alpha_{\mathrm{PU,PN}},$$

$$\alpha^*_{\mathrm{NU,PN}} = \lim_{n_{\mathrm{P}},n_{\mathrm{N}},n_{\mathrm{U}} \to \infty} \alpha_{\mathrm{NU,PN}}.$$

If $\alpha^*_{\mathrm{PU,PN}} < 1$, (5.11) implies $\alpha^*_{\mathrm{NU,PN}} > 1$ and thus it holds that

$$R(\widehat{g}_{\mathrm{PU}}) < R(\widehat{g}_{\mathrm{PN}}) < R(\widehat{g}_{\mathrm{NU}}).$$

Therefore, PU classification is more promising than PN classification in the above sense. Similarly, if $\alpha^*_{\mathrm{PU,PN}} > 1$, (5.11) implies $\alpha^*_{\mathrm{NU,PN}} < 1$ and thus it holds that

$$R(\widehat{g}_{\mathrm{NU}}) < R(\widehat{g}_{\mathrm{PN}}) < R(\widehat{g}_{\mathrm{PU}}).$$

Therefore, NU classification is more promising than PN classification in the above sense.

In real-world applications, we may not be able to know whether the number of unlabeled samples is sufficiently large or not. Therefore, in practice, it would be promising to combine the best methods in both the finite-sample and asymptotic cases. In fact, PNU classification is the combination of the best methods in both cases (i.e., PU+PN and NU+PN), but PU+NU classification is not. Therefore, PNU classification can be expected to work better than PU+NU classification. Indeed, in section 5.6.3, we will see that this theoretical result well agrees with experimental tendency.

5.5.3 Theoretical Analysis

Here, we investigate theoretical properties of the empirical risk estimators for PU+PN and NU+PN classification.

5.5.3.1   Estimation error bounds

As a function class, we consider the linear-in-parameter function class $\mathcal{G}_{\mathrm{LIP}}$ in section 3.1.2.4 for the sake of simplicity, but results can be extended to other function classes as long as the empirical Rademacher complexity is bounded. We then assume that the ramp loss function satisfies the symmetric condition in (4.4) in section 4.3.1.

For convenience, let us define

$$\chi(c_{\mathrm{P}}, c_{\mathrm{N}}, c_{\mathrm{U}}) := c_{\mathrm{P}}\pi_{\mathrm{P}}/\sqrt{n_{\mathrm{P}}} + c_{\mathrm{N}}\pi_{\mathrm{N}}\sqrt{n_{\mathrm{N}}} + c_{\mathrm{U}}/\sqrt{n_{\mathrm{U}}}.$$

For the empirical risk estimators for PU+PN, NU+PN, and PU+NU classification, we first have the following lemma:

Lemma 5.2   *For any $\delta > 0$, the following inequalities hold separately with probability at least $1 - \delta$:*

$$\sup_{g \in \mathcal{G}_{\mathrm{LIP}}} \left| \widehat{R}^{\gamma}_{\mathrm{PU+PN}}(g) - R(g) \right| \leq \frac{1}{2} C_{\alpha,\phi,\delta} \cdot \chi(1+\gamma, 1-\gamma, \gamma), \tag{5.12}$$

$$\sup_{g \in \mathcal{G}_{\mathrm{LIP}}} \left| \widehat{R}^{\gamma}_{\mathrm{NU+PN}}(g) - R(g) \right| \leq \frac{1}{2} C_{\alpha,\phi,\delta} \cdot \chi(1-\gamma, 1+\gamma, \gamma), \tag{5.13}$$

$$\sup_{g \in \mathcal{G}_{\mathrm{LIP}}} \left| \widehat{R}^{\gamma}_{\mathrm{PU+NU}}(g) - R(g) \right| \leq \frac{1}{2} C_{\alpha,\phi,\delta} \cdot \chi(2-2\gamma, 2\gamma, |2\gamma - 1|), \tag{5.14}$$

where $C_{\alpha,\phi,\delta} = 2 C_{\phi} C_{\alpha} + \sqrt{2\ln(6/\delta)}$.

Proof   Assume the linear-in-parameter function class in (3.6) and the margin-based ramp loss

$$\ell(z) = \frac{1}{2} \min\{2, \max\{0, 1-z\}\}$$

with $z = yg(x)$, which satisfies the symmetric condition (4.4). Based on the Rademacher analysis (see section 3.1.2.4), for any $\delta > 0$, with probability at least $1 - \delta/6$, the following inequalities hold separately for any $g \in \mathcal{G}_{\mathrm{LIP}}$:

$$R_{\mathrm{P}}(g) - \widehat{R}_{\mathrm{P}}(g) \leq 2\mathfrak{R}_{n_{\mathrm{P}},p_{\mathrm{P}}}(\ell \circ \mathcal{G}_{\mathrm{LIP}}) + \sqrt{\frac{\ln(6/\delta)}{2n_{\mathrm{P}}}},$$

$$R_{\mathrm{N}}(g) - \widehat{R}_{\mathrm{N}}(g) \leq 2\mathfrak{R}_{n_{\mathrm{P}},p_{\mathrm{P}}}(\ell \circ \mathcal{G}_{\mathrm{LIP}}) + \sqrt{\frac{\ln(6/\delta)}{2n_{\mathrm{N}}}},$$

$$R_{\mathrm{U}}(g) - \widehat{R}_{\mathrm{U,N}}(g) \leq 2\mathfrak{R}_{n_{\mathrm{U}},p_{\mathrm{U}}}(\ell \circ \mathcal{G}_{\mathrm{LIP}}) + \sqrt{\frac{\ln(6/\delta)}{2n_{\mathrm{U}}}},$$

$$R_{\mathrm{U}}(g) - \widehat{R}_{\mathrm{U,P}}(g) \leq 2\mathfrak{R}_{n_{\mathrm{U}},p_{\mathrm{U}}}(\ell \circ \mathcal{G}_{\mathrm{LIP}}) + \sqrt{\frac{\ln(6/\delta)}{2n_{\mathrm{U}}}}.$$

The ramp loss is a $1/2$-Lipschitz function. By the contraction lemma (see, e.g., Mohri et al., 2012) and theorem 3.2, we obtain

$$\mathfrak{R}_{n_P, p_P}(\ell \circ \mathcal{G}_{\mathrm{LIP}}) \le C_\phi C_\alpha / (2\sqrt{n_P}),$$

$$\mathfrak{R}_{n_N, p_N}(\ell \circ \mathcal{G}_{\mathrm{LIP}}) \le C_\phi C_\alpha / (2\sqrt{n_N}),$$

$$\mathfrak{R}_{n_U, p_U}(\ell \circ \mathcal{G}_{\mathrm{LIP}}) \le C_\phi C_\alpha / (2\sqrt{n_U}).$$

As shown in section 4.3.2, if the loss function satisfies the symmetric condition (4.4), we have

$$\widehat{R}_{\mathrm{PU}}(g) = 2\pi_P \widehat{R}_P(g) + \widehat{R}_{\mathrm{U,N}}(g) - \pi_P,$$

$$\widehat{R}_{\mathrm{NU}}(g) = 2\pi_N \widehat{R}_N(g) + \widehat{R}_{\mathrm{U,P}}(g) - \pi_N.$$

The PU+PN, NU+PN, and PU+NU risk estimators are expressed as

$$\widehat{R}^\gamma_{\mathrm{PU+PN}}(g) = (1+\gamma)\pi_P \widehat{R}_P(g) + (1-\gamma)\pi_N \widehat{R}_N(g) + \gamma \widehat{R}_{\mathrm{U,N}}(g) + \mathrm{Const.},$$

$$\widehat{R}^\gamma_{\mathrm{NU+PN}}(g) = (1-\gamma)\pi_P \widehat{R}_P(g) + (1+\gamma)\pi_N \widehat{R}_N(g) + \gamma \widehat{R}_{\mathrm{U,P}}(g) + \mathrm{Const.},$$

$$\widehat{R}^\gamma_{\mathrm{PU+NU}}(g) = (2-2\gamma)\pi_P \widehat{R}_P(g) + 2\gamma \pi_N \widehat{R}_N(g) + \gamma \widehat{R}_{\mathrm{U,P}}(g)$$

$$+ (1-\gamma)\widehat{R}_{\mathrm{U,N}}(g) + \mathrm{Const.}$$

For $\widehat{R}_{\mathrm{PU+NU}}(g)$, $\widehat{R}_{\mathrm{U,P}}(g) + \widehat{R}_{\mathrm{U,N}}(g) = 1$ holds because of the symmetric condition. We thus have

$$(1-\gamma)\widehat{R}_{\mathrm{U,P}}(g) + \gamma \widehat{R}_{\mathrm{U,N}}(g) = \begin{cases} (2\gamma - 1)\widehat{R}_{\mathrm{U,P}}(g) + \mathrm{Const.} & (\gamma \ge 1/2), \\ (1-2\gamma)\widehat{R}_{\mathrm{U,N}}(g) + \mathrm{Const.} & (\gamma < 1/2). \end{cases}$$

By the union bound, for any $\delta > 0$, with probability at least $1 - \delta/2$, the following inequalities hold separately for any $g \in \mathcal{G}_{\mathrm{LIP}}$:

$$R(g) - \widehat{R}^\gamma_{\mathrm{PU+PN}}(g) \le \frac{1}{2} C_{\alpha,\phi,\delta} \left(\frac{(1+\gamma)\pi_P}{\sqrt{n_P}} + \frac{(1-\gamma)\pi_N}{\sqrt{n_N}} + \frac{\gamma}{\sqrt{n_U}} \right),$$

$$R(g) - \widehat{R}^\gamma_{\mathrm{NU+PN}}(g) \le \frac{1}{2} C_{\alpha,\phi,\delta} \left(\frac{(1-\gamma)\pi_P}{\sqrt{n_P}} + \frac{(1+\gamma)\pi_N}{\sqrt{n_N}} + \frac{\gamma}{\sqrt{n_U}} \right),$$

$$R(g) - \widehat{R}^\gamma_{\mathrm{PU+PN}}(g) \le \frac{1}{2} C_{\alpha,\phi,\delta} \left(\frac{(2-2\gamma)\pi_P}{\sqrt{n_P}} + \frac{2\gamma \pi_N}{\sqrt{n_N}} + \frac{|2\gamma - 1|}{\sqrt{n_U}} \right),$$

where $C_{\alpha,\phi,\delta} := 2C_\phi C_\alpha + \sqrt{2\ln(6/\delta)}$. Note that with $I(g) \le 2R(g)$ of the ramp loss, we can obtain the generalization bounds in Sakai et al. (2017).

Similarly, the other direction of bounds is obtained with probability at least $1 - \delta/2$. Thus, the two directions hold simultaneously with probability at least $1 - \delta$, which completes the proof. $\qquad\square$

Let us define

$$\widehat{g}^{\gamma}_{\text{PU+PN}} := \underset{g \in \mathcal{G}_{\text{LIP}}}{\text{argmin}} \; \widehat{R}^{\gamma}_{\text{PU+PN}}(g),$$

$$\widehat{g}^{\gamma}_{\text{NU+PN}} := \underset{g \in \mathcal{G}_{\text{LIP}}}{\text{argmin}} \; \widehat{R}^{\gamma}_{\text{NU+PN}}(g),$$

$$\widehat{g}^{\gamma}_{\text{PU+NU}} := \underset{g \in \mathcal{G}_{\text{LIP}}}{\text{argmin}} \; \widehat{R}^{\gamma}_{\text{PU+NU}}(g).$$

With lemma 5.2, we have the following theorem:

Theorem 5.3 *For any $\delta > 0$, the following inequalities hold separately with probability at least $1 - \delta$:*

$$R(\widehat{g}^{\gamma}_{\text{PU+PN}}) - R(g^*) \leq C_{\alpha,\phi,\delta} \cdot \chi(1 + \gamma, 1 - \gamma, \gamma),$$

$$R(\widehat{g}^{\gamma}_{\text{NU+PN}}) - R(g^*) \leq C_{\alpha,\phi,\delta} \cdot \chi(1 - \gamma, 1 + \gamma, \gamma),$$

$$R(\widehat{g}^{\gamma}_{\text{PU+NU}}) - R(g^*) \leq C_{\alpha,\phi,\delta} \cdot \chi(2 - 2\gamma, 2\gamma, |2\gamma - 1|),$$

where $C_{\alpha,\phi,\delta} := 2C_\phi C_\alpha + \sqrt{2\ln(6/\delta)}$.

Proof Similarly to the proof of theorem 3.6 (the estimation error bound in PN classification), we have

$$R(\widehat{g}^{\gamma}_{\text{PU+PN}}) - R(g^*) \leq 2 \sup_{g \in \mathcal{G}_{\text{LIP}}} \left| \widehat{R}^{\gamma}_{\text{PU+PN}}(g) - R(g) \right|.$$

With lemma 5.2, we obtain the estimation error bound for PU+PN classification. Likewise, we prove the estimation error bounds for NU+PN and PU+NU classification. $\qquad\square$

Theorem 5.3 guarantees that the estimation errors vanish asymptotically with high probability. Since n_{P}, n_{N}, and n_{U} can increase independently in the current setup, the obtained convergence rate

$$\mathcal{O}_p(1/\sqrt{n_{\text{P}}} + 1/\sqrt{n_{\text{N}}} + 1/\sqrt{n_{\text{U}}})$$

is the optimal without any additional assumption (Vapnik, 1998; Mendelson, 2008).

5.5.3.2 Variance reduction

As we saw, the empirical risk estimators are all unbiased. Therefore, among unbiased estimators, the one having a smaller variance is more preferable. Below, we elucidate when and how much the variance of empirical risk estimators are smaller than that of $\widehat{R}_{\text{PN}}(g)$. In other words, we will see that how much $\mathcal{X}_{\text{U}}$ can help reduce the variance in estimating $R(g)$. In the following analyses, $n_{\text{U}} \to \infty$ is assumed to illustrate the maximum variance reduction that can be achieved. Also, we assume that a loss function satisfies the symmetric condition in (4.4).

Similarly to $R_P(g)$ and $R_N(g)$, let $\sigma_P^2(g)$ and $\sigma_N^2(g)$ be the corresponding variances:

$$\sigma_P^2(g) := \mathbb{V}_P[\ell(g(\boldsymbol{x}), +1)] \quad \text{and} \quad \sigma_N^2(g) := \mathbb{V}_N[\ell(g(\boldsymbol{x}), -1)],$$

where $\mathbb{V}_P$ and $\mathbb{V}_N$ denote the variances over $p_P(\boldsymbol{x})$ and $p_N(\boldsymbol{x})$, respectively. Moreover, denote

$$\psi_P := \pi_P^2 \sigma_P^2(g)/n_P \quad \text{and} \quad \psi_N := \pi_N^2 \sigma_N^2(g)/n_N$$

for short, and let $\mathbb{V}$ be the variance over $p_P(\boldsymbol{x}_1^P) \cdots p_P(\boldsymbol{x}_{n_P}^P) \cdot p_N(\boldsymbol{x}_1^N) \cdots p_N(\boldsymbol{x}_{n_N}^N) \cdot p(\boldsymbol{x}_1^U) \cdots p(\boldsymbol{x}_{n_U}^U)$.

We first investigate the variances of the PU+PN and NU+PN risk estimators and obtain the following theorem:

Theorem 5.4   *(Sakai et al., 2017) Assume $n_U \to \infty$. For any fixed[4] g, let*

$$\gamma_{\text{PU+PN}} = \underset{g}{\arg\min} \, \mathbb{V}\big[\widehat{R}^{\gamma}_{\text{PU+PN}}(g)\big] = \frac{\psi_N - \psi_P}{\psi_P + \psi_N}, \tag{5.15}$$

$$\gamma_{\text{NU+PN}} = \underset{g}{\arg\min} \, \mathbb{V}\big[\widehat{R}^{\gamma}_{\text{NU+PN}}(g)\big] = \frac{\psi_P - \psi_N}{\psi_P + \psi_N}. \tag{5.16}$$

Then, we have

$$\gamma_{\text{PU+PN}} \in [0, 1] \text{ if } \psi_P \leq \psi_N,$$

$$\gamma_{\text{NU+PN}} \in [0, 1] \text{ if } \psi_N \leq \psi_P.$$

Furthermore,

$$\mathbb{V}\big[\widehat{R}^{\gamma}_{\text{PU+PN}}(g)\big] < \mathbb{V}\big[\widehat{R}_{\text{PN}}(g)\big] \text{ for all } \gamma \in (0, 2\gamma_{\text{PU+PN}}) \text{ if } \psi_P < \psi_N,$$

$$\mathbb{V}\big[\widehat{R}^{\gamma}_{\text{NU+PN}}(g)\big] < \mathbb{V}\big[\widehat{R}_{\text{PN}}(g)\big] \text{ for all } \gamma \in (0, 2\gamma_{\text{NU+PN}}) \text{ if } \psi_N < \psi_P.$$

Proof   Notice that g is independent of the data for evaluating $\widehat{R}^{\gamma}_{\text{PU+PN}}(g)$, since g is fixed in the evaluation. As shown in section 4.3.2,

$$\widehat{R}_{\text{PU}}(g) = 2\pi_P \widehat{R}_P^+(g) + \widehat{R}_U^-(g) - \pi_P$$

if the loss function satisfies the symmetric condition. The PU+PN and NU+PN risk estimators are then expressed as

$$\widehat{R}^{\gamma}_{\text{PU+PN}}(g) = (1+\gamma)\pi_P \widehat{R}_P^+(g) + (1-\gamma)\pi_N \widehat{R}_N^-(g) + \gamma \widehat{R}_U^-(g) - \gamma\pi_P,$$

$$\widehat{R}^{\gamma}_{\text{NU+PN}}(g) = (1-\gamma)\pi_P \widehat{R}_P^+(g) + (1+\gamma)\pi_N \widehat{R}_N^-(g) + \gamma \widehat{R}_U^+(g) - \gamma\pi_N.$$

When $n_U \to \infty$,

$$\mathbb{V}\big[\widehat{R}^{\gamma}_{\mathrm{PU+PN}}(g)\big] = (1+\gamma)^2 \pi_{\mathrm{P}}^2 \mathbb{V}_{\mathrm{P}}\big[\widehat{R}^{+}_{\mathrm{P}}(g)\big] + (1-\gamma)^2 \pi_{\mathrm{N}}^2 \mathbb{V}_{\mathrm{N}}\big[\widehat{R}^{-}_{\mathrm{N}}(g)\big]$$

$$= (1+\gamma)^2 \psi_{\mathrm{P}} + (1-\gamma)^2 \psi_{\mathrm{N}},$$

$$\mathbb{V}\big[\widehat{R}^{\gamma}_{\mathrm{NU+PN}}(g)\big] = (1-\gamma)^2 \pi_{\mathrm{P}}^2 \mathbb{V}_{\mathrm{P}}\big[\widehat{R}^{+}_{\mathrm{P}}(g)\big] + (1+\gamma)^2 \pi_{\mathrm{N}}^2 \mathbb{V}_{\mathrm{N}}\big[\widehat{R}^{-}_{\mathrm{N}}(g)\big]$$

$$= (1+\gamma)^2 \psi_{\mathrm{P}} + (1-\gamma)^2 \psi_{\mathrm{N}},$$

where we used $\mathbb{V}_{\mathrm{P}}[\widehat{R}^{+}_{\mathrm{P}}(g)] = \sigma_{\mathrm{P}}^2(g)/n_{\mathrm{P}}$ and $\mathbb{V}_{\mathrm{N}}[\widehat{R}^{-}_{\mathrm{N}}(g)] = \sigma_{\mathrm{N}}^2(g)/n_{\mathrm{N}}$. This shows that $\gamma_{\mathrm{PU+PN}} \geq 0$ if $\psi_{\mathrm{P}} \leq \psi_{\mathrm{N}}$ or $\gamma_{\mathrm{NU+PN}} \geq 0$ if $\psi_{\mathrm{P}} \geq \psi_{\mathrm{N}}$. Also, $\mathbb{V}\big[\widehat{R}_{\mathrm{PN}}(g)\big] = \mathbb{V}\big[\widehat{R}^{\gamma}_{\mathrm{PU+PN}}(g)\big] = \mathbb{V}\big[\widehat{R}^{\gamma}_{\mathrm{NU+PN}}(g)\big]$ if $\gamma = 0$. By completing the square, we have

$$\mathbb{V}\big[\widehat{R}^{\gamma}_{\mathrm{PU+PN}}(g)\big] = (\psi_{\mathrm{P}} + \psi_{\mathrm{N}})\Big(\gamma - \frac{\psi_{\mathrm{N}} - \psi_{\mathrm{P}}}{\psi_{\mathrm{P}} + \psi_{\mathrm{N}}}\Big)^2 + \frac{4\psi_{\mathrm{P}}\psi_{\mathrm{N}}}{\psi_{\mathrm{P}} + \psi_{\mathrm{N}}},$$

showing that $\mathbb{V}\big[\widehat{R}^{\gamma}_{\mathrm{PU+PN}}(g)\big] < \mathbb{V}\big[\widehat{R}_{\mathrm{PN}}(g)\big]$ for all $\gamma \in (0, 2\gamma_{\mathrm{PU+PN}})$ if $\psi_{\mathrm{P}} < \psi_{\mathrm{N}}$. Similarly, we can show that $\mathbb{V}\big[\widehat{R}^{\gamma}_{\mathrm{NU+PN}}(g)\big] < \mathbb{V}\big[\widehat{R}_{\mathrm{PN}}(g)\big]$ for all $\gamma \in (0, 2\gamma_{\mathrm{NU+PN}})$ if $\psi_{\mathrm{N}} < \psi_{\mathrm{P}}$. $\square$

Theorem 5.4 shows that thanks to U data, the PU+PN and NU+PN risk estimators can be more stable than the PN risk estimator in terms of the variance.

We next analyze the variance of the PU+NU risk estimators.

Theorem 5.5 *(Sakai et al., 2017) Assume $n_{\mathrm{U}} \to \infty$. For any fixed g, let*

$$\gamma_{\mathrm{PU+NU}} = \underset{g}{\arg\min}\ \mathbb{V}[\widehat{R}^{\gamma}_{\mathrm{PU+NU}}(g)] = \frac{\psi_{\mathrm{P}}}{\psi_{\mathrm{P}} + \psi_{\mathrm{N}}}.$$

Then, we have $\gamma_{\mathrm{PU+NU}} \in [0, 1]$. Additionally,

$$\mathbb{V}[\widehat{R}^{\gamma}_{\mathrm{PU+NU}}(g)] < \mathbb{V}[\widehat{R}_{\mathrm{PN}}(g)]$$

for all $\gamma \in (2\gamma_{\mathrm{PU+NU}} - 1/2, 1/2)$ if $\psi_{\mathrm{P}} < \psi_{\mathrm{N}}$, or for all $\gamma \in (1/2, 2\gamma_{\mathrm{PU+NU}} - 1/2)$ if $\psi_{\mathrm{N}} < \psi_{\mathrm{P}}$.

Proof If the symmetric loss is used,

$$\widehat{R}^{\gamma}_{\mathrm{PU+NU}}(g) = 2(1-\gamma)\pi_{\mathrm{P}}\widehat{R}_{\mathrm{P}}(g) + 2\gamma\pi_{\mathrm{N}}\widehat{R}_{\mathrm{N}}(g) + (1-\gamma)\widehat{R}^{-}_{\mathrm{U}}(g)$$

$$+ \gamma\widehat{R}^{+}_{\mathrm{U}}(g) - (1-\gamma)\pi_{\mathrm{P}} - \gamma\pi_{\mathrm{N}}.$$

When $n_{\mathrm{U}} \to \infty$,

$$\mathbb{V}\big[\widehat{R}^{\gamma}_{\mathrm{PU+NU}}(g)\big] = 4(1-\gamma)^2 \pi_{\mathrm{P}}^2 \mathbb{V}_{\mathrm{P}}\big[\widehat{R}^{+}_{\mathrm{P}}(g)\big] + 4\gamma^2 \pi_{\mathrm{N}}^2 \mathbb{V}_{\mathrm{N}}\big[\widehat{R}^{-}_{\mathrm{N}}(g)\big]$$

$$= 4(1-\gamma)^2 \psi_{\mathrm{P}} + 4\gamma^2 \psi_{\mathrm{N}}$$

$$= 4(\psi_{\mathrm{P}} + \psi_{\mathrm{N}})\Big(\gamma - \frac{\psi_{\mathrm{P}}}{\psi_{\mathrm{P}} + \psi_{\mathrm{N}}}\Big)^2 + \frac{4\psi_{\mathrm{P}}\psi_{\mathrm{N}}}{\psi_{\mathrm{P}} + \psi_{\mathrm{N}}}.$$

The second equation shows that $\mathbb{V}\left[\widehat{R}^{\gamma}_{\mathrm{PU+NU}}(g)\right] = \mathbb{V}\left[\widehat{R}_{\mathrm{PN}}(g)\right]$ if $\gamma = 1/2$. Besides, since $\mathbb{V}\left[\widehat{R}^{\gamma}_{\mathrm{PU+NU}}(g)\right]$ is quadratic in γ, $\mathbb{V}[\widehat{R}^{\gamma}_{\mathrm{PU+NU}}(g)] < \mathbb{V}[\widehat{R}_{\mathrm{PN}}(g)]$ for all $\gamma \in (2\gamma_{\mathrm{PU+NU}} - 1/2, 1/2)$ if $\psi_{\mathrm{P}} < \psi_{\mathrm{N}}$, or for all $\gamma \in (1/2, 2\gamma_{\mathrm{PU+NU}} - 1/2)$ if $\psi_{\mathrm{N}} < \psi_{\mathrm{P}}$. $\qquad\square$

Theorems 5.4 and 5.5 indicate that the minimum variance reduction achievable by $\widehat{R}^{\gamma}_{\mathrm{PU+PN}}$, $\widehat{R}^{\gamma}_{\mathrm{NU+PN}}$, and $\widehat{R}^{\gamma}_{\mathrm{PU+NU}}$ at their optima $\gamma_{\mathrm{PU+PN}}$, $\gamma_{\mathrm{NU+PN}}$, and $\gamma_{\mathrm{PU+NU}}$ is exactly the same—namely, $4\psi_{\mathrm{P}}\psi_{\mathrm{N}}/(\psi_{\mathrm{P}} + \psi_{\mathrm{N}})$. However, in terms of the range of good γ values, $\widehat{R}^{\gamma}_{\mathrm{PU+PN}}$ and $\widehat{R}^{\gamma}_{\mathrm{NU+PN}}$ are more preferable to $\widehat{R}^{\gamma}_{\mathrm{PU+NU}}$, which is summarized as the next corollary (its proof is straightforward from the proofs of theorems 5.4 and 5.5).

Corollary 5.6 *Assume g is fixed. Let $l_{\mathrm{PU+PN}}$, $l_{\mathrm{NU+PN}}$, and $l_{\mathrm{PU+NU}}$ be the length of the range of γ such that*

$$\mathbb{V}\left[\widehat{R}^{\gamma}_{\mathrm{PU+PN}}(g)\right] < \mathbb{V}\left[\widehat{R}_{\mathrm{PN}}(g)\right],$$

$$\mathbb{V}\left[\widehat{R}^{\gamma}_{\mathrm{NU+PN}}(g)\right] < \mathbb{V}\left[\widehat{R}_{\mathrm{PN}}(g)\right],$$

$$\mathbb{V}\left[\widehat{R}^{\gamma}_{\mathrm{PU+NU}}(g)\right] < \mathbb{V}\left[\widehat{R}_{\mathrm{PN}}(g)\right],$$

respectively. If $\psi_{\mathrm{P}} < \psi_{\mathrm{N}}$,

$$l_{\mathrm{PU+PN}} = \min\{2(\psi_{\mathrm{N}} - \psi_{\mathrm{P}})/(\psi_{\mathrm{P}} + \psi_{\mathrm{N}}), 1\},$$

$$l_{\mathrm{NU+PN}} = 0,$$

$$l_{\mathrm{PU+NU}} = \min\{(\psi_{\mathrm{N}} - \psi_{\mathrm{P}})/(\psi_{\mathrm{P}} + \psi_{\mathrm{N}}), 1/2\}.$$

If $\psi_{\mathrm{N}} < \psi_{\mathrm{P}}$,

$$l_{\mathrm{PU+PN}} = 0,$$

$$l_{\mathrm{NU+PN}} = \min\{2(\psi_{\mathrm{P}} - \psi_{\mathrm{N}})/(\psi_{\mathrm{P}} + \psi_{\mathrm{N}}), 1\},$$

$$l_{\mathrm{PU+NU}} = \min\{(\psi_{\mathrm{P}} - \psi_{\mathrm{N}})/(\psi_{\mathrm{P}} + \psi_{\mathrm{N}}), 1/2\}.$$

In particular, if $3\psi_{\mathrm{P}} \le \psi_{\mathrm{N}}$ (resp. $3\psi_{\mathrm{N}} \le \psi_{\mathrm{P}}$), $\widehat{R}^{\gamma}_{\mathrm{PU+PN}}(g)$ (resp. $\widehat{R}^{\gamma}_{\mathrm{NU+PN}}(g)$) given any $\gamma \in (0, 1)$ can reduce the variance, i.e., $l_{\mathrm{PU+PN}} = 1$ (resp. $l_{\mathrm{NU+PN}} = 1$).

Corollary 5.6 indicates that $\widehat{R}^{\gamma}_{\mathrm{PU+PN}}(g)$ and $\widehat{R}^{\gamma}_{\mathrm{NU+PN}}(g)$ have twice the length of the range of good γ values than $\widehat{R}^{\gamma}_{\mathrm{PU+NU}}(g)$, which is a preferable property in practice. Since γ may be picked from several candidates, a wider length gives us a better chance to pick a better γ for $\widehat{R}^{\gamma}_{\mathrm{PU+PN}}(g)$ and $\widehat{R}^{\gamma}_{\mathrm{NU+PN}}(g)$. If we further assume $\sigma_{\mathrm{P}}(g) = \sigma_{\mathrm{N}}(g)$, the condition in theorems 5.4 and 5.5 as to whether $\psi_{\mathrm{P}} \le \psi_{\mathrm{N}}$ or $\psi_{\mathrm{P}} \ge \psi_{\mathrm{N}}$ will be independent of g.

A final remark is that learning is not involved in theorems 5.4 and 5.5. Therefore, $\ell(m)$ can be any loss that satisfies the symmetric condition (4.5), and g can be any fixed decision function. For instance, we may adopt the zero-one loss $\ell_{0\text{-}1}(m)$, define $\widehat{I}_{\mathrm{PN}}(g)$ and $\widehat{I}^{\eta}_{\mathrm{PNU}}(g)$

as the risk estimators with $\ell_{0\text{-}1}(m)$, respectively, and pick some g resulted from some other classification methods. As a consequence, the variance of $\widehat{I}_{\mathrm{PNU}}(g)$ over the validation data can be smaller than that of $\widehat{I}_{\mathrm{PN}}(g)$, and then the cross-validation should be more stable, given that n_{U} is sufficiently large. Therefore, even without being minimized, the PNU and PU+NU risk estimators are themselves of practical importance.

5.6   Experiments

In this section, we experimentally compare the performance of semi-supervised classification methods.

5.6.1   Datasets

We used 16 benchmark datasets taken from the *UCI Machine Learning Repository* (Lichman, 2013), *Semi-Supervised Learning* (Chapelle et al., 2006), the *LIBSVM* (Chang and Lin, 2011), and the *ELENA Project*.[5] Each feature was rescaled into $[0, 1]$.

5.6.2   PNU Risk for Validation

In the first experiments, we investigated the validity of the PNU risk given in section 5.5.
 As a classifier, we used the Gaussian kernel model:

$$g(\boldsymbol{x}) = \sum_{i=1}^{n} w_i \exp\left(-\frac{\|\boldsymbol{x} - \boldsymbol{x}_i\|^2}{2\sigma^2}\right),$$

where $n = n_{\mathrm{P}} + n_{\mathrm{N}}$, $\{w_i\}_{i=1}^{n}$ are the parameters, $\{\boldsymbol{x}_i\}_{i=1}^{n} = \mathscr{X}_{\mathrm{P}} \cup \mathscr{X}_{\mathrm{N}}$, and $\sigma > 0$ is the Gaussian bandwidth. The classifier trained by minimizing the empirical PN risk estimator is denoted by $\widehat{g}_{\mathrm{PN}}$. The number of labeled samples for training was set to 20, where the class-prior was fixed at 0.5. In all experiments, we used the squared loss for training. Note that the class-prior of test data was the same as that of unlabeled data.
 As discussed in section 5.5.3.2, the empirical PNU risk is expected to be a reliable validation score thanks to its smaller variance than the empirical PN risk estimator. We show here that the empirical PNU risk estimator is a promising alternative to an ordinary validation score.
 To focus on the effect of validation scores only, we trained two classifiers by using the same risk, e.g., $\widehat{R}_{\mathrm{PN}}$. We then selected the Gaussian bandwidth using the empirical PN and PNU risks from $\{1/8, 1/4, 1/2, 1, 3/2, 2\} \times \mathrm{median}(\|\boldsymbol{x}_i - \boldsymbol{x}_j\|_{i,j=1}^{n})$. The obtained classifiers are denoted by $\widehat{g}_{\mathrm{PN}}^{\mathrm{PN}}$ and $\widehat{g}_{\mathrm{PN}}^{\mathrm{PNU}}$, respectively. For validation, we used additional $n_{\mathrm{P}}^{\mathrm{V}}$ positive samples, $n_{\mathrm{N}}^{\mathrm{V}}$ negative samples, and $n_{\mathrm{U}}^{\mathrm{V}}$ unlabeled samples, where we fixed $n_{\mathrm{P}}^{\mathrm{V}} = n_{\mathrm{N}}^{\mathrm{V}} = 10$ and changed $n_{\mathrm{U}}^{\mathrm{V}}$. The additional samples were used also for approximating $\widehat{\sigma}_{\mathrm{P}}(\widehat{g}_{\mathrm{PN}})$ and $\widehat{\sigma}_{\mathrm{N}}(\widehat{g}_{\mathrm{PN}})$ to compute η, i.e., γ in (5.15) and (5.16).

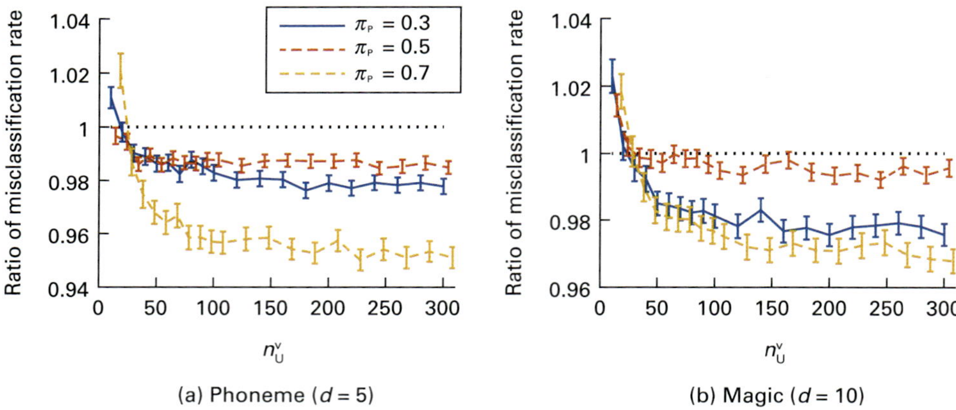

Figure 5.3
Average and standard error of the ratio between the misclassification rates of $\widehat{g}_{\mathrm{PN}}^{\mathrm{PNU}}$ and $\widehat{g}_{\mathrm{PN}}^{\mathrm{PN}}$ as a function of un-labeled samples over 1,000 trials. In many cases, the ratio becomes less than 1, implying that the empirical PNU risk estimator is a promising alternative to the standard empirical PN risk estimator in validation if unlabeled data are available.

Figure 5.3 shows the ratio between the misclassification rates of $\widehat{g}_{\mathrm{PN}}^{\mathrm{PNU}}$ and $\widehat{g}_{\mathrm{PN}}^{\mathrm{PN}}$. The number of unlabeled samples, $n_{\mathrm{U}}^{\mathrm{V}}$, for validation was increased from 10 to 300. With a rather small number of unlabeled samples, the ratio becomes less than 1, i.e., $\widehat{g}_{\mathrm{PN}}^{\mathrm{PNU}}$ achieves better performance than $\widehat{g}_{\mathrm{PN}}^{\mathrm{PN}}$. In particular, when $\pi_{\mathrm{P}} = 0.3$ and 0.7, $\widehat{g}_{\mathrm{PN}}^{\mathrm{PNU}}$ improves the performance substantially. This result shows that the use of the empirical PNU risk estimator for validation improves the classification performance even with a relatively small number of unlabeled data.

5.6.3   Comparison with Other Methods

We next compare the performance of the following methods:

- PNU classification

- PU+NU classification

- *Entropy regularization* (ER) (Grandvalet and Bengio, 2005)

- The *Laplacian support vector machine* (LapSVM) (Belkin et al., 2006; Melacci and Belkin, 2011)

- *Squared-loss mutual information regularization* (SMIR) (Niu et al., 2013)

- The *weakly labeled support vector machine* (WellSVM) (Li et al., 2013)

- The *safe semi-supervised support vector machine* (S4VM) (Li and Zhou, 2015).

We used the Gaussian kernel model for all methods. The training data is $\{x_i\}_{i=1}^{n} = \mathcal{X}_P \cup \mathcal{X}_N \cup \mathcal{X}_U$, where $n = n_P + n_N + n_U$. We selected all hyper-parameters with validation samples of size 20 ($n_P^V = n_N^V = 10$). For training, we drew n_{labeled} labeled and $n_U = 300$ unlabeled samples. The class-prior of labeled data was set at 0.7 and that of unlabeled data was set at $\pi_P = 0.5$ that were assumed to be known. In practice, the class-prior, π_P, can be estimated by class-prior estimation methods (see chapter 12 for details).

Table 5.1 lists the average and standard error of the misclassification rates over 50 trials and the number of best/comparable performances of each method in the bottom row. The superior performance of PNU classification over PU+NU classification agrees well with the discussion in section 5.5.2. With the g50c dataset, which well satisfies the low-density separation principle, the WellSVM achieved the best performance. However, in the Banana dataset, where the two classes are highly overlapped, the performance of WellSVM was worse than the other methods. In contrast, PNU classification achieved consistently better/comparable performance and its performance did not degenerate considerably across datasets. These results show that the idea of using PU classification in semi-supervised classification is promising.

5.7 Extensions

As theoretically and experimentally shown, semi-supervised classification based on PU classification called PNU classification is promising. In addition to good performance, it is advantageous in that the idea is quite general and can be applied to various applications (Sakai et al., 2018; Hayashi et al., 2018; Tsuchiya et al., 2019). In this section, we show three extensions of PNU classification: multi-class classification, AUC optimization, and matrix imputation. Note that later in sections 7.5 and 9.4, we will further see promising applications of the combination idea.

5.7.1 Multi-Class Extension

In section 5.5, we described the PNU classification method solely for binary classification. However, its idea can be naturally extended to multi-class classification. Let c be the number of classes, let $g \colon \mathbb{R}^d \to \mathbb{R}^c$ be a c-class classifier, and let $\mathcal{L} \colon \mathbb{R}^c \times \mathcal{Y} \to \mathbb{R}_{\geq 0}$ be a multi-class surrogate loss (see section 2.2 for details).

Recall the classification risk for multi-class (MC) classification:

$$\widehat{R}_{\text{MC}}(g) = \sum_{y=1}^{c} \frac{\pi_y}{n_y} \sum_{i=1}^{n_y} \mathcal{L}\big(g(x_i^{(y)}), y\big). \tag{5.17}$$

Similar to PU classification, the risk estimator from labeled data without class k and unlabeled data ($\text{MC}^{\backslash k}$) (Xu et al., 2017) is defined as[6]

$$\widehat{R}_{\text{MC}^{\backslash k}}(g) := \sum_{y \neq k}^{c} \frac{\pi_y}{n_y} \sum_{i=1}^{n_y} \widetilde{\mathcal{L}}^{\backslash k}\big(g(x_i^{(y)}), y\big) + \frac{1}{n_{\text{unlabeled}}} \sum_{i=1}^{n_{\text{unlabeled}}} \mathcal{L}\big(g(x_i), k\big),$$

Table 5.1
Average and standard error of the misclassification rates of each method over 50 trials for benchmark datasets. Boldfaced numbers denote the best and comparable methods in terms of the average misclassification rate according to the t-test at significance level 5%. The bottom row gives the number of best/comparable cases of each method.

Dataset	n_{labeled}	PNU	PU+NU	ER	LapSVM	SMIR	WellSVM	S4VM
Banana	10	**30.1 (1.0)**	**32.1 (1.1)**	35.8 (1.0)	36.9 (1.0)	37.7 (1.1)	41.8 (0.6)	45.3 (1.0)
$d=2$	50	**19.0 (0.6)**	26.4 (1.2)	20.6 (0.7)	21.3 (0.7)	21.1 (1.0)	42.6 (0.5)	38.7 (0.9)
Phoneme	10	32.5 (0.8)	33.5 (1.0)	33.4 (1.2)	36.5 (1.5)	36.4 (1.2)	**28.4 (0.6)**	33.7 (1.4)
$d=5$	50	28.1 (0.5)	32.8 (0.9)	27.8 (0.6)	27.0 (0.8)	28.6 (1.0)	26.8 (0.4)	**25.1 (0.2)**
Magic	10	**31.7 (0.8)**	34.1 (0.9)	34.2 (1.1)	37.9 (1.3)	36.0 (1.2)	**30.1 (0.8)**	33.3 (0.9)
$d=10$	50	**29.9 (0.8)**	33.4 (0.9)	30.9 (0.5)	31.0 (0.9)	**30.8 (0.9)**	**28.8 (0.8)**	29.2 (0.4)
Image	10	**29.8 (0.9)**	**31.7 (0.8)**	33.7 (1.1)	36.6 (1.2)	36.7 (1.2)	34.7 (1.1)	35.9 (1.0)
$d=18$	50	**20.7 (0.8)**	26.6 (1.1)	**20.8 (0.8)**	**20.3 (1.0)**	20.9 (0.9)	27.2 (1.0)	23.2 (0.7)
Susy	10	**44.6 (0.6)**	**45.0 (0.6)**	47.7 (0.4)	48.2 (0.4)	**45.1 (0.7)**	48.0 (0.3)	46.8 (0.3)
$d=18$	50	**38.9 (0.6)**	41.5 (0.6)	**37.9 (0.7)**	43.1 (0.6)	43.9 (0.8)	43.8 (0.7)	42.1 (0.4)
German	10	**40.8 (0.9)**	**42.4 (0.7)**	43.6 (0.9)	45.9 (0.7)	46.2 (0.8)	**42.4 (0.8)**	**42.0 (0.7)**
$d=20$	50	**36.2 (0.8)**	39.0 (0.8)	38.9 (0.6)	40.6 (0.6)	38.4 (1.1)	38.5 (1.0)	**34.9 (0.5)**
Waveform	10	**17.4 (0.6)**	**18.0 (0.9)**	18.5 (0.6)	24.9 (1.4)	**18.0 (1.0)**	**16.7 (0.6)**	20.8 (0.8)
$d=21$	50	16.3 (0.6)	23.7 (1.2)	**14.2 (0.4)**	18.1 (0.8)	**15.4 (0.6)**	15.5 (0.5)	15.3 (0.3)
ijcnn1	10	43.6 (0.6)	**40.3 (1.0)**	49.7 (0.1)	49.2 (0.3)	44.0 (1.0)	45.9 (0.7)	49.3 (0.8)
$d=22$	50	**34.5 (0.8)**	37.1 (0.9)	**35.5 (0.8)**	**33.4 (1.1)**	49.4 (0.3)	46.2 (0.8)	48.6 (0.4)
g50c	10	11.4 (0.6)	12.5 (0.6)	23.3 (2.3)	39.8 (1.6)	21.9 (1.3)	**6.6 (0.4)**	27.0 (1.4)
$d=50$	50	12.5 (1.1)	10.1 (0.6)	8.7 (0.4)	22.5 (1.5)	10.6 (0.6)	**7.4 (0.4)**	12.1 (0.5)
covtype	10	**46.2 (0.4)**	**46.0 (0.4)**	**46.0 (0.5)**	47.1 (0.5)	47.9 (0.5)	**46.9 (0.6)**	**46.4 (0.4)**
$d=54$	50	**41.3 (0.5)**	42.3 (0.5)	**41.0 (0.4)**	**41.5 (0.5)**	46.2 (0.8)	43.6 (0.6)	**40.8 (0.4)**
Spambase	10	27.2 (0.9)	28.1 (1.1)	31.8 (1.4)	39.7 (1.4)	30.9 (1.3)	**23.8 (0.8)**	36.1 (1.5)
$d=57$	50	23.4 (1.0)	26.6 (1.0)	22.1 (0.7)	28.5 (1.3)	20.9 (0.5)	**19.1 (0.4)**	24.5 (0.9)
Splice	10	**38.3 (0.8)**	**39.3 (0.8)**	43.9 (0.8)	47.9 (0.5)	41.6 (0.7)	42.0 (1.0)	42.4 (0.6)
$d=60$	50	**30.6 (0.8)**	34.7 (0.9)	**30.9 (0.8)**	38.8 (1.0)	**30.6 (0.9)**	40.9 (0.8)	35.9 (0.7)
phishing	10	**24.2 (1.2)**	**25.8 (1.0)**	**27.3 (1.6)**	37.2 (1.6)	27.6 (1.6)	27.5 (1.4)	31.7 (1.3)
$d=68$	50	**15.8 (0.6)**	18.3 (0.8)	**15.4 (0.5)**	21.1 (1.3)	**14.7 (0.8)**	17.2 (0.7)	16.7 (0.8)
a9a	10	**31.4 (0.9)**	**31.3 (1.0)**	34.3 (1.2)	41.0 (1.1)	37.3 (1.3)	**33.1 (1.2)**	34.3 (1.2)
$d=83$	50	27.9 (0.6)	29.9 (0.8)	28.6 (0.7)	33.3 (1.0)	**26.9 (0.7)**	28.9 (0.8)	**26.2 (0.4)**
Coil2	10	**38.7 (0.8)**	**40.1 (0.8)**	42.8 (0.7)	43.9 (0.8)	43.2 (0.8)	**39.1 (0.9)**	44.0 (0.8)
$d=241$	50	**23.2 (0.6)**	30.5 (0.9)	23.6 (0.9)	**22.8 (0.9)**	25.1 (0.9)	**22.6 (0.8)**	25.4 (0.8)
w8a	10	35.9 (0.9)	**33.6 (1.0)**	41.6 (1.0)	46.6 (0.8)	39.4 (0.9)	42.1 (0.8)	43.0 (0.8)
$d=300$	50	**28.1 (0.7)**	**27.6 (0.6)**	27.0 (0.9)	38.7 (0.8)	**28.0 (0.9)**	33.7 (0.8)	35.2 (1.0)
#Best/Comp.		23	13	11	4	9	13	7

where $\widetilde{\mathscr{L}}^{\setminus k}(g(\boldsymbol{x}), y) := \mathscr{L}(g(\boldsymbol{x}), y) - \mathscr{L}(g(\boldsymbol{x}), k)$. The MC$^{\setminus k}$+MC classification risk estimator is then defined as

$$\widehat{R}^{\gamma}_{\mathrm{MC}^{\setminus k}+\mathrm{MC}}(g) := \gamma \widehat{R}^{\setminus k}_{\mathrm{MC}}(g) + (1 - \gamma)\widehat{R}_{\mathrm{MC}}(g),$$

where $0 \leq \gamma \leq 1$. Varying k, we obtain different multi-class classification risk estimators. Finally, by combining these risk estimators, we obtain a semi-supervised classification risk estimator, similar to PNU classification.

5.7.2 AUC Maximization

In real-world applications, data is sometimes imbalanced, e.g., the size of positive samples is quite smaller than that of negative samples. To handle such imbalanced data, a standard approach is to train a classifier by maximizing the *area under the receiver operating characteristic curve* (AUC) (Cortes and Mohri, 2004; Herschtal and Raskutti, 2004).

Let

$$f(\boldsymbol{x}, \boldsymbol{x}') := g(\boldsymbol{x}) - g(\boldsymbol{x}').$$

In supervised classification, we use the following PN-AUC risk estimator:

$$\widehat{R}_{\mathrm{PN\text{-}AUC}}(f) := \frac{1}{n_{\mathrm{P}}n_{\mathrm{N}}} \sum_{i=1}^{n_{\mathrm{P}}} \sum_{j=1}^{n_{\mathrm{N}}} \ell\big(f(\boldsymbol{x}_i^{\mathrm{P}}, \boldsymbol{x}_j^{\mathrm{N}}), +1\big).$$

For PU data, we use the PU-AUC risk estimator defined as

$$\widehat{R}_{\mathrm{PU\text{-}AUC}}(f) := \frac{\pi_{\mathrm{N}}^{-1}}{n_{\mathrm{P}}n_{\mathrm{U}}} \sum_{i=1}^{n_{\mathrm{P}}} \sum_{j=1}^{n_{\mathrm{U}}} \ell\big(f(\boldsymbol{x}_i^{\mathrm{P}}, \boldsymbol{x}_j^{\mathrm{U}}), 1\big) - \frac{\pi_{\mathrm{P}}\pi_{\mathrm{N}}^{-1}}{n_{\mathrm{P}}(n_{\mathrm{P}} - 1)} \sum_{i=1}^{n_{\mathrm{P}}} \sum_{i' \neq i}^{n_{\mathrm{P}}} \ell\big(f(\boldsymbol{x}_i^{\mathrm{P}}, \boldsymbol{x}_{i'}^{\mathrm{P}}), 1\big).$$

Similar to PU classification, $\mathbb{E}[\widehat{R}_{\mathrm{PN\text{-}AUC}}(f)] = \mathbb{E}[\widehat{R}_{\mathrm{PU\text{-}AUC}}(f)]$ holds. We then define the PU+PN AUC risk estimator as

$$\widehat{R}^{\gamma}_{\mathrm{PU+PN\text{-}AUC}}(f) := \gamma \widehat{R}_{\mathrm{PN\text{-}AUC}}(f) + (1 - \gamma)\widehat{R}_{\mathrm{PU\text{-}AUC}}(f),$$

where $0 \leq \gamma \leq 1$. We can also consider the NU+PN AUC risk estimator and obtain the PNU-AUC risk estimator. See Sakai et al. (2018) for more details.

5.7.3 Matrix Imputation

The idea of PNU classification can be applied to *matrix imputation* (MI) (also known as matrix completion) (Candès and Recht, 2009; Davenport and Romberg, 2016).

Let $y_{i,j}$ be the (i,j)-th element of underlying matrix $\boldsymbol{Y} \in \{\pm 1\}^{I \times J}$, where I and J denote the size of rows and columns, respectively. Let $a_{i,j} \in \{\pm 1, 0\}$ be the (i,j)-th element of observation matrix $\boldsymbol{A}$, where 0 denotes an unobserved entry. The goal of matrix imputation is to output a matrix G that is as close to Y as possible. As an observation process, Hayashi et al. (2018) assumed

$$p(a = \pm 1 \,|\, y = \pm 1) = 1 - \rho,$$

$$p(a = 0 \,|\, y = \pm 1) = \rho, \tag{5.18}$$

$$p(a = \mp 1 \,|\, y = \pm 1) = 0,$$

where $0 \leq \rho < 1/2$ is the perturbation rate. It means that some of the entries in Y are not observed with probability ρ and there is no label flipping.

Let $\widetilde{\ell}_{\mathrm{PN}}$ be a loss function for supervised MI (PN-MI). In supervised MI, unobserved entries are naturally ignored, meaning that $\widetilde{\ell}_{\mathrm{PN}}(g, a) = 0$ if $a = 0$. Specifically, $\widetilde{\ell}$ is defined as

$$\widetilde{\ell}_{\mathrm{PN}}(g, a) := \begin{cases} \frac{1}{1-\rho}\ell(g, +1) & (a = +1), \\ 0 & (a = 0), \\ \frac{1}{1-\rho}\ell(g, -1) & (a = -1). \end{cases}$$

Under this observation process, we have $\mathbb{E}_a[\widetilde{\ell}_{\mathrm{PN}}(g, a)] = \ell(g, y)$, which can be confirmed by, e.g., if $y = +1$,

$$\mathbb{E}_a[\widetilde{\ell}_{\mathrm{PN}}(g, a)] = p(a = +1 \,|\, y = +1)\widetilde{\ell}_{\mathrm{PN}}(g, +1) + p(a = -1 \,|\, y = -1)\widetilde{\ell}_{\mathrm{PN}}(g, -1)$$

$$= \ell(g, +1).$$

The PN-MI risk estimator is then defined as

$$\widehat{R}_{\mathrm{PN\text{-}MI}}(G) := \sum_{i=1}^{I} \sum_{j=1}^{J} \widetilde{\ell}_{\mathrm{PN}}(g_{i,j}, a_{i,j}),$$

where $g_{i,j}$ is the (i,j)-th element of G.

On the other hand, in PU-MI (Hsieh et al., 2015), the loss function of PU matrix imputation is defined as

$$\widetilde{\ell}_{\mathrm{PU}}(g, a) := \begin{cases} \frac{1}{1-\rho}\left(\ell(g, +1) - \rho\ell(g, -1)\right) & (a = +1), \\ \ell(g, -1) & (a = 0), \\ \ell(g, -1) & (a = -1), \end{cases}$$

where $\ell(g, -1)$ is applied for both $a = 0$ and $a = -1$. This means that while the PN-MI risk estimator ignores unobserved entries, PU-MI utilizes unobserved entries. PU-MI still ensures $\mathbb{E}_a[\widetilde{\ell}_{\mathrm{PU}}(g, a)] = \ell(g, y)$, which is similar to *classification under noisy labels* (Natarajan et al., 2013). The PU-MI risk estimator is then defined as

$$\widehat{R}_{\mathrm{PU\text{-}MI}}(G) := \sum_{i=1}^{I} \sum_{j=1}^{J} \widetilde{\ell}_{\mathrm{PU}}(g_{i,j}, a_{i,j}).$$

By combining the PU-MI and PN-MI risk estimators, we can use all positive, negative, and *unobserved entries* in MI. The PU+PN MI risk estimator is defined as

$$\widehat{R}^{\gamma}_{\text{PU+PN-MI}}(\boldsymbol{G}) := \gamma \widehat{R}_{\text{PU-MI}}(\boldsymbol{G}) + (1 - \gamma)\widehat{R}_{\text{PN-MI}}(\boldsymbol{G}),$$

where $0 \leq \gamma \leq 1$. For the detailed discussions about combinations, we refer the readers to Hayashi et al. (2018).

6   Positive-Confidence (Pconf) Classification

Can we learn a binary classifier from only positive data, without any negative data? In chapter 4, we already discussed positive-unlabeled (PU) classification using additional unlabeled data to learn a binary classifier in this situation. However, what if even unlabeled data is not available?

In this chapter, we consider *strengthening the supervision* from positive data. Specifically, we demonstrate that if one can equip positive data with confidence (positive-confidence), one can successfully learn a binary classifier without negative or unlabeled data (Ishida et al., 2018). We name this problem *positive-confidence (Pconf) classification*.

6.1   Introduction

At a glance, being restricted from collecting negative or even unlabeled data but having access to confidence information may seem peculiar. However, the Pconf classification scenario we are discussing in this chapter is conceivable in various real-world problems. For example, in purchase prediction, we can easily collect customer data from our own company (positive data) but not from rival companies (negative data). Oftentimes, customers are asked to answer questionnaires/surveys on how strong their buying intention is in choosing one product over rival products. This may be transformed into a probability between 0 and 1 by preprocessing, and then it can be used as a Pconf value, which is all we need for Pconf classification.

Another example is a common task for app developers, where they need to predict whether app users will continue using the app or unsubscribe it in the future. The critical issue is that depending on the privacy/opt-out policy or data regulation, they need to fully discard the unsubscribed user's data. Hence, developers will not have access to users who quit using their services, but they can associate a Pconf score with each remaining user according to, for example, how actively they use the app.

Shinoda et al. (2021) applied Pconf classification to a real-world problem where they considered the classification task of predicting whether drivers are alert or drowsy. Waiting until drivers start feeling drowsy in a driving simulator can be extremely time-consuming, and it would be easier to annotate the alert samples with confidence instead.

In these potential and real-world applications, as long as Pconf data can be collected, Pconf classification allows us to obtain a classifier that discriminates between positive and negative data.

6.2   Related Works

Pconf classification is related to *one-class classification*, which is aimed at "describing" the positive class typically from hard-labeled positive data without confidence. To the best of our knowledge, previous one-class methods are motivated geometrically (Tax and Duin, 2004; Schölkopf et al., 2001), by information theory (Sugiyama et al., 2014), or by density estimation (Breunig et al., 2000). However, because of the descriptive nature of all previous methods, there is no systematic way to tune hyper-parameters to "classify" positive and negative data. In the conceptual example in figure 6.1, one-class methods do not have any knowledge of the negative distribution, such that the negative distribution is in the lower right of the positive distribution. Therefore, even if we have infinite positive data, one-class methods will still require regularization to have a tight boundary in all directions, wherever the positive-posterior probability becomes low. Note that even if we knew that the negative distribution lies in the lower right of the positive distribution, it is still impossible to find the decision boundary because we still need to know the degree of overlap between the two distributions and the class-prior. One-class methods are designed for and work well for anomaly detection which is aimed at *ranking* samples according to their positivity. However, they have critical limitations if the problem of interest is classification, which is aimed at *deciding* whether samples belong to the positive or negative class.

Pconf classification that we are discussing in this chapter falls into the category of classification. Therefore, it constructs a discriminative classifier, and thus hyper-parameters can be objectively chosen to discriminate between the positive and negative classes. We will later see that Pconf classification relies on the key ingredient recurring throughout the book, which is the empirical risk minimization (ERM; Vapnik [1998]) and this makes it suitable for binary classification.

Pconf classification is also related to PU classification discussed in chapter 4, which uses hard-labeled positive data and additional unlabeled data for constructing a binary classifier. A practical advantage of the Pconf classification method introduced in this chapter over typical PU classification methods is that the Pconf method does not involve estimation of the *class-prior probability*. As we will see in chapter 12, class-prior estimation is a challenging task. This is enabled by the additional confidence information that indirectly includes the information of the class-prior probability, bridging the class-conditional probabilities and class-posterior probabilities.

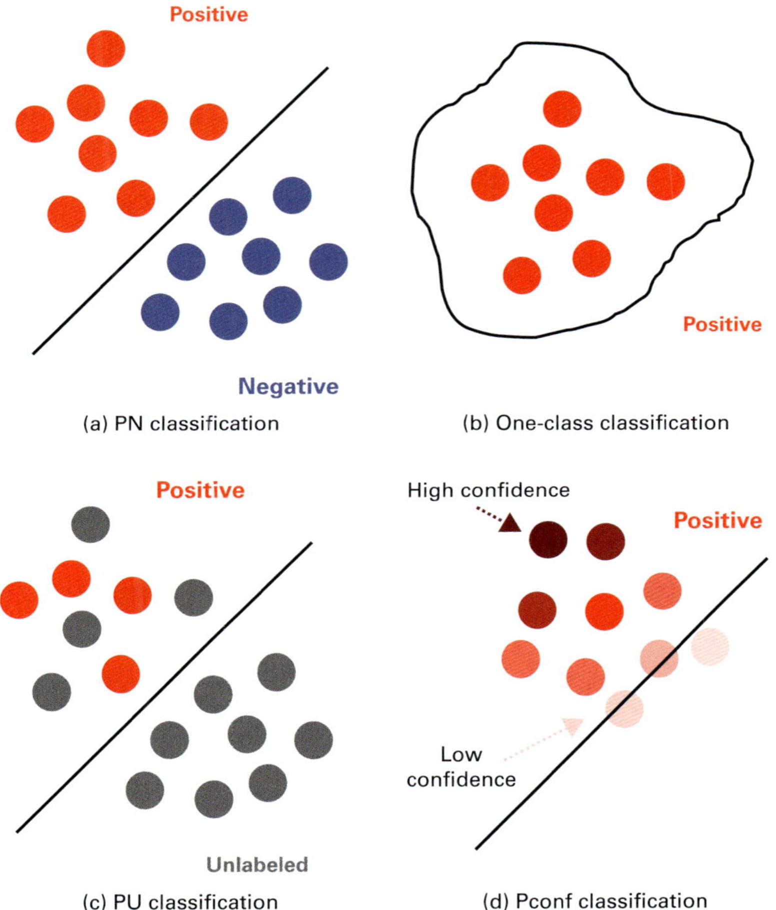

Figure 6.1
Illustrations of the *Pconf classification* and other related classification settings. Red points are positive data, blue points are negative data, and gray points are unlabeled data. The dark/light red colors in (d) show high/low confidence values for positive data.

6.3   Problem Formulation

In this section, we formulate the Pconf classification problem (Ishida et al., 2018). Suppose that a pair of d-dimensional pattern $\boldsymbol{x} \in \mathbb{R}^d$ and its class label $y \in \{+1, -1\}$ follow an unknown probability distribution with density $p(\boldsymbol{x}, y)$. Our goal is to train a binary classifier $g(\boldsymbol{x}) : \mathbb{R}^d \to \mathbb{R}$ so that the classification risk $R(g)$ is minimized:

$$R(g) = \mathbb{E}_{p(\boldsymbol{x}, y)}[\ell(g(\boldsymbol{x}), y)], \tag{6.1}$$

where $\mathbb{E}_{p(x,y)}$ denotes the expectation over $p(x,y)$, and $\ell : \mathbb{R} \times \{+1,-1\} \to \mathbb{R}$ is a loss function. Since $p(x,y)$ is unknown, the ordinary ERM approach (Vapnik, 1998) replaces the expectation with the average over training data drawn independently from $p(x,y)$.

However, in the Pconf classification scenario, we are only given positive data equipped with the confidence

$$\mathscr{X} := \{(x_i, r_i)\}_{i=1}^{n},$$

where x_i is a positive pattern drawn independently from $p(x|y=+1)$ and r_i is the positive confidence given by $r_i := p(y=+1|x_i)$. Since we have no access to negative data in the Pconf classification scenario, we cannot directly employ the standard ERM approach. In the next section, we show how the classification risk can be estimated only from Pconf data.

6.4   Empirical Risk Minimization (ERM) Framework

Let $\pi_+ := p(y=+1)$ and $r(x) := p(y=+1|x)$, and let $\mathbb{E}_+$ denote the expectation over $p(x|y=+1)$. Then the following theorem holds, which forms the basis of our approach.

Theorem 6.1   *(Ishida et al., 2018) If $p(y=+1|x) > 0$ for all x sampled from $p(x)$, the classification risk (6.1) can be expressed as*

$$R(g) = \pi_+ \mathbb{E}_+ \left[\ell\big(g(x),+1\big) + \frac{1-r(x)}{r(x)} \ell\big(g(x),-1\big) \right]. \tag{6.2}$$

Proof   The classification risk (6.1) can be expressed and decomposed as

$$R(g) = \sum_{y=\pm 1} \int \ell\big(g(x),y\big) p(x|y) p(y) dx$$

$$= \int \ell\big(g(x),+1\big) p(x|y=+1) p(y=+1) dx$$

$$+ \int \ell\big(g(x),-1\big) p(x|y=-1) p(y=-1) dx$$

$$= \pi_+ \mathbb{E}_+[\ell(g(x),+1)] + \pi_- \mathbb{E}_-[\ell(g(x),-1)], \tag{6.3}$$

where $\pi_- := p(y=-1)$ and $\mathbb{E}_-$ denotes the expectation over $p(x|y=-1)$. We can derive

$$\pi_+ p(x|y=+1) + \pi_- p(x|y=-1) = p(x,y=+1) + p(x,y=-1)$$

$$= p(x) = \frac{p(x,y=+1)}{p(y=+1|x)} = \frac{\pi_+ p(x|y=+1)}{r(x)},$$

where the third equality requires the assumption of $p(y=+1|x) > 0$ stated in theorem 6.1. We then have

$$\pi_- p(x|y=-1) = \pi_+ p(x|y=+1) \left(\frac{1-r(x)}{r(x)} \right).$$

The second term in (6.3) can be expressed as

$$\pi_- \mathbb{E}_- [\ell(g(x), -1)] = \int \pi_- p(x|y=-1) \ell(g(x), -1) dx$$

$$= \int \pi_+ p(x|y=+1) \left(\frac{1-r(x)}{r(x)} \right) \ell(g(x), -1) dx$$

$$= \pi_+ \mathbb{E}_+ \left[\frac{1-r(x)}{r(x)} \ell(g(x), -1) \right],$$

which concludes the proof.   □

Equation (6.2) does not include the expectation over negative data; it only includes the expectation over positive data and the data's confidence values. Furthermore, when (6.2) is minimized with respect to g, unknown π_+ is a proportional constant and thus can be safely ignored. Conceptually, the assumption of $p(y=+1|x) > 0$ implies that the support of the negative distribution is the same as, or is included in, the support of the positive distribution.

Based on the above theorem, we can immediately have the following ERM framework for Pconf classification:

$$\min_g \sum_{i=1}^{n} \left[\ell(g(x_i), +1) + \frac{1-r_i}{r_i} \ell(g(x_i), -1) \right]. \tag{6.4}$$

It might be tempting to consider this similar but more intuitively straightforward empirical formulation

$$\min_g \sum_{i=1}^{n} \left[r_i \ell(g(x_i), +1) + (1 - r_i) \ell(g(x_i), -1) \right], \tag{6.5}$$

in which the positive loss is weighted with Pconf score r_i and the negative loss is weighted with negative-confidence score $1 - r_i$. However, if we simply consider the population version of the objective function of (6.5), we have

$$\mathbb{E}_+ \left[r(x)\ell(g(x), +1) + (1 - r(x))\ell(g(x), -1) \right]$$

$$= \mathbb{E}_+ \left[p(y=+1|x)\ell(g(x), +1) + p(y=-1|x)\ell(g(x), -1) \right]$$

$$= \mathbb{E}_+ \left[\sum_{y \in \{\pm 1\}} p(y|x)\ell\big(g(x),y\big) \right]$$

$$= \mathbb{E}_+ \left[\mathbb{E}_{p(y|x)}\big[\ell\big(g(x),y\big)\big] \right], \tag{6.6}$$

which is *not* equivalent to the classification risk $R(g)$ defined by (6.1). If the outer expectation was over $p(x)$ instead of $p(x|y=+1)$ in (6.6), then it would be equal to (6.1). This implies that if we had a different problem setting of having Pconf equipped for x sampled from $p(x)$, this would be trivially solved by the above naive weighting idea.

From this viewpoint, (6.4) can be regarded as an application of *importance sampling* (Fishman, 1996; Sugiyama and Kawanabe, 2012) to (6.5) to cope with the distribution difference between $p(x)$ and $p(x|y=+1)$, but with the advantage of *not* requiring training data from the test distribution $p(x)$.

6.5   Theoretical Analysis

Based on the mathematical tools detailed in section 3.1.2, next we derive an estimation error bound for the above Pconf classification method.

To begin with, let $\mathscr{G}$ be our function class for ERM. Assume there exists $C_g > 0$ such that $\sup_{g \in \mathscr{G}} \|g\|_\infty \le C_g$ and $C_\ell > 0$ such that $\sup_{|m| \le C_g} \ell(m) \le C_\ell$, where $\ell(m) = \ell(g(x),y)$ is the margin-based loss with $m = yg(x)$. The existence of C_ℓ may be guaranteed for any reasonable ℓ given a reasonable $\mathscr{G}$ in the sense that C_g exists. We assume that $\ell(m)$ is Lipschitz continuous for all $|m| \le C_g$ with a (not necessarily optimal) Lipschitz constant L_ℓ.

Denote by $\widehat{R}(g)$ the objective function of (6.4) times π_+, which is unbiased in estimating $R(g)$ in (6.1) according to theorem 6.1. Subsequently, let $g^* := \text{argmin}_{g \in \mathscr{G}} R(g)$ be the true risk minimizer and $\hat{g} := \text{argmin}_{g \in \mathscr{G}} \widehat{R}(g)$ be the empirical risk minimizer, respectively. The estimation error is defined as $R(\hat{g}) - R(g^*)$, and we are going to bound it from above.

In theorem 6.1, $(1 - r(x))/r(x)$ is playing a role inside the expectation, for the fact that

$$r(x) = p(y = +1 \,|\, x) > 0 \text{ for } x \sim p(x|y=+1).$$

In order to derive any error bound based on statistical learning theory, we should ensure that $r(x)$ could never be too close to zero. To this end, assume there is $C_r > 0$ such that $r(x) \ge C_r$ almost surely. We may trim $r(x)$ and then analyze the bounded but biased version of $\widehat{R}(g)$ alternatively. For simplicity, only the unbiased version is involved after assuming C_r exists.

Lemma 6.2   *(Ishida et al., 2018) For any $\delta > 0$, the following uniform deviation bound holds with probability at least $1 - \delta$ over repeated sampling of data for evaluating $\widehat{R}(g)$:*

$$\sup_{g\in\mathcal{G}} |\widehat{R}(g) - R(g)| \leq 2\pi_+ \left(L_\ell + \frac{L_\ell}{C_r}\right) \mathfrak{R}_n(\mathcal{G}) + \pi_+ \frac{C_\ell}{C_r} \sqrt{\frac{\ln(2/\delta)}{2n}}, \qquad (6.7)$$

where $\mathfrak{R}_n(\mathcal{G})$ is the Rademacher complexity of $\mathcal{G}$ for $\mathcal{X}$ of size n drawn from $p(x\,|\,y=+1)$.[1]

Proof By assumption, it holds almost surely that

$$\frac{1-r(x)}{r(x)} \leq \frac{1}{C_r} - 1.$$

Because of the existence of C_ℓ, the change of $\widehat{R}(g)$ will be no more than $(C_\ell/C_r)/n$ if some x_i is replaced with x_i'.

Consider a single direction of the uniform deviation: $\sup_{g\in\mathcal{G}} \widehat{R}(g) - R(g)$. Note that the change of $\sup_{g\in\mathcal{G}} \widehat{R}(g) - R(g)$ shares the same upper bound with the change of $\widehat{R}(g)$, and *McDiarmid's inequality* (McDiarmid, 1989) implies that

$$\Pr\left\{ \sup_{g\in\mathcal{G}} \widehat{R}(g) - R(g) - \mathbb{E}_{\mathcal{X}}\left[\sup_{g\in\mathcal{G}} \widehat{R}(g) - R(g) \right] \geq \epsilon \right\}$$

$$\leq \exp\left(-\frac{2\epsilon^2 n}{(C_\ell/C_r)^2} \right),$$

or equivalently, with probability at least $1 - \delta/2$,

$$\sup_{g\in\mathcal{G}} \widehat{R}(g) - R(g) \leq \mathbb{E}_{\mathcal{X}}\left[\sup_{g\in\mathcal{G}} \widehat{R}(g) - R(g) \right] + \frac{C_\ell}{C_r} \sqrt{\frac{\ln(2/\delta)}{2n}}.$$

Since $\widehat{R}(g)$ is unbiased, we can show that

$$\mathbb{E}_{\mathcal{X}}\left[\sup_{g\in\mathcal{G}} \widehat{R}(g) - R(g) \right] \leq 2\mathfrak{R}_n\left(\left(1 + \frac{1-r}{r}\right) \circ \ell \circ \mathcal{G} \right)$$

$$\leq 2\left(1 + \frac{1}{C_r}\right) \mathfrak{R}_n(\ell \circ \mathcal{G})$$

$$\leq 2\left(L_\ell + \frac{L_\ell}{C_r}\right) \mathfrak{R}_n(\mathcal{G}),$$

which proves this direction.

The other direction $\sup_{g\in\mathcal{G}} R(g) - \widehat{R}(g)$ can be proved similarly. $\square$

Lemma 6.2 guarantees that with high probability $\widehat{R}(g)$ concentrates around $R(g)$ for all $g \in \mathcal{G}$, and the degree of such concentration is controlled by $\mathfrak{R}_n(\mathcal{G})$. Based on this lemma, we are able to establish an estimation error bound as follows:

Theorem 6.3 *(Ishida et al., 2018) For any $\delta > 0$, with probability at least $1 - \delta$ over repeated sampling of data for training $\hat{g}$, we have*

$$R(\hat{g}) - R(g^*) \leq 4\pi_+ \left(L_\ell + \frac{L_\ell}{C_r} \right) \mathfrak{R}_n(\mathcal{G}) + 2\pi_+ \frac{C_\ell}{C_r} \sqrt{\frac{\ln(2/\delta)}{2n}}. \tag{6.8}$$

Proof Based on lemma 6.2, the estimation error bound (6.8) is proved through

$$R(\hat{g}) - R(g^*) = \left(\widehat{R}(\hat{g}) - \widehat{R}(g^*) \right) + \left(R(\hat{g}) - \widehat{R}(\hat{g}) \right) + \left(\widehat{R}(g^*) - R(g^*) \right)$$

$$\leq 0 + 2 \sup_{g \in \mathcal{G}} |\widehat{R}(g) - R(g)|$$

$$\leq 4 \left(L_\ell + \frac{L_\ell}{C_r} \right) \mathfrak{R}_n(\mathcal{G}) + 2 \frac{C_\ell}{C_r} \sqrt{\frac{\ln(2/\delta)}{2n}},$$

where $\widehat{R}(\hat{g}) \leq \widehat{R}(g^*)$ by the definition of $\widehat{R}$. $\qquad\qquad\square$

Theorem 6.3 guarantees that classification with (6.4) is consistent (Ledoux and Talagrand, 1991), i.e., $n \to \infty$ always means $R(\hat{g}) \to R(g^*)$. Consider linear-in-parameter models defined by

$$\mathcal{G} = \{ g(\boldsymbol{x}) = \langle w, \phi(\boldsymbol{x}) \rangle_{\mathcal{H}} \mid \|w\|_{\mathcal{H}} \leq C_w, \|\phi(\boldsymbol{x})\|_{\mathcal{H}} \leq C_\phi \},$$

where $\mathcal{H}$ is a Hilbert space, $\langle \cdot, \cdot \rangle_{\mathcal{H}}$ is the inner product in $\mathcal{H}$, $w \in \mathcal{H}$ is the normal, $\phi : \boldsymbol{R}^d \to \mathcal{H}$ is a feature map, and $C_w > 0$ and $C_\phi > 0$ are constants (Schölkopf and Smola, 2002). It is known that $\mathfrak{R}_n(\mathcal{G}) \leq C_w C_\phi / \sqrt{n}$ (Mohri et al., 2012), and thus $R(\hat{g}) \to R(g^*)$ in $\mathcal{O}_p(1/\sqrt{n})$, where $\mathcal{O}_p$ denotes the order in probability. This order is already the optimal parametric rate and cannot be improved without additional strong assumptions on $p(\boldsymbol{x}, y)$, ℓ and $\mathcal{G}$ jointly (Mendelson, 2008). Additionally, if ℓ is strictly convex, we have $\hat{g} \to g^*$, and if the aforementioned $\mathcal{G}$ is used, $\hat{g} \to g^*$ in $\mathcal{O}_p(1/\sqrt{n})$ (Boyd and Vandenberghe, 2004).

At first glance, classification with (6.5) is numerically more stable; however, it is generally inconsistent, especially when g is linear in parameters and ℓ is strictly convex. Denote by $\widehat{R}'(g)$ the objective function of (6.5) times π_+, which is unbiased to $R'(g) = \pi_+ \mathbb{E}_+ \mathbb{E}_{p(y|x)} [\ell(g(\boldsymbol{x}), y)]$ rather than $R(g)$. By using the same technique for proving (6.7) and (6.8), it is not difficult to show that with probability at least $1 - \delta$,

$$\sup_{g \in \mathcal{G}} |\widehat{R}'(g) - R'(g)| \leq 4\pi_+ L_\ell \mathfrak{R}_n(\mathcal{G}) + 2\pi_+ C_\ell \sqrt{\frac{\ln(2/\delta)}{2n}},$$

and hence

$$R'(\hat{g}') - R'(g'^*) \leq 8\pi_+ L_\ell \mathfrak{R}_n(\mathcal{G}) + 4\pi_+ C_\ell \sqrt{\frac{\ln(2/\delta)}{2n}},$$

where

$$g'^* = \mathrm{argmin}_{g \in \mathcal{G}}\, R'(g) \quad \text{and} \quad \hat{g}' = \mathrm{argmin}_{g \in \mathcal{G}}\, \widehat{R}'(g).$$

As a result, when the strict convexity of $R'(g)$ and $\widehat{R}'(g)$ is also met, we have $\hat{g}' \to g'^*$. This demonstrates the inconsistency of classification with (6.5), since $R'(g) \neq R(g)$, which leads to $g'^* \neq g^*$ given any reasonable $\mathcal{G}$.

6.6   Implementation

Finally, we give examples of implementations of the Pconf classification method. As a classifier g, let us consider a linear-in-parameter model $g(x) = \alpha^\top \phi(x)$, where $^\top$ denotes the transpose, $\phi(x)$ is a vector of basis functions, and α is a parameter vector (see section 2.1.2.2). Then, from (6.4), the ℓ_2-regularized ERM is formulated as

$$\min_\alpha \sum_{i=1}^n \left[\ell\big(\alpha^\top \phi(x_i), +1\big) + \frac{1 - r_i}{r_i} \ell\big(-\alpha^\top \phi(x_i), -1\big) \right] + \frac{\lambda}{2} \alpha^\top R \alpha,$$

where λ is a non-negative constant and R is a positive semi-definite matrix. In practice, we can use any margin-based loss functions such as squared loss $\ell_S(z) = (z - 1)^2$, hinge loss $\ell_H(z) = \max\{0, 1 - z\}$, and ramp loss $\ell_R(z) = \min\{1, \max\{0, 1 - z\}\}$ (see section 2.1.3). In the experiments in section 6.7, we use the logistic loss $\ell_L(z) = \log(1 + e^{-z})$, which yields

$$\min_\alpha \sum_{i=1}^n \left[\log\big(1 + e^{-\alpha^\top \phi(x_i)}\big) + \frac{1 - r_i}{r_i} \log\big(1 + e^{\alpha^\top \phi(x_i)}\big) \right] + \frac{\lambda}{2} \alpha^\top R \alpha. \tag{6.9}$$

The above objective function is continuous and differentiable, and therefore optimization can be efficiently performed, for example, by quasi-Newton (Nocedal and Wright, 1999) or stochastic gradient methods (Shalev-Shwartz and Ben-David, 2014).

6.7   Experiments

Next, we numerically illustrate the behavior of the Pconf classification method on synthetic datasets for linear models. We further demonstrate the usefulness of the Pconf method on benchmark datasets for deep neural networks that are highly nonlinear. The implementation is based on *PyTorch* (Paszke et al., 2019) and *Sklearn* (Pedregosa et al., 2011).[2]

6.7.1   Synthetic Experiments with Linear Models

Setup: We used two-dimensional Gaussian distributions with means $\mu_+ = [0, 0]^\top$ and μ_- and covariance matrices Σ_+ and Σ_-, for $p(x|y = +1)$ and $p(x|y = -1)$, respectively. For

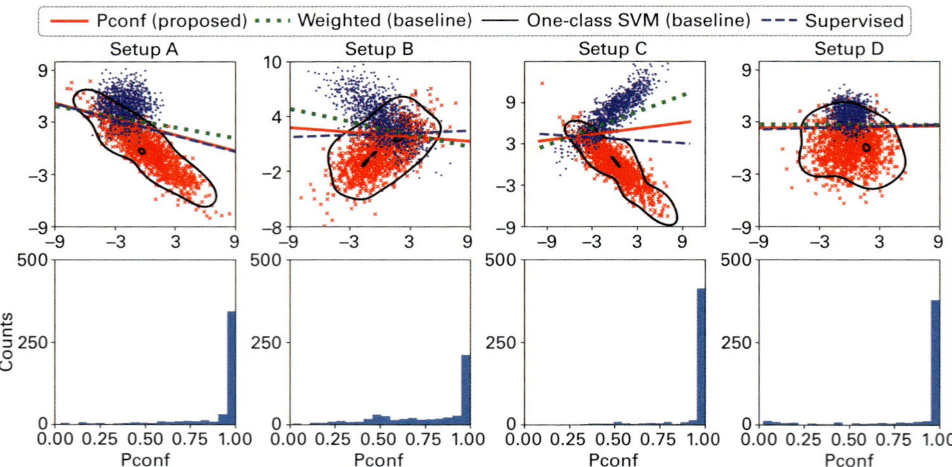

Figure 6.2
Illustrations based on a single trail of the four setups used in experiments with various Gaussian distributions. The red and green lines are decision boundaries obtained by Pconf and weighted classification, respectively, where only positive data with confidence are used (no negative data). The black boundary is obtained by O-SVM (one-class support vector machine), which uses only hard-labeled positive data. The blue boundary is obtained by the fully supervised method using data from both classes. Histograms of confidence of positive data are shown below.

these parameters, we tried various combinations (visually shown in figure 6.2). The specific parameters used for each setup are as follows:

- Setup A: $\boldsymbol{\mu}_- = [-2, 5]^\top$, $\boldsymbol{\Sigma}_+ = \begin{bmatrix} 7 & -6 \\ -6 & 7 \end{bmatrix}$, $\boldsymbol{\Sigma}_- = \begin{bmatrix} 2 & 0 \\ 0 & 2 \end{bmatrix}$.

- Setup B: $\boldsymbol{\mu}_- = [0, 4]^\top$, $\boldsymbol{\Sigma}_+ = \begin{bmatrix} 5 & 3 \\ 3 & 5 \end{bmatrix}$, $\boldsymbol{\Sigma}_- = \begin{bmatrix} 5 & -3 \\ -3 & 5 \end{bmatrix}$.

- Setup C: $\boldsymbol{\mu}_- = [0, 8]^\top$, $\boldsymbol{\Sigma}_+ = \begin{bmatrix} 7 & -6 \\ -6 & 7 \end{bmatrix}$, $\boldsymbol{\Sigma}_- = \begin{bmatrix} 7 & 6 \\ 6 & 7 \end{bmatrix}$.

- Setup D: $\boldsymbol{\mu}_- = [0, 4]^\top$, $\boldsymbol{\Sigma}_+ = \begin{bmatrix} 4 & 0 \\ 0 & 4 \end{bmatrix}$, $\boldsymbol{\Sigma}_- = \begin{bmatrix} 1 & 0 \\ 0 & 1 \end{bmatrix}$.

Using two Gaussian distributions, $p(y = +1|\boldsymbol{x}) > 0$ is satisfied for any $\boldsymbol{x}$ sampled from $p(\boldsymbol{x})$, which is a necessary condition for applying theorem 6.1. In this case, 500 positive data points and 500 negative data points were generated independently of each distribution for training.[3] Similarly, 1,000 positive and 1,000 negative data points were generated for testing.

    We compared the following methods:

- The Pconf classification method (6.4) with the weighted classification method (6.5).

- The regression-based method, which predicts the confidence value and then thresholds it at 0.5 for binarization.

- The one-class support vector machine (O-SVM) with the Gaussian kernel (Schölkopf et al., 2001).

- The fully supervised method based on the empirical version of (6.1).

Note that the Pconf classification method, weighted method, and regression-based method only use Pconf data, O-SVM only uses (hard-labeled) positive data, and the fully supervised method uses both positive and negative data.

In the Pconf classification, weighted, fully supervised methods, linear-in-input model $g(x) = \alpha^\top x + b$ and the logistic loss were commonly used, and *vanilla gradient descent* with 5, 000 epochs (full-batch size), and learning rate 0.001 was used for optimization. For the regression-based method, we used the squared loss and analytical solution (Hastie et al., 2001). For the purpose of clear comparison of the risk, we did not use regularization in this toy experiment. The exception was O-SVM, where the user is required to subjectively pre-specify regularization parameter ν and Gaussian bandwidth γ. We set them at $\nu = 0.05$ and $\gamma = 0.1$.[4]

Analysis with true Pconf   First, we assume that the true Pconf values are known; we can analytically compute them from the Gaussian densities and give them to positive samples. The results in table 6.1 show that the Pconf method is significantly better than other methods in all four cases. In most cases, the Pconf method's accuracy is similar to the fully supervised method, excluding setup C, where there is a few percent loss. Note that the naive weighted method is consistent if the model is correctly specified, but it becomes inconsistent if the model is misspecified (Sugiyama and Kawanabe, 2012).

Analysis with noisy Pconf   In the above toy experiments, we assumed that true positive confidence $r(x) = p(y = +1|x)$ is exactly accessible. However, this can be unrealistic in practice. To investigate the influence of noise in Pconf, we conducted experiments with noisy Pconf data. More specifically, we added zero-mean Gaussian noise with standard

Table 6.1
Comparison on toy datasets with varying degrees of overlap between the positive and negative distributions. The mean and standard deviation of the classification accuracy over 20 trials are reported. The best and equivalent methods based on the t-test with significance level 5 percent were shown in boldface, excluding the fully supervised method and O-SVM, whose settings are different from the others, thus making comparison unfair.

Setup	Pconf	Weighted	Regression	O-SVM	Supervised
A	**89.7 ± 0.6**	88.7 ± 1.2	68.4 ± 6.5	76.0 ± 3.5	89.8 ± 0.7
B	**81.2 ± 1.1**	78.1 ± 1.8	73.2 ± 3.2	71.3 ± 2.3	81.4 ± 1.0
C	**90.2 ± 9.1**	82.7 ± 13.1	50.5 ± 1.7	90.8 ± 1.2	93.6 ± 0.5
D	**91.5 ± 0.5**	90.8 ± 0.7	64.6 ± 5.3	57.1 ± 4.8	91.4 ± 0.5

Table 6.2
Mean and standard deviation of the classification accuracy with noisy positive confidence. The experimental setup is the same as table 6.1, except that Pconf scores for positive data are noisy. "Std." denotes the standard deviation of Gaussian noise.

Setup A			Setup B		
Std.	Pconf	Weighted	Std.	Pconf	Weighted
0.01	**89.8 ± 0.6**	88.8 ± 0.9	0.01	**81.2 ± 0.9**	78.2 ± 1.4
0.05	**89.7 ± 0.6**	88.3 ± 1.1	0.05	**80.7 ± 2.3**	78.1 ± 1.4
0.10	**89.2 ± 0.7**	87.6 ± 1.4	0.10	**80.8 ± 1.2**	77.8 ± 1.5
0.20	**85.9 ± 2.5**	85.8 ± 2.5	0.20	77.8 ± 1.4	**77.2 ± 1.9**

Setup C			Setup D		
Std.	Pconf	Weighted	Std.	Pconf	Weighted
0.01	**92.4 ± 1.7**	84.0 ± 8.2	0.01	**91.6 ± 0.5**	90.6 ± 0.9
0.05	**92.2 ± 3.3**	78.5 ± 11.3	0.05	**91.5 ± 0.5**	89.9 ± 1.2
0.10	**90.8 ± 9.5**	72.6 ± 12.9	0.10	**90.8 ± 0.7**	88.7 ± 1.8
0.20	**88.0 ± 9.5**	65.5 ± 13.1	0.20	**87.7 ± 0.8**	85.5 ± 3.7

deviation chosen from $\{0.01, 0.05, 0.1, 0.2\}$ to the true Pconf values. As the standard deviation gets larger, more noise will be incorporated into generated Pconf values. When generated Pconf values were beyond 1 or below 0.01, we clipped it to 1 or rounded it up to 0.01, respectively.

The results are shown in table 6.2. As expected, the performance tends to deteriorate as the confidence becomes more noisy (i.e., as the standard deviation of Gaussian noise is larger). However, the Pconf method still works reasonably well in almost all cases.

6.7.2   Benchmark Experiments with Neural Network Models

In the next experiments, we used more realistic benchmark datasets and more flexible neural network models.

Fashion-MNIST: This dataset[5] consists of 70,000 examples where each sample is a 28×28 gray-scale image (input dimension is 784) associated with a label from 10 fashion item classes. We standardized the Fashion-MINST data to have zero mean and unit variance.

First, we chose "T-shirt/top" as the positive class and another item for the negative class. The binary dataset was then divided into four sub-datasets: a training set, a validation set, a test set, and a dataset for learning a probabilistic classifier to estimate Pconf initially. Note that in real-world Pconf classification, positive-confidence values may be obtained from labelers, but we generated them with a pretrained probabilistic classifier here. We used logistic regression with the same network architecture as the probabilistic classifier to generate confidence. Note that both positive and negative data were used to train the probabilistic classifier to estimate Pconf, but they were separated from any other process of experiments. However, instead of weight decay, we used *dropout* (Srivastava et al., 2014) with rate 50 percent after each fully connected layer and with 20 epochs. We trained the network for a small number of epochs since softmax output of flexible neural networks tends to be extremely close to 0 or 1 (Goodfellow et al., 2016); thus a network trained for a large number of epochs is not suitable as a Pconf estimator. Furthermore, we rounded up Pconf values less than 1 percent to 1 percent, to stabilize the optimization process.

We compared the following methods:

- Pconf classification (6.4)

- Weighted classification (6.5)

- Fully supervised classification based on the empirical version of (6.1)

- Auto-encoder (Hinton and Salakhutdinov, 2006) as a one-class classification method

For the first three methods we used the logistic loss and a fully connected neural network of three hidden layers (d-100-100-100-1) with *rectified linear units* (ReLU) (Nair and Hinton, 2010) as the activation functions. Weight decay candidates were chosen from $\{10^{-7}, 10^{-4}, 10^{-1}\}$, and *Adam* (Kingma and Ba, 2015) was used for optimization with 200 epochs and mini-batch size 100.

To select hyper-parameters with validation data, we used the zero-one loss versions of (6.4) and (6.5) for Pconf classification and weighted classification, respectively, since no negative data was available in the validation process, and thus we could not directly use the classification accuracy. On the other hand, the classification accuracy was directly used for hyper-parameter tuning of the fully supervised method, which is extremely advantageous. We reported the test accuracy of the model with the best validation score out of all epochs.

The auto-encoder was trained with (hard-labeled) positive data. We classify test data into the positive class if the mean squared error (MSE) is below a threshold of 70 percent quantile and into the negative class otherwise. Since we have no negative data for validating hyper-parameters, we sort the MSEs of training positive data in ascending order. We set the weight decay to 10^{-4}. The architecture is d-100-100-100-100 for encoding and the reversed version for decoding, with *ReLU* after hidden layers and *Tanh* after the final layer.

CIFAR-10:   This dataset[6] consists of 10 classes, with 5,000 images in each class. Each image is given in a $32 \times 32 \times 3$ format. We chose "airplane" as the positive class and one of the other classes as the negative class in order to construct a dataset for binary classification. We used the following neural network architecture:

- convolution (3 in- /18 out-channels, kernel size 5)

- max-pooling (kernel size 2, stride 2)

- convolution (18 in- /48 out-channels, kernel size 5)

- max-pooling (kernel size 2, stride 2)

- fully connected (800 units) with ReLU

- fully connected (400 units) with ReLU

- fully connected (1 unit)

For the probabilistic classifier, the same architecture as that for Fashion-MNIST was used (except *dropout* with rate 50 percent added after the first two fully connected layers). For the auto-encoder, the MSE threshold was set to 80 percent quantile, and we used the following architecture:

Table 6.3
Mean and standard deviation of the classification accuracy over 20 trials for the Fashion-MNIST dataset with *T-shirt* as the positive class and different choices for the negative class. Fully connected three hidden-layer neural networks were used for Pconf classification, weighted classification, and fully supervised classification. The best and equivalent methods between Pconf classification and weighted classification are shown in boldface based on the t-test at significance level 5 percent.

P / N	Pconf	Weighted	Auto-encoder	Supervised
T-shirt / trouser	**92.14 ± 4.06**	85.30 ± 9.07	71.06 ± 1.00	98.98 ± 0.16
T-shirt / pullover	**96.00 ± 0.29**	**96.08 ± 1.05**	70.27 ± 1.22	96.17 ± 0.34
T-shirt / dress	**91.52 ± 1.14**	89.31 ± 1.08	53.82 ± 0.93	96.56 ± 0.34
T-shirt / coat	**98.12 ± 0.33**	**98.13 ± 1.12**	68.74 ± 0.98	98.44 ± 0.13
T-shirt / sandal	**99.55 ± 0.22**	87.83 ± 18.79	82.02 ± 0.49	99.93 ± 0.09
T-shirt / shirt	**83.70 ± 0.46**	**83.60 ± 0.65**	57.76 ± 0.55	85.57 ± 0.69
T-shirt / sneaker	**89.86 ± 13.32**	58.26 ± 14.27	83.70 ± 0.26	100.00 ± 0.00
T-shirt / bag	**97.56 ± 0.99**	95.34 ± 1.00	82.79 ± 0.70	99.02 ± 0.29
T-shirt / ankle boot	**98.84 ± 1.43**	88.87 ± 7.86	85.07 ± 0.37	99.76 ± 0.07

Table 6.4
Mean and standard deviation of the classification accuracy over 20 trials for the CIFAR-10 dataset with *airplane* as the positive class and different choices for the negative class. Convolutional neural networks were used for Pconf classification, weighted classification, and fully supervised classification. The best and equivalent methods between Pconf classification and weighted classification are shown in boldface based on the t-test at significance level 5 percent.

P / N	Pconf	Weighted	Auto-encoder	Supervised
airplane / automobile	**82.68 ± 1.89**	76.21 ± 2.43	75.13 ± 0.42	93.96 ± 0.58
airplane / bird	**82.23 ± 1.21**	80.66 ± 1.60	54.83 ± 0.39	87.76 ± 4.97
airplane / cat	85.18 ± 1.35	**89.60 ± 0.92**	61.03 ± 0.59	92.90 ± 0.58
airplane / deer	**87.68 ± 1.36**	**87.24 ± 1.58**	55.60 ± 0.53	93.35 ± 0.77
airplane / dog	**89.91 ± 0.85**	**89.08 ± 1.95**	62.64 ± 0.63	94.61 ± 0.45
airplane / frog	**90.80 ± 0.98**	81.84 ± 3.92	62.52 ± 0.68	95.95 ± 0.40
airplane / horse	**89.82 ± 1.07**	85.10 ± 2.61	67.55 ± 0.73	95.65 ± 0.37
airplane / ship	69.71 ± 2.37	**70.68 ± 1.45**	52.09 ± 0.42	81.45 ± 8.87
airplane / truck	81.76 ± 2.09	**86.74 ± 0.85**	73.74 ± 0.38	92.10 ± 0.82

- convolution (3 in- /18 out-channels, kernel size 5, stride 1) with ReLU

- max-pooling (kernel size 2, stride 2)

- convolutional layer (18 in- /48 out-channels, kernel size 5, stride 1) with ReLU

- max-pooling (kernel size 2, stride 2)

- deconvolution (48 in- /18 out-channels, kernel size 5, stride 2) with ReLU

- deconvolution (18 in- /5 out-channels, kernel size 5, stride 2)

- deconvolution (5 in- /3 out-channels, kernel size 4, stride 1) with Tanh

Other details such as the loss function and weight-decay follow the same setup as the Fashion-MNIST experiments.

Results: The results in table 6.3 and table 6.4 show that in most cases, Pconf classification either outperforms or is comparable to weighted classification, outperforms Auto-encoder, and is even comparable to the fully supervised method in some cases.

7   Pairwise-Constraint Classification

In the previous chapters, we have explored classification from partially labeled data such as PU classification, PNU classification, and Pconf classification. In this chapter, we consider classification only from pairwise constraints. Specifically, we consider *pairwise similar (S) data* and *pairwise dissimilar (D) data*, which indicates that a pair of unlabeled samples belong to the same class or different classes, but none of the samples have explicit class labels such as P or N. We first show that it is possible to train a classifier only from S and U data, which we name *similar-unlabeled (SU) classification*. Then we extend the discussion and derive *similar-dissimilar (SD) classification* and *dissimilar-unlabeled (DU) classification*. Finally, we give *similar-dissimilar-unlabeled (SDU) classification* for the case where all S, D, and U data are available, following the same construction as PNU classification in chapter 5.

7.1   Introduction

Labeling each data sample explicitly in the absolute sense is often a difficult task. For example, picking a single label tends to be biased because of the human nature of decision making (Thurstone, 1927) and social desirability (Fisher, 1993). In contrast, giving an implicit and relative label to a *pair* of data samples has been shown to be substantially easier in several problem domains (Eric et al., 2008; Jamieson and Nowak, 2011). In the context of classification, we may collect such pairwise comparison data in the form of pairwise similar data (two data points belong to the same class) and pairwise dissimilar data (two data points belong to the different classes). Since it is essentially impossible to manually label a huge number of training data in many real-world problems, it is favorable to train a classifier only with such pairwise comparison data. For example, pairwise comparison data is available and useful in the following scenarios:

• In verification tasks of fingerprints (Baldi and Chauvin, 1993) and signatures (Bromley et al., 1994), systems are asked to discriminate patterns based on their similarity. Many pairs of similar and dissimilar patterns are available.

- In contrastive representation learning (Saunshi et al., 2019), useful data representations for downstream tasks can be learned by making similar data points closer to each other and far from *negative* samples. This approach has been widely used in domains where semantically similar data is easy to collect, such as natural language (Mikolov et al., 2013; Logeswaran and Lee, 2018), computer vision (Yan et al., 2006; Wang and Gupta, 2015; Noroozi and Favaro, 2016), and social networks (Tang et al., 2015).

- In protein function prediction (Klein et al., 2002), knowledge about similar/dissimilar proteins can be obtained via experiments.

- When we attempt to predict people's opinions on sensitive matters such as religion, politics, and racial issues, people often hesitate to give explicit answers. Instead, we may consider indirect questions that might be easier to answer: "Which person do you have the same belief as?"

Semi-supervised clustering is another approach to deal with pairwise comparison from the clustering viewpoint, where pairwise similarity and dissimilarity (also known as must-link and cannot-link constraints) are used to guide unsupervised clustering to a desired solution. Common approaches are:

- *Constrained clustering* (Wagstaff et al., 2001; Basu et al., 2002, 2004; Li and Liu, 2009), which uses pairwise links as constraints on clustering solutions

- *Metric learning* (Xing et al., 2002; Bilenko et al., 2004; Weinberger et al., 2005; Davis et al., 2007; Li et al., 2008; Niu et al., 2012), which performs (typically k-means) clustering with a metric that reflects the pairwise similarity and dissimilarity

- *Matrix completion* (Yi et al., 2013; Chiang et al., 2015), which recovers unknown entries in a similarity matrix

Semi-supervised clustering and weakly supervised classification are similar in that they do not use fully supervised data. However, semi-supervised clustering is aimed at assigning cluster membership only to given training data. In contrast, weakly supervised classification learns an *inductive model*, which provides a decision function and thus can be applied for out-of-sample prediction (i.e., prediction of labels for unseen test data—see section 1.3.4).

7.2   Formulation

In this section, we formulate the problem of weakly supervised classification from pairwise constraints (Bao et al., 2018).

First, let us briefly review the formulation and notation of standard binary classification shown in chapter 2. Let $\mathscr{X} \subset \mathbb{R}^d$ be a d-dimensional example space and $\mathscr{Y} = \{+1, -1\}$ be the binary label space. The goal of binary classification is to obtain a classifier $g : \mathscr{X} \to \mathbb{R}$ that minimizes the classification *risk* defined as

$$R(g) := \mathbb{E}_{p(\boldsymbol{x},y)}\left[\ell(g(\boldsymbol{x}),y)\right], \tag{7.1}$$

where $\mathbb{E}_{p(\boldsymbol{x},y)}[\cdot]$ denotes the expectation over the joint distribution with density $p(\boldsymbol{x},y)$ and $\ell : \mathbb{R} \times \mathcal{Y} \to \mathbb{R}$ is a loss function. Let $\pi_{\mathrm{P}} := p(y=+1)$ and $\pi_{\mathrm{N}} := p(y=-1)$ such that $\pi_{\mathrm{P}} + \pi_{\mathrm{N}} = 1$. In addition, let $p_{\mathrm{P}}(\boldsymbol{x}) := p(\boldsymbol{x}\,|\,y=+1)$ and $p_{\mathrm{N}}(\boldsymbol{x}) := p(\boldsymbol{x}\,|\,y=-1)$.

As discussed in chapter 3, in standard supervised classification scenarios, we are given positive and negative training data independently following $p(\boldsymbol{x},y)$. Then, based on the training data, the classification risk (7.1) is empirically approximated and the empirical risk minimizer is obtained. The goal in this chapter is to train a binary classifier only from pairwise constraints and unlabeled data, which are cheaper to collect than fully labeled data.

To formulate the pairwise-constraint classification, we first discuss the underlying distributions of pairwise similarity, dissimilarity, and unlabeled data. We distinguish two different formulations, one-sample and two-sample cases, as in section 3.1.1. We assume that pairs of two data points $(\boldsymbol{x},y)$ and $(\boldsymbol{x}',y')$ are drawn independently from the distribution with density $p(\boldsymbol{x},\boldsymbol{x}',y,y') = p(\boldsymbol{x},y)p(\boldsymbol{x}',y')$.

7.2.1   One-Sample Case

Here, we observe pairwise similarity (S) and dissimilarity (D) as a set of labeled data pairs:

$$\mathcal{X}_{\mathrm{SD}} := \{(\boldsymbol{x}_i^{\mathrm{SD}}, \boldsymbol{x}_i^{\mathrm{SD}'}, \tau_i)\}_{i=1}^{n_{\mathrm{SD}}} \overset{\text{i.i.d.}}{\sim} p(\boldsymbol{x},\boldsymbol{x}',\tau), \tag{7.2}$$

where τ is a random variable indicating similarity or dissimilarity such as

$$\tau = \begin{cases} +1 & (y=y'), \\ -1 & (y\neq y'). \end{cases}$$

We can see that the S and D data are sampled together from the pairwise marginal distribution

$$p(\boldsymbol{x},\boldsymbol{x}') = \underbrace{p(\tau=+1)}_{:=\pi_{\mathrm{S}}}\underbrace{p(\boldsymbol{x},\boldsymbol{x}'\,|\,\tau=+1)}_{:=p_{\mathrm{S}}(\boldsymbol{x},\boldsymbol{x}')} + \underbrace{p(\tau=-1)}_{:=\pi_{\mathrm{D}}}\underbrace{p(\boldsymbol{x},\boldsymbol{x}'\,|\,\tau=-1)}_{:=p_{\mathrm{D}}(\boldsymbol{x},\boldsymbol{x}')},$$

where

$$\pi_{\mathrm{S}} = p(y=+1)p(y'=+1) + p(y=-1)p(y'=-1) = \pi_{\mathrm{P}}^2 + \pi_{\mathrm{N}}^2,$$

$$\pi_{\mathrm{D}} = p(y=+1)p(y'=-1) + p(y=-1)p(y'=+1) = 2\pi_{\mathrm{P}}\pi_{\mathrm{N}},$$

$$p_{\mathrm{S}}(\boldsymbol{x},\boldsymbol{x}') := p(\boldsymbol{x},\boldsymbol{x}'\,|\,y=y')$$

$$= \frac{p(\boldsymbol{x},\boldsymbol{x}',y=y'=+1) + p(\boldsymbol{x},\boldsymbol{x}',y=y'=-1)}{\pi_{\mathrm{S}}}$$

$$= \frac{\pi_{\mathrm{P}}^2 p_{\mathrm{P}}(\boldsymbol{x})p_{\mathrm{P}}(\boldsymbol{x}') + \pi_{\mathrm{N}}^2 p_{\mathrm{N}}(\boldsymbol{x})p_{\mathrm{N}}(\boldsymbol{x}')}{\pi_{\mathrm{S}}}, \tag{7.3}$$

and

$$p_D(x, x') := p(x, x' \mid y \neq y')$$

$$= \frac{p(x, x', y = +1, y' = -1) + p(x, x', y = -1, y' = +1)}{\pi_D}$$

$$= \frac{p_P(x)p_N(x') + p_N(x)p_P(x')}{2}. \tag{7.4}$$

We occasionally treat the S and D data independently with the following notation:

$$\mathcal{X}_S = \{(x_i^S, x_i^{S'})\}_{i=1}^{n_S} := \{(x, x') \mid (x, x', \tau = +1) \in \mathcal{X}_{SD}\},$$

$$\mathcal{X}_D = \{(x_i^D, x_i^{D'})\}_{i=1}^{n_D} := \{(x, x') \mid (x, x', \tau = -1) \in \mathcal{X}_{SD}\}.$$

In addition, we assume that unlabeled (U) data points are available:

$$\mathcal{X}_U := \{x_i^U\}_{i=1}^{n_U} \overset{\text{i.i.d.}}{\sim} p_U(x), \quad \text{where} \quad p_U(x) := \pi_P p_P(x) + \pi_N p_N(x). \tag{7.5}$$

Importantly, the sample sizes n_S and n_D are random variables dependent on n_{SD}.

7.2.2 Two-Sample Case

In the second case, we observe sets of S, D, and U data separately:

$$\mathcal{X}_S := \{(x_i^S, x_i^{S'})\}_{i=1}^{n_S} \overset{\text{i.i.d.}}{\sim} p_S(x, x'),$$

$$\mathcal{X}_D := \{(x_i^D, x_i^{D'})\}_{i=1}^{n_D} \overset{\text{i.i.d.}}{\sim} p_D(x, x'),$$

$$\mathcal{X}_U := \{x_i^U\}_{i=1}^{n_U} \overset{\text{i.i.d.}}{\sim} p_U(x).$$

Here, we no longer observe the similarity label τ, and the sample sizes n_S and n_D are not random variables but fixed budgets for data collection in advance. In the two-sample case, we write $n_{SD} := n_S + n_D$.

7.2.3 Comparison of Sampling Schemes

As discussed in section 3.1.1, the two-sample case is more general, therefore we mostly focus on this scheme in our formulation risk estimation. The two-sample case is assumed unless otherwise stated. Nevertheless, we briefly shed light on the one-sample case for the two reasons:

• Intuitive interpretation of risk estimators: As we will see in sections 7.3 and 7.4, the classification risk can be estimated only with SU, DU, or SD data. However, it is not straightforward to see the underlying mechanisms of these risk estimators. In section 7.4.2, we will interpret the SD risk estimator with the aid of the one-sample data.

- Natural data collection process: When we collect pairwise constraints, we would often request human annotators to compare whether two data points are similar or dissimilar. In this case, it is natural to consider a data collection strategy such that the maximal budget n_{SD} is fixed first, and then we let annotators provide the similarity labels τ. This scenario fits into the one-sample case. In addition, we will compare SU, DU, and SD estimators under this scenario in section 7.6.2.

7.2.4 Pairwise Constraints as Pointwise Data

Although the S and D data are generated in the pairwise fashion, they can be *decoupled* and regarded as if they were drawn from the pointwise distributions.

Assuming pairwise independence $x \perp\!\!\!\perp x'$, we may marginalize out x' in $p_{\mathrm{S}}(x, x')$:

$$\int p_{\mathrm{S}}(x, x')\mathrm{d}x' = \frac{\pi_{\mathrm{P}}^2}{\pi_{\mathrm{S}}}p_{\mathrm{P}}(x)\int p_{\mathrm{P}}(x')\mathrm{d}x' + \frac{\pi_{\mathrm{N}}^2}{\pi_{\mathrm{S}}}p_{\mathrm{N}}(x)\int p_{\mathrm{N}}(x')\mathrm{d}x'$$

$$= \frac{\pi_{\mathrm{P}}^2}{\pi_{\mathrm{S}}}p_{\mathrm{P}}(x) + \frac{\pi_{\mathrm{N}}^2}{\pi_{\mathrm{S}}}p_{\mathrm{N}}(x)$$

$$:= \widetilde{p}_{\mathrm{S}}(x).$$

Hence, we may regard the decoupled S data

$$\widetilde{\mathcal{X}}_{\mathrm{S}} = \{\widetilde{x}_i^{\mathrm{S}}\}_{i=1}^{2n_{\mathrm{S}}} := \{x_1^{\mathrm{S}}, x_1^{\mathrm{S}'}, \ldots, x_{n_{\mathrm{S}}}^{\mathrm{S}}, x_{n_{\mathrm{S}}}^{\mathrm{S}'}\}$$

as being drawn from the pointwise distribution such that $\widetilde{\mathcal{X}}_{\mathrm{S}} \overset{\text{i.i.d.}}{\sim} \widetilde{p}_{\mathrm{S}}(x)$.

In the same manner, we can treat the pointwise dissimilar data

$$\widetilde{\mathcal{X}}_{\mathrm{D}} = \{\widetilde{x}_i^{\mathrm{D}}\}_{i=1}^{2n_{\mathrm{D}}} := \{x_1^{\mathrm{D}}, x_1^{\mathrm{D}'}, \ldots, x_{n_{\mathrm{D}}}^{\mathrm{D}}, x_{n_{\mathrm{D}}}^{\mathrm{D}'}\}$$

as if it were independent and identically drawn from the following pointwise distribution:

$$\widetilde{p}_{\mathrm{D}}(x) := \frac{1}{2}p_{\mathrm{P}}(x) + \frac{1}{2}p_{\mathrm{N}}(x).$$

This perspective is important when we analyze the variance of the risk estimator in section 7.3.2 and estimate the class-prior probability in section 12.7.[1]

7.3 Similar-Unlabeled (SU) Classification

In this section, we first show how to train a classifier only from pairwise similarity and unlabeled data.

7.3.1　Classification Risk Estimation

Let us express the classification risk (7.1) only in terms of SU data. Assume $\pi_{\mathrm{P}} \neq \frac{1}{2}$ and let

$$\ell_\sharp(\hat{y}, y) := \frac{\pi_{\mathrm{P}}}{2\pi_{\mathrm{P}} - 1}\ell(\hat{y}, y) - \frac{\pi_{\mathrm{N}}}{2\pi_{\mathrm{P}} - 1}\ell(\hat{y}, -y), \tag{7.6}$$

$$\widetilde{\ell}_\sharp(\hat{y}) := \ell_\sharp(\hat{y}, +1) - \ell_\sharp(\hat{y}, -1). \tag{7.7}$$

Then we have the following theorem.

Theorem 7.1　*(Bao et al., 2018) Suppose $\pi_{\mathrm{P}} \neq \frac{1}{2}$. Then the classification risk (7.1) can be equivalently expressed as*

$$R(g) = R_{\mathrm{SU}}(g) := \pi_{\mathrm{S}}\, \mathbb{E}_{\mathrm{S}}\left[\frac{\widetilde{\ell}_\sharp(g(\boldsymbol{x})) + \widetilde{\ell}_\sharp(g(\boldsymbol{x}'))}{2}\right] + \mathbb{E}_{\mathrm{U}}\left[\ell_\sharp(g(\boldsymbol{x}), -1)\right],$$

where $\mathbb{E}_{\mathrm{S}}$ and $\mathbb{E}_{\mathrm{U}}$ denote the expectations over $p_{\mathrm{S}}(\boldsymbol{x}, \boldsymbol{x}')$ and $p_{\mathrm{U}}(\boldsymbol{x})$, respectively:

$$\mathbb{E}_{\mathrm{S}}[\ \cdot\ ] := \int \ \cdot\ p_{\mathrm{S}}(\boldsymbol{x}, \boldsymbol{x}')\mathrm{d}\boldsymbol{x}\mathrm{d}\boldsymbol{x}',$$

$$\mathbb{E}_{\mathrm{U}}[\ \cdot\ ] := \int \ \cdot\ p_{\mathrm{U}}(\boldsymbol{x})\mathrm{d}\boldsymbol{x}.$$

Theorem 7.1 can be proved as a special case of *unlabeled-unlabeled (UU) classification* (discussed in chapter 8). Indeed, theorem 7.1 can be obtained by setting $(\theta, \theta') = (\pi_{\mathrm{P}}^2/\pi_{\mathrm{S}}, \pi_{\mathrm{P}})$ in theorem 8.4, by noting that the S data can be decoupled into pointwise data (see section 7.2.4).

From theorem 7.1, we can immediately obtain an unbiased estimator of the classification risk (7.1):

$$\widehat{R}_{\mathrm{SU}}(g) = \frac{\pi_{\mathrm{S}}}{n_{\mathrm{S}}}\sum_{i=1}^{n_{\mathrm{S}}}\frac{\widetilde{\ell}_\sharp(g(\boldsymbol{x}_i^{\mathrm{S}})) + \widetilde{\ell}_\sharp(g(\boldsymbol{x}_i^{\mathrm{S}'}))}{2} + \frac{1}{n_{\mathrm{U}}}\sum_{i=1}^{n_{\mathrm{U}}}\ell_\sharp(g(\boldsymbol{x}_i^{\mathrm{U}}), -1)$$

$$= \frac{\pi_{\mathrm{S}}}{2n_{\mathrm{S}}}\sum_{i=1}^{2n_{\mathrm{S}}}\widetilde{\ell}_\sharp(f(\widetilde{\boldsymbol{x}}_i^{\mathrm{S}})) + \frac{1}{n_{\mathrm{U}}}\sum_{i=1}^{n_{\mathrm{U}}}\ell_\sharp(g(\boldsymbol{x}_i^{\mathrm{U}}), -1), \tag{7.8}$$

where in the last line we used the decomposed version of similar pairs $\widetilde{\mathscr{X}}_{\mathrm{S}}$ instead of $\mathscr{X}_{\mathrm{S}}$.

7.3.2　Minimum-Variance Risk Estimation

While (7.8) is one of the unbiased SU risk estimator, we can express the first term of $R_{\mathrm{SU}}(g)$ with arbitrary $\beta \in \mathbb{R}$, due to the symmetry of $(\boldsymbol{x}, \boldsymbol{x}') \sim p_{\mathrm{S}}(\boldsymbol{x}, \boldsymbol{x}')$:

$$\mathbb{E}_{\mathrm{S}}\left[\frac{\widetilde{\ell}_\sharp(g(\boldsymbol{x})) + \widetilde{\ell}_\sharp(g(\boldsymbol{x}'))}{2}\right] \tag{7.9}$$

$$= \mathbb{E}_{\mathrm{S}} \left[\widetilde{\ell}_\sharp(g(\boldsymbol{x})) \right]$$

$$= \mathbb{E}_{\mathrm{S}} \left[\beta \widetilde{\ell}_\sharp(g(\boldsymbol{x})) + (1 - \beta)\widetilde{\ell}_\sharp(g(\boldsymbol{x})) \right]$$

$$= \mathbb{E}_{\mathrm{S}} \left[\beta \widetilde{\ell}_\sharp(g(\boldsymbol{x})) \right] + \mathbb{E}_{\mathrm{S}} \left[(1 - \beta)\widetilde{\ell}_\sharp(g(\boldsymbol{x}')) \right]$$

$$= \mathbb{E}_{\mathrm{S}} \left[\beta \widetilde{\ell}_\sharp(g(\boldsymbol{x})) + (1 - \beta)\widetilde{\ell}_\sharp(g(\boldsymbol{x}')) \right].$$

Hence,

$$\frac{\pi_{\mathrm{S}}}{n_{\mathrm{S}}} \sum_{i=1}^{n_{\mathrm{S}}} \left\{ \beta \widetilde{\ell}_\sharp(g(\boldsymbol{x}_i^{\mathrm{S}})) + (1 - \beta)\widetilde{\ell}_\sharp(g(\boldsymbol{x}_i^{\mathrm{S}'})) \right\} \tag{7.10}$$

is also an unbiased estimator of (7.9). Then, a natural question arises: *Is the risk estimator* (7.8) *best among all β?* This question is answered by the following theorem.

Theorem 7.2   *(Bao et al., 2018) The estimator*

$$\frac{\pi_{\mathrm{S}}}{n_{\mathrm{S}}} \sum_{i=1}^{n_{\mathrm{S}}} \frac{\widetilde{\ell}_\sharp(g(\boldsymbol{x}_i^{\mathrm{S}})) + \widetilde{\ell}_\sharp(g(\boldsymbol{x}_i^{\mathrm{S}'}))}{2} \tag{7.11}$$

has the minimum variance among all estimators of the form (7.10) *with respect to $\beta \in \mathbb{R}$.*

Proof   For a fixed g, we show that (7.11) is the minimizer of the variance with respect to $\beta \in \mathbb{R}$:

$$S(g; \beta) := \frac{1}{n_{\mathrm{S}}} \sum_{i=1}^{n_{\mathrm{S}}} \left\{ \beta \widetilde{\ell}_\sharp(g(\boldsymbol{x}_i^{\mathrm{S}})) + (1 - \beta)\widetilde{\ell}_\sharp(g(\boldsymbol{x}_i^{\mathrm{S}'})) \right\}.$$

Let

$$\mu_1 := \mathbb{E}_{\mathscr{X}_{\mathrm{S}}}[S(g; \beta)],$$

$$\widetilde{\mu}_1 := \mathbb{E}_{\widetilde{\mathscr{X}}_{\mathrm{S}}} \left[\frac{1}{n_{\mathrm{S}}} \sum_{i=1}^{n_{\mathrm{S}}} \widetilde{\ell}_\sharp(g(\boldsymbol{x}_i^{\mathrm{S}})) \right],$$

$$\widetilde{\mu}_2 := \mathbb{E}_{\widetilde{\mathscr{X}}_{\mathrm{S}}} \left[\left(\frac{1}{n_{\mathrm{S}}} \sum_{i=1}^{n_{\mathrm{S}}} \widetilde{\ell}_\sharp(g(\boldsymbol{x}_i^{\mathrm{S}})) \right)^2 \right],$$

where $\mathbb{E}_{\mathscr{X}_{\mathrm{S}}}$ denotes the expectation over $\{(\boldsymbol{x}_i^{\mathrm{S}}, \boldsymbol{x}_i^{\mathrm{S}'})\}_{i=1}^{n_{\mathrm{S}}}$ drawn independently from $p_{\mathrm{S}}(\boldsymbol{x}, \boldsymbol{x}')$, and $\mathbb{E}_{\widetilde{\mathscr{X}}_{\mathrm{S}}}$ denotes the expectation over $\{\boldsymbol{x}_i^{\mathrm{S}}\}_{i=1}^{n_{\mathrm{S}}}$ drawn independently from $\widetilde{p}_{\mathrm{S}}(\boldsymbol{x})$. Then, the variance of $S(g; \beta)$, $\mathbb{V}[S(g; \beta)]$, is

$$\mathbb{V}[S(g; \beta)] = \mathbb{E}_{\mathscr{X}_{\mathrm{S}}} \left[(S(g; \beta) - \mu_1)^2 \right]$$

$$= \mathbb{E}_{\mathscr{X}_{\mathrm{S}}} \left[S(g; \beta)^2 \right] - \mu_1^2$$

$$= \widetilde{\mu}_2\beta^2 + 2\widetilde{\mu}_1^2\beta(1-\beta) + \widetilde{\mu}_2(1-\beta)^2 - \mu_1^2$$

$$= 2(\widetilde{\mu}_2 - \widetilde{\mu}_1^2)\left\{\left(\beta - \frac{1}{2}\right)^2 - \frac{1}{4}\right\} + \widetilde{\mu}_2 - \mu_1^2.$$

Since $\widetilde{\mu}_2 - \widetilde{\mu}_1^2$ is the variance of $\frac{1}{n_S}\sum_i \widetilde{\ell}_\sharp(g(x_i^S))$, it holds that $\widetilde{\mu}_2 - \widetilde{\mu}_1^2 \geq 0$. Thus, $\mathbb{V}[S(g;\beta)]$ is minimized when $\beta = \frac{1}{2}$. $\qquad\qquad\square$

Thus, the variance minimality of the risk estimator (7.8) is guaranteed by theorem 7.2. We use this risk estimator in the following sections.

7.3.3   Convex Formulation

Here, we investigate the objective function when the linear-in-parameter model $g(x) = \boldsymbol{\alpha}^\top\boldsymbol{\phi}(x)$ is employed as a classifier, where $\boldsymbol{\alpha} \in \mathbb{R}^b$ are parameters, $\boldsymbol{\phi}: \mathbb{R}^d \to \mathbb{R}^b$ are basis functions, and b is the number of parameters (see section 2.1.2.2). We formulate SU classification as the following empirical risk minimization (ERM) problem using (7.8) together with the ridge regularization (see section 2.1.5):

$$\widehat{\boldsymbol{\alpha}} = \underset{\boldsymbol{\alpha}\in\mathbb{R}^b}{\mathrm{argmin}}\ \widehat{J}_\ell(\boldsymbol{\alpha}), \tag{7.12}$$

where

$$\widehat{J}_\ell(\boldsymbol{\alpha}) := \frac{\pi_S}{2n_S}\sum_{i=1}^{2n_S}\widetilde{\ell}_\sharp(\boldsymbol{\alpha}^\top\boldsymbol{\phi}(\widetilde{x}_i^S)) + \frac{1}{n_U}\sum_{i=1}^{n_U}\ell_\sharp(\boldsymbol{\alpha}^\top\boldsymbol{\phi}(x_i^U), -1) + \frac{\lambda}{2}\|\boldsymbol{\alpha}\|^2, \tag{7.13}$$

and $\lambda > 0$ is the regularization parameter. When we solve this optimization problem, we need to know the class-prior π_P (included in π_S, $\ell_\sharp$, and $\widetilde{\ell}_\sharp$). We discuss how to estimate π_P in section 12.7.

In general, our objective function (7.13) is non-convex even if a convex loss function is used for ℓ. On the other hand, if the loss satisfies the linearity condition (4.5) such that

$$\ell(\widehat{y}, +1) - \ell(\widehat{y}, -1) = -\widehat{y} \quad \text{for all } \widehat{y}\in\mathbb{R},$$

then $\widehat{J}_\ell(\boldsymbol{\alpha})$ is convex in $\boldsymbol{\alpha}$ (Bao et al., 2018). This is similar to the convex approach of PU classification in section 4.3.3. If the loss ℓ is margin-based, we denote $\ell(\widehat{y}, y) = \ell(y\widehat{y})$ with a slight abuse of notation. Then, the linearity condition reduces to $\ell(\widehat{y}) - \ell(-\widehat{y}) = -\widehat{y}$, resulting in $\widetilde{\ell}_\sharp(\widehat{y}) = -\widehat{y}$ and $\ell_\sharp(\widehat{y}, y) = \frac{1}{2\pi_P - 1}\ell(\widehat{y}, y) - \frac{\pi_P}{2\pi_P - 1}$. Examples of margin-based loss functions satisfying the linearity condition are listed in section 4.3.3. Below, we show the objective functions for the squared loss and the double-hinge loss as special cases.

Squared loss   Substituting squared loss $\ell(m) = \frac{1}{4}(1-m)^2$ into (7.13), we can express the objective function as

$$\widehat{J}(\boldsymbol{\alpha}) = \boldsymbol{\alpha}^{\top}\left(\frac{1}{4n_{\mathrm{U}}}\boldsymbol{\Phi}_{\mathrm{U}}^{\top}\boldsymbol{\Phi}_{\mathrm{U}} + \frac{\lambda}{2}I\right)\boldsymbol{\alpha}$$

$$+ \frac{1}{2\pi_{\mathrm{P}}-1}\left(-\frac{\pi_{\mathrm{S}}}{2n_{\mathrm{S}}}\mathbf{1}^{\top}\boldsymbol{\Phi}_{\mathrm{S}} + \frac{1}{2n_{\mathrm{U}}}\mathbf{1}^{\top}\boldsymbol{\Phi}_{\mathrm{U}}\right)\boldsymbol{\alpha} + \mathrm{Const.},$$

where $\mathbf{1}$ is the vector whose elements are all ones, I is the identity matrix, and $\boldsymbol{\Phi}_{\mathrm{S}}$ and $\boldsymbol{\Phi}_{\mathrm{U}}$ are the $2n_{\mathrm{S}} \times b$ and $n_{\mathrm{U}} \times b$ matrices whose (i,j)-entries are given, respectively, by

$$[\boldsymbol{\Phi}_{\mathrm{S}}]_{i,j} := \phi_j(\widetilde{\boldsymbol{x}}_i^{\mathrm{S}}),$$

$$[\boldsymbol{\Phi}_{\mathrm{U}}]_{i,j} := \phi_j(\boldsymbol{x}_i^{\mathrm{U}}).$$

The minimizer of this objective function can be obtained analytically as

$$\boldsymbol{\alpha} = \frac{n_{\mathrm{U}}}{2\pi_{\mathrm{P}}-1}\left(\boldsymbol{\Phi}_{\mathrm{U}}^{\top}\boldsymbol{\Phi}_{\mathrm{U}} + 2\lambda n_{\mathrm{U}}I\right)^{-1}\left(\frac{\pi_{\mathrm{S}}}{n_{\mathrm{S}}}\boldsymbol{\Phi}_{\mathrm{S}}^{\top}\mathbf{1} - \frac{1}{n_{\mathrm{U}}}\boldsymbol{\Phi}_{\mathrm{U}}^{\top}\mathbf{1}\right).$$

Thus, we can obtain the global minimizer uniquely and the optimization problem can be easily implemented and solved highly efficiently if the number of basis functions is not so large.

Double-hinge loss   The double-hinge loss $\ell(m) = \max\{-m, \frac{1}{2}\max\{0, 1-m\}\}$ was proposed in du Plessis et al. (2015) as an alternative to the hinge loss $\ell(m) = \max\{0, 1-m\}$ to satisfy the linearity condition. With the double-hinge loss, the optimization problem (7.13) can be expressed as follows (Bao et al., 2018):

$$\min_{\boldsymbol{\alpha},\boldsymbol{\xi},\boldsymbol{\eta}} -\frac{\pi_{\mathrm{S}}}{2n_{\mathrm{S}}(2\pi_{\mathrm{P}}-1)}\mathbf{1}^{\top}\boldsymbol{\Phi}_{\mathrm{S}}\boldsymbol{\alpha} - \frac{\pi_{\mathrm{N}}}{n_{\mathrm{S}}(2\pi_{\mathrm{P}}-1)}\mathbf{1}^{\top}\boldsymbol{\xi}$$

$$+ \frac{\pi_{\mathrm{P}}}{n_{\mathrm{U}}(2\pi_{\mathrm{P}}-1)}\mathbf{1}^{\top}\boldsymbol{\eta} + \frac{\lambda}{2}\boldsymbol{\alpha}^{\top}\boldsymbol{\alpha}$$

$$\text{s.t.}\quad \boldsymbol{\xi} \geq \mathbf{0}, \quad \boldsymbol{\xi} \geq \frac{1}{2}\mathbf{1} + \frac{1}{2}\boldsymbol{\Phi}_{\mathrm{U}}\boldsymbol{\alpha}, \quad \boldsymbol{\xi} \geq \boldsymbol{\Phi}_{\mathrm{U}}\boldsymbol{\alpha},$$

$$\boldsymbol{\eta} \geq \mathbf{0}, \quad \boldsymbol{\eta} \geq \frac{1}{2}\mathbf{1} - \frac{1}{2}\boldsymbol{\Phi}_{\mathrm{U}}\boldsymbol{\alpha}, \quad \boldsymbol{\eta} \geq -\boldsymbol{\Phi}_{\mathrm{U}}\boldsymbol{\alpha},$$

where $\geq$ for vectors denotes the element-wise inequality and $\mathbf{0}$ is the vector whose elements are all zeros. This optimization problem is a *quadratic program* (QP) and thus can be solved rather efficiently by a standard solver.

7.3.4   Class-Priors in SU Classification

When class-prior π_{P} is provided as prior knowledge, we can directly solve the optimization problem (7.12). In this case, the solution of SU classification does not only separate data but also *identifies classes*, i.e., we can determine which class is positive or negative, even though we do not use any single explicit label.

Table 7.1
Behavior of SU classification with respect to the prior knowledge
on class-priors. "Separation" means that it is possible only to separate two classes, while "identification" further gives which class is
positive or negative.

Case	Prior knowledge	Identification	Separation
1	exact π_P	✓	✓
2	nothing	✗	✓
3	$\mathrm{sign}(\pi_P - \pi_N)$	✓	✓

However, when no prior knowledge on π_P is available, we need to estimate π_P before
solving (7.12). In section 12.7, we will give a method to estimate the class-prior only from
SU data, which gives an estimate of the class-prior always larger than 0.5. This means that
we cannot identify which class is positive or negative in general.

On the other hand, if we can know which class has a larger class-prior, i.e., $\mathrm{sign}(\pi_P - \pi_N)$,
it is possible to estimate π_P (and π_N) by the method given in section 12.7. Therefore, class
identification is still possible. This discussion is summarized in table 7.1.

7.4   Similar-Dissimilar (SD) and Dissimilar-Unlabeled (DU) Classification

Next, SU classification is extended to similar-dissimilar (SD) classification and dissimilar-
unlabeled (DU) classification (Shimada et al., 2021).

7.4.1   Classification Risk Estimation

First, we show that the classification risk can be estimated only from DU data or SD data.
Recall that

$$\ell_\sharp(\hat{y}, y) = \frac{\pi_P}{2\pi_P - 1}\ell(\hat{y}, y) - \frac{\pi_N}{2\pi_P - 1}\ell(\hat{y}, -y), \quad \widetilde{\ell}_\sharp(\hat{y}) = \ell_\sharp(\hat{y}, +1) - \ell_\sharp(\hat{y}, -1).$$

Theorem 7.3   *(Shimada et al., 2021) Suppose $\pi_P \neq \frac{1}{2}$. Then classification risk (7.1) can
be equivalently represented as*

$$R(g) = R_{\mathrm{DU}}(g) := \pi_D \, \mathbb{E}_D\left[-\frac{\widetilde{\ell}_\sharp(g(x)) + \widetilde{\ell}_\sharp(g(x'))}{2}\right] + \mathbb{E}_U\left[\ell_\sharp(g(x), +1)\right],$$

$$R(g) = R_{\mathrm{SD}}(g) := \pi_S \, \mathbb{E}_S\left[\frac{\ell_\sharp(g(x), +1) + \ell_\sharp(g(x'), +1)}{2}\right]$$

$$+ \pi_D \, \mathbb{E}_D\left[\frac{\ell_\sharp(g(x), -1) + \ell_\sharp(g(x'), -1)}{2}\right],$$

where $\mathbb{E}_D$ *denotes the expectation over* $p_D(\boldsymbol{x}, \boldsymbol{x}')$,

$$\mathbb{E}_D[\,\cdot\,] := \int \cdot\, p_D(\boldsymbol{x}, \boldsymbol{x}')\mathrm{d}\boldsymbol{x}\mathrm{d}\boldsymbol{x}',$$

and $\ell_\sharp(\hat{y}, y)$ *and* $\widetilde{\ell}_\sharp(\hat{y})$ *are defined in* (7.6) *and* (7.7), *respectively.*

Like theorem 7.1, theorem 7.3 can also be proved as special cases of UU classification (shown in chapter 8). Indeed, theorem 7.3 can be obtained by setting (θ, θ') in theorem 8.4 as

$$(\theta, \theta') = \begin{cases} \left(\frac{1}{2}, \pi_P\right) & \text{(DU classification),} \\[2ex] \left(\frac{\pi_P^2}{\pi_S}, \frac{1}{2}\right) & \text{(SD classification).} \end{cases}$$

Theorem 7.3 immediately gives unbiased risk estimators for DU classification and SD classification. However, similar to SU classification, the class-prior probability π_P (and also π_S in SD classification) is needed to perform ERM. In the case of DU classification, one may follow the same strategy as SU classification (discussed in section 7.3.4) to handle π_P (see section 12.7 for the detail of class-prior estimation). In the case of SD classification, one may estimate π_S directly by $n_S/(n_S + n_D)$, then estimate π_P by

$$\frac{\sqrt{2\pi_S - 1} + 1}{2}, \tag{7.14}$$

which is obtained by the solution of $\pi_P^2 + \pi_N^2 = \pi_S.^2$

7.4.2   Interpretation of SD Risk

The SD risk is not only an equivalent expression of the classification risk, but it can also be interpreted as a binary classification risk that aims to predict the "similar" and "dissimilar" labels.

To show this, we turn to the one-sample case and recall that a binary random variable τ indicates similarity ($\tau = +1$) and dissimilarity ($\tau = -1$) (see section 7.2). Then we can rewrite the SD risk as

$$\begin{aligned} R_{SD}(g) = {}&\pi_S\, \mathbb{E}_{p_S(\boldsymbol{x}, \boldsymbol{x}')}\left[\frac{\ell_\sharp(g(\boldsymbol{x}), +1) + \ell_\sharp(g(\boldsymbol{x}'), +1)}{2}\right] \\[2ex] &+ \pi_D\, \mathbb{E}_{p_D(\boldsymbol{x}, \boldsymbol{x}')}\left[\frac{\ell_\sharp(g(\boldsymbol{x}), -1) + \ell_\sharp(g(\boldsymbol{x}'), -1)}{2}\right] \\[2ex] = {}&\iint p(\boldsymbol{x}, \boldsymbol{x}', \tau = +1)\left\{\frac{\ell_\sharp(g(\boldsymbol{x}), +1) + \ell_\sharp(g(\boldsymbol{x}'), +1)}{2}\right\}\mathrm{d}\boldsymbol{x}\mathrm{d}\boldsymbol{x}' \\[2ex] &+ \iint p(\boldsymbol{x}, \boldsymbol{x}', \tau = -1)\left\{\frac{\ell_\sharp(g(\boldsymbol{x}), -1) + \ell_\sharp(g(\boldsymbol{x}'), -1)}{2}\right\}\mathrm{d}\boldsymbol{x}\mathrm{d}\boldsymbol{x}' \end{aligned}$$

$$= \iint \sum_{\tau \in \{+1,-1\}} p(\boldsymbol{x},\boldsymbol{x}',\tau) \left\{ \frac{\ell_\sharp(g(\boldsymbol{x}),\tau) + \ell_\sharp(g(\boldsymbol{x}'),\tau)}{2} \right\} \mathrm{d}\boldsymbol{x}\mathrm{d}\boldsymbol{x}'$$

$$= \mathbb{E}_{p(\boldsymbol{x},\boldsymbol{x}',\tau)} \left[\frac{\ell_\sharp(g(\boldsymbol{x}),\tau) + \ell_\sharp(g(\boldsymbol{x}'),\tau)}{2} \right].$$

To interpret this expression of the risk from a different perspective, we visualize the landscape of loss $\ell_\sharp$ in figure 7.1, with ℓ set to several margin-based losses. As can be seen in these figures, $\ell_\sharp(\hat{y},y)$ has a similar profile to $\ell(\hat{y},y)$ when $\pi_{\mathrm{P}} > \frac{1}{2}$; otherwise, $\ell_\sharp(\hat{y},y)$ is similar to $\ell(\hat{y},-y)$. This enables us to give another interpretation of SD classification as binary classification with loss function $\ell_\sharp$, where classifier g takes input $\boldsymbol{x}$ and predicts its associated pairwise label τ.

From this point of view, the relationship among SD, SU, and DU classification actually corresponds to that among PN, PU, and NU classification. The main idea of PU and NU classification is to complement missing N and P data, respectively, with U data. (see chapter 4 for details). In other words, the classification risk can be represented by PU data as

$$R(g) = \mathbb{E}_{p(\boldsymbol{x},y)} \left[\ell(g(\boldsymbol{x}),y) \right]$$

$$= \pi_{\mathrm{P}} \, \mathbb{E}_{p_{\mathrm{P}}(\boldsymbol{x})} \left[\ell(g(\boldsymbol{x}),+1) \right] + \pi_{\mathrm{N}} \, \mathbb{E}_{p_{\mathrm{N}}(\boldsymbol{x})} \left[\ell(g(\boldsymbol{x}),-1) \right]$$

$$= \pi_{\mathrm{P}} \, \mathbb{E}_{p_{\mathrm{P}}(\boldsymbol{x})} \left[\ell(g(\boldsymbol{x}),+1) \right]$$

$$+ \underbrace{ \left\{ \mathbb{E}_{p_{\mathrm{U}}(\boldsymbol{x})} \left[\ell(g(\boldsymbol{x}),-1) \right] - \pi_{\mathrm{P}} \, \mathbb{E}_{p_{\mathrm{P}}(\boldsymbol{x})} \left[\ell(g(\boldsymbol{x}),-1) \right] \right\} }_{\text{since } \pi_{\mathrm{N}} p(\boldsymbol{x}) = p_{\mathrm{U}}(\boldsymbol{x}) - \pi_{\mathrm{P}} p_{\mathrm{P}}(\boldsymbol{x})}$$

$$= \pi_{\mathrm{P}} \, \mathbb{E}_{p_{\mathrm{P}}(\boldsymbol{x})} \left[\ell(g(\boldsymbol{x}),+1) - \ell(g(\boldsymbol{x}),-1) \right] + \mathbb{E}_{p_{\mathrm{U}}(\boldsymbol{x})} \left[\ell(g(\boldsymbol{x}),-1) \right]. \tag{7.15}$$

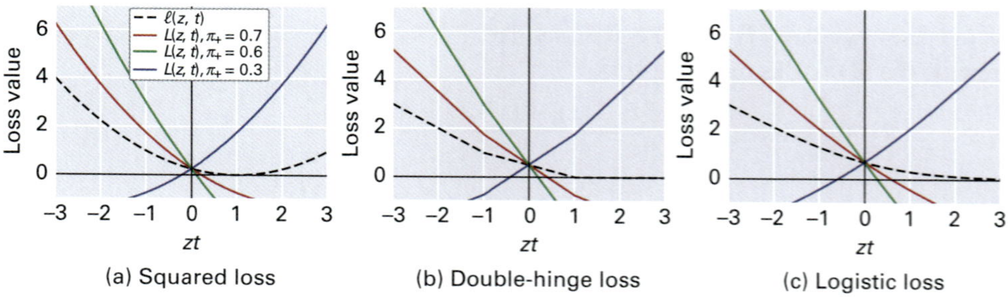

Figure 7.1

Visualization of loss $\ell_\sharp$ defined in (7.6). ℓ is set to the squared, double-hinge, and logistic loss. Each line with $L(z,t)$ in the figures represents $\ell_\sharp$. As shown in the graphs, $\ell_\sharp(\hat{y},y)$ approaches $\ell(\hat{y},y)$ as π_{P} gets larger, and $\ell_\sharp(\hat{y},y)$ approaches $\ell(\hat{y},-y)$ as π_{P} gets smaller.

Furthermore, when loss ℓ is symmetric (Ghosh et al., 2015; Charoenphakdee et al., 2019), i.e., for some $K \in \mathbb{R}$, $\ell(z, +1) + \ell(z, -1) = K$, then the following relationship holds.

Corollary 7.4 *Assume $\pi_{\mathrm{P}} \neq \frac{1}{2}$. Define Q_{SD} by replacing $\ell_\sharp$ in R_{SD} with ℓ as*

$$Q_{\mathrm{SD}}(g) := \mathbb{E}_{p(\boldsymbol{x},\boldsymbol{x}',\tau)} \left[\frac{\ell(g(\boldsymbol{x}), \tau) + \ell(g(\boldsymbol{x}'), \tau)}{2} \right].$$

Suppose that ℓ is a symmetric loss. Then R_{SD} and Q_{SD} share the optimal solution as

$$\operatorname*{argmin}_{g \in \mathscr{G}} R_{\mathrm{SD}}(g) = \begin{cases} \operatorname{argmin}_{g \in \mathscr{G}} Q_{\mathrm{SD}}(g) & (\pi_{\mathrm{P}} > \frac{1}{2}), \\ \operatorname{argmax}_{g \in \mathscr{G}} Q_{\mathrm{SD}}(g) & (\pi_{\mathrm{P}} < \frac{1}{2}). \end{cases} \tag{7.16}$$

Proof We show that $\ell_\sharp(\hat{y}, y)$ is a linear function of $\ell(\hat{y}, y)$ as follows:

$$\begin{aligned} \ell_\sharp(\hat{y}, y) &= \frac{\pi_{\mathrm{P}}}{\pi_{\mathrm{P}} - \pi_{\mathrm{N}}} \ell(\hat{y}, y) - \frac{\pi_{\mathrm{N}}}{\pi_{\mathrm{P}} - \pi_{\mathrm{N}}} \ell(\hat{y}, -y) \\ &= \frac{\pi_{\mathrm{P}}}{\pi_{\mathrm{P}} - \pi_{\mathrm{N}}} \ell(\hat{y}, y) - \frac{\pi_{\mathrm{N}}}{\pi_{\mathrm{P}} - \pi_{\mathrm{N}}} \left\{ K - \ell(\hat{y}, y) \right\} \\ &= \frac{1}{\pi_{\mathrm{P}} - \pi_{\mathrm{N}}} \ell(\hat{y}, y) - \frac{\pi_{\mathrm{N}}}{\pi_{\mathrm{P}} - \pi_{\mathrm{N}}} K. \end{aligned}$$

By using the above relationship, we obtain

$$R_{\mathrm{SD}}(g) = \frac{1}{\pi_{\mathrm{P}} - \pi_{\mathrm{N}}} Q_{\mathrm{SD}}(g) - \frac{\pi_{\mathrm{N}}}{\pi_{\mathrm{P}} - \pi_{\mathrm{N}}} K,$$

which results in (7.16). $\square$

If we use a symmetric loss in the classification risk, corollary 7.4 gives us a practical advantage. Specifically, when writing a program code of a certain classification algorithm, we do not have to implement loss $\ell_\sharp$ by ourselves. Instead, we can treat the S and D labels as the P and N labels associated with each point in a pair, and we can use any standard binary classification algorithm with a loss function ℓ.

Actually, the relationship in corollary 7.4 holds for R_{SU} and R_{DU} as well. The alternative objective functions Q_{SU} and Q_{DU} are defined as follows:

$$Q_{\mathrm{SU}}(g) := \pi_{\mathrm{S}} \, \mathbb{E}_{p_{\mathrm{S}}(\boldsymbol{x},\boldsymbol{x}')} \left[\frac{\widetilde{\ell}(g(\boldsymbol{x})) + \widetilde{\ell}(g(\boldsymbol{x}'))}{2} \right] + \mathbb{E}_{p_{\mathrm{U}}(\boldsymbol{x})} \left[\ell(g(\boldsymbol{x}), -1) \right],$$

$$Q_{\mathrm{DU}}(g) := \pi_{\mathrm{D}} \, \mathbb{E}_{p_{\mathrm{D}}(\boldsymbol{x},\boldsymbol{x}')} \left[-\frac{\widetilde{\ell}(g(\boldsymbol{x})) + \widetilde{\ell}(g(\boldsymbol{x}'))}{2} \right] + \mathbb{E}_{p_{\mathrm{U}}(\boldsymbol{x})} \left[\ell(g(\boldsymbol{x}), +1) \right],$$

where

$$\widetilde{\ell}(\hat{y}) := \ell(\hat{y}, +1) - \ell(\hat{y}, -1).$$

Similarly to (7.16), we observe that Q_{SU} and Q_{DU} have the same optimal solutions as R_{SU} and R_{DU}, respectively. Moreover, we can confirm that Q_{SU} corresponds to the PU risk in (7.15) with a different coefficient π_{S} from π_{P}—Q_{DU} corresponds to the NU risk as well— which gives us an intuitive interpretation of the SU (respectively, DU) classification method as the PU (respectively, NU) classification method.

7.5   Similar-Dissimilar-Unlabeled (SDU) Classification

This classification method (Shimada et al., 2021) incorporates all S, D, and U data into the ERM framework.

The main idea is to combine the risks computed from SU, DU, and SD data in a similar manner to PNU classification in chapter 5. Since each of $R_{\mathrm{SU}}(g)$, $R_{\mathrm{DU}}(g)$, and $R_{\mathrm{SD}}(g)$ is an equivalent expression of the true classification risk $R(g)$, their convex combination is still equivalent to $R(g)$:

$$R_{\mathrm{SDU}}^{\gamma}(g) := \gamma_1 R_{\mathrm{SU}}(g) + \gamma_2 R_{\mathrm{DU}}(g) + \gamma_3 R_{\mathrm{SD}}(g),$$

where $\gamma = (\gamma_1, \gamma_2, \gamma_3)$ is the hyper-parameter that satisfies $\gamma_1, \gamma_2, \gamma_3 \geq 0$ and $\gamma_1 + \gamma_2 + \gamma_3 = 1$.

In section 7.6, theoretical analyses of SU, DU and SD classification will be conducted, where the DU and SD classification methods are more favorable than the SU classification method. Based on the discussion there, we practically propose to fix $\gamma_1 = 0$, i.e., SU classification is completely dropped from the combination. This reduction of the number of hyper-parameters will contribute to both improving the classification performance and reducing the computation time. We refer to SDU classification without SU classification as *SD+DU classification*.

7.6   Theoretical Analysis

Now, let us establish estimation error bounds for SU, DU, and SD classification and investigate their relations.

7.6.1   Derivation of Estimation Error Bounds

Hereafter, let $\mathcal{G} \subset \mathbb{R}^{\mathcal{X}}$ be a function class of a specific model. Define the Rademacher complexity of $\mathcal{G}$ for the sample size n with respect to $q(x)$ as

$$\mathfrak{R}_{n,q}(\mathcal{G}) := \mathbb{E}_{x_1,\ldots,x_n} \mathbb{E}_{\sigma_1,\ldots,\sigma_n} \left[\sup_{g \in \mathcal{G}} \frac{1}{n} \sum_{i=1}^{n} \sigma_i g(x_i) \right],$$

where $\sigma_1, \ldots, \sigma_n$ are the Rademacher variables, i.e., random variables taking $+1$ and -1 with the same probabilities (see section 3.1.2.3). For any probability density $q(x)$, we assume that our model class $\mathscr{G}$ satisfies

$$\mathfrak{R}_{n,q}(\mathscr{G}) \leq \frac{C_{\mathscr{G}}}{\sqrt{n}}$$

for some constant $C_{\mathscr{G}} > 0$. This covers linear-in-parameter models with bounded normals and feature maps as a special case:

$$\mathscr{G} = \{g(x) = \boldsymbol{\alpha}^\top \boldsymbol{\phi}(x) \mid \|\boldsymbol{\alpha}\| \leq C_\alpha, \|\boldsymbol{\phi}\|_\infty \leq C_\phi\} \quad \text{for some } C_\alpha, C_\phi > 0.$$

More details on the Rademacher complexity bounds can be found in section 3.1.2.4.

We have estimation error bounds for the SU, DU and SD classification methods as follows.

Theorem 7.5   *(Bao et al., 2018; Shimada et al., 2021) Let $R(g) = \mathbb{E}_{p(x,y)}[\ell(g(x),y)]$ be the classification risk for function g, let $g^* \in \mathscr{G}$ be its minimizer, and $\widehat{g}_{\mathrm{SU}}$, $\widehat{g}_{\mathrm{DU}}$, and let $\widehat{g}_{\mathrm{SD}}$ be minimizers of the empirical SU, DU, and SD risks in $\mathscr{G}$, respectively. Assume that $\pi_\mathrm{P} \neq \frac{1}{2}$, the loss function $\ell(\cdot, y)$ is L_ℓ-Lipschitz continuous for any $y \in \mathscr{Y}$ $(0 < L_\ell < \infty)$, and all functions in the model class $\mathscr{G}$ are bounded, i.e., there exists a constant C_g such that $\|g\|_\infty \leq C_\mathrm{g}$ for any $g \in \mathscr{G}$. Let $C_\ell := \sup_{y \in \{\pm 1\}} \ell(C_\mathrm{g}, y)$. For any $\delta > 0$, each of the following inequalities holds independently with probability at least $1 - \delta$:*

$$R(\widehat{g}_{\mathrm{SU}}) - R(g^*) \leq C_{\mathscr{G},\ell,\delta} \left(\frac{2\pi_\mathrm{S}}{\sqrt{2n_\mathrm{S}}} + \frac{1}{\sqrt{n_\mathrm{U}}} \right), \tag{7.17}$$

$$R(\widehat{g}_{\mathrm{DU}}) - R(g^*) \leq C_{\mathscr{G},\ell,\delta} \left(\frac{2\pi_\mathrm{D}}{\sqrt{2n_\mathrm{D}}} + \frac{1}{\sqrt{n_\mathrm{U}}} \right), \tag{7.18}$$

$$R(\widehat{g}_{\mathrm{SD}}) - R(g^*) \leq C_{\mathscr{G},\ell,\delta} \left(\frac{\pi_\mathrm{S}}{\sqrt{2n_\mathrm{S}}} + \frac{\pi_\mathrm{D}}{\sqrt{2n_\mathrm{D}}} \right), \tag{7.19}$$

where

$$C_{\mathscr{G},\ell,\delta} = \frac{1}{|\pi_\mathrm{P} - \pi_\mathrm{N}|} \left(4L_\ell C_{\mathscr{G}} + \sqrt{2C_\ell^2 \ln \frac{8}{\delta}} \right).$$

Theorem 7.5 can be proved in the same manner as theorem 3.6 in the PN classification case, so its proof is omitted. This theorem shows that if we have π_P in advance, the SU, DU, and SD classification methods are consistent, meaning that the estimation errors vanish asymptotically with high probability. The convergence rates are also the same as the PN classification case; for example, the convergence rate of SU classification is $\mathcal{O}_p(1/\sqrt{n_\mathrm{S}} + 1/\sqrt{n_\mathrm{U}})$, where $\mathcal{O}_p$ denotes the order in probability.

7.6.2   Comparison of Estimation Error Bounds

Here, we compare the SU, DU, and SD classification methods from the perspective of their estimation error bounds. We assume the one-sample case in this analysis in order to understand the behavior of the SU, DU, and SD risk estimators with sufficiently large sample size n_{SD}.

Theorem 7.6　*(Shimada et al., 2021) Suppose $\pi_{\mathrm{P}} \neq \frac{1}{2}$ and similar and dissimilar data pairs follow the generation process in (7.2). We denote the right-hand sides of (7.17), (7.18), and (7.19) by V_{SU}, V_{DU}, and V_{SD}, respectively. Then,*

$$\Pr\{V_{\mathrm{DU}} < V_{\mathrm{SU}} \wedge V_{\mathrm{SD}} < V_{\mathrm{SU}}\} \geq 1 - \exp(-cn_{\mathrm{SD}})$$

for some constant $c > 0$.

Proof　We will show that the events $V_{\mathrm{DU}} < V_{\mathrm{SU}}$ and $V_{\mathrm{SD}} < V_{\mathrm{SU}}$ hold simultaneously with high probability. First, if $\pi_{\mathrm{S}}/\sqrt{2n_{\mathrm{S}}} > \pi_{\mathrm{D}}/\sqrt{2n_{\mathrm{D}}}$ holds, then we have

$$\frac{V_{\mathrm{SU}} - C_{\mathscr{G},\ell,\delta}/\sqrt{n_{\mathrm{U}}}}{V_{\mathrm{DU}} - C_{\mathscr{G},\ell,\delta}/\sqrt{n_{\mathrm{U}}}} = \frac{\pi_{\mathrm{S}}/\sqrt{2n_{\mathrm{S}}}}{\pi_{\mathrm{D}}/\sqrt{2n_{\mathrm{D}}}} > 1,$$

and

$$V_{\mathrm{SU}} - V_{\mathrm{SD}} = C_{\mathscr{G},\ell,\delta} \left(\frac{\pi_{\mathrm{S}}}{\sqrt{2n_{\mathrm{S}}}} - \frac{\pi_{\mathrm{D}}}{\sqrt{2n_{\mathrm{D}}}} + \frac{1}{\sqrt{n_{\mathrm{U}}}} \right) > \frac{C_{\mathscr{G},\ell,\delta}}{\sqrt{n_{\mathrm{U}}}} > 0.$$

These two inequalities imply $V_{\mathrm{DU}} < V_{\mathrm{SU}}$ and $V_{\mathrm{SD}} < V_{\mathrm{SU}}$. Hence, it is sufficient to show that $\pi_{\mathrm{S}}/\sqrt{2n_{\mathrm{S}}} > \pi_{\mathrm{D}}/\sqrt{2n_{\mathrm{D}}}$ holds with high probability in order to show the high probability bound of $V_{\mathrm{DU}} < V_{\mathrm{SU}}$ and $V_{\mathrm{SD}} < V_{\mathrm{SU}}$.

Since we assumed the data generation process in (7.2), the class of each data pair (i.e., similar or dissimilar) follows a Bernoulli distribution. Therefore, the number of pairs in each class follows a binomial distribution—namely, $n_{\mathrm{D}} \sim \mathrm{Binomial}(n_{\mathrm{SD}}, \pi_{\mathrm{D}})$ and $n_{\mathrm{S}} = n_{\mathrm{SD}} - n_{\mathrm{D}}$. By using the upper tail probability bound of a binomial random variable derived from Hoeffding's inequality, we have

$$\Pr\left\{ \frac{\pi_{\mathrm{S}}}{\sqrt{2n_{\mathrm{S}}}} > \frac{\pi_{\mathrm{D}}}{\sqrt{2n_{\mathrm{D}}}} \right\} = \Pr\left\{ n_{\mathrm{D}} > \frac{n_{\mathrm{SD}}\pi_{\mathrm{D}}^2}{\pi_{\mathrm{S}}^2 + \pi_{\mathrm{D}}^2} \right\}$$

$$= 1 - \Pr\left\{ n_{\mathrm{D}} \leq \frac{n_{\mathrm{SD}}\pi_{\mathrm{D}}^2}{\pi_{\mathrm{S}}^2 + \pi_{\mathrm{D}}^2} \right\}$$

$$\geq 1 - \exp\left(-\frac{n_{\mathrm{SD}}\pi_{\mathrm{D}}}{2(1 - \pi_{\mathrm{D}})} \left(1 - \frac{\pi_{\mathrm{D}}}{\pi_{\mathrm{S}}^2 + \pi_{\mathrm{D}}^2} \right)^2 \right).$$

Finally, we obtain

$$\Pr\{V_{\mathrm{DU}} \leq V_{\mathrm{SU}} \wedge V_{\mathrm{SD}} \leq V_{\mathrm{SU}}\} \geq 1 - \exp\left(-\frac{n_{\mathrm{SD}}\pi_{\mathrm{D}}}{2(1 - \pi_{\mathrm{D}})} \left(1 - \frac{\pi_{\mathrm{D}}}{\pi_{\mathrm{S}}^2 + \pi_{\mathrm{D}}^2} \right)^2 \right).$$

When $\pi_P \neq \frac{1}{2}$, this probability bound is not vacuous, i.e., the right-hand side is strictly larger than the zero. $\square$

Theorem 7.6 states that $\max\{V_{SD}, V_{DU}\} < V_{SU}$ holds with high probability when n_{SD} is sufficiently large.[3] This suggests that both the SD and DU classification methods are likely to outperform the SU classification method when all of pairwise similarities/dissimilarities and unlabeled data are given in advance. Inspired by this result, SU classification was dropped from the SDU classification in section 7.5.

7.7 Experiments

In this section, pairwise-constraint classification is experimentally evaluated.

7.7.1 Setup

Experiments were conducted on benchmark datasets obtained from the *UCI Machine Learning Repository* (Lichman, 2013), the *LIBSVM* (Chang and Lin, 2011), and the *ELENA* project.[4] The original datasets were randomly subsampled so that S data pairs consist of P and N data pairs with the ratio of π_P^2 to π_N^2 (7.3), while D data pairs with the ratio of 1 to 1 (7.4). The ratios of U and test data are π_P to π_N (7.5).

To implement SU classification, we used the linear-in-input model $g(x) = \alpha^\top x$. In section 7.7.2, the squared loss was used, and π_P was assumed to be known (which corresponds to case 1 in table 7.1). In section 7.7.3, the squared loss and the double-hinge loss were used, and the class-prior was estimated (see section 12.7 for details) with KM2 (Ramaswamy et al., 2016) (which corresponds to case 2 in table 7.1). The regularization parameter λ was chosen from $\{10^{-1}, 10^{-4}, 10^{-7}\}$.

To implement SDU and SD+DU classification, the linear-in-input model was used as well. In section 7.7.4, the double-hinge loss was used. The class-prior was estimated by (7.14). The regularization parameter λ was chosen from $\{10^{-1}, 10^{-4}, 10^{-7}\}$. For SDU classification, the combination parameter γ was chosen from $\{0.0, 0.2, 0.4, \ldots, 1.0\}$.

To choose hyper-parameters, fivefold cross-validation was used. Since we do not have any labeled data in the training phase, the validation error cannot be computed directly. Instead, (7.8) equipped with the zero-one loss $\ell_{0\text{-}1}(\cdot) = \frac{1}{2}(1 - \text{sign}(\cdot))$ was used as a proxy to estimate the validation error. In each experimental trial, the model with minimum validation error was chosen.

7.7.2 Illustration of SU Classification

The number of S data pairs was fixed to 200, while the number of U data, n_U, was changed as 200, 400, 800, and 1,600.

The experimental results are shown in figure 7.2. It indicates that the classification error decreases as n_U grows, which well agrees with the theoretical analysis in theorem 7.5.

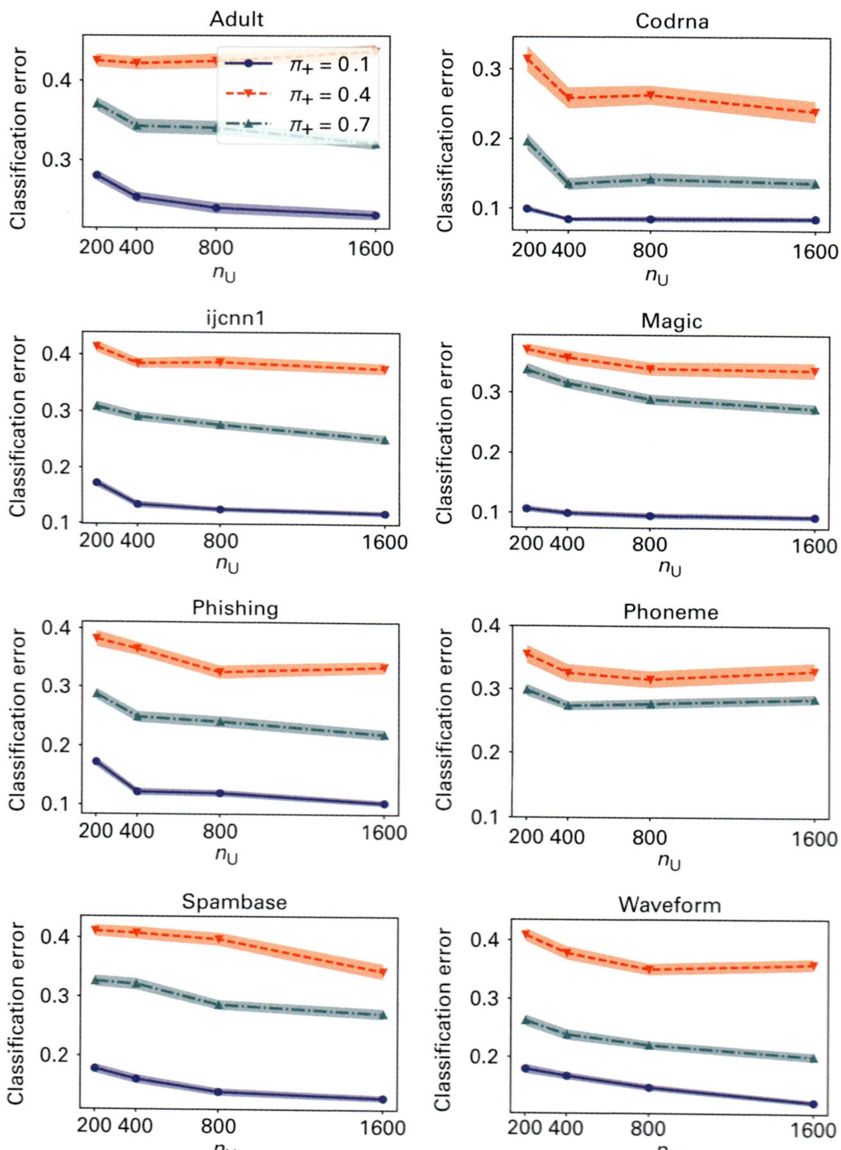

Figure 7.2
The average (vertical axes) and standard error (shaded areas) of the classification error over 50 trials. Different $n_U \in \{200, 400, 800, 1600\}$ were tested, while n_S was fixed to 200. For each dataset, results with different class-priors ($\pi_P \in \{0.1, 0.4, 0.7\}$) are plotted, which was assumed to be known in this experiment. Dataset "phoneme" does not have a plot for $\pi_P = 0.1$ because the number of data points in the original dataset is insufficient to subsample an SU dataset with $\pi_P = 0.1$.

Furthermore, we can observe a tendency that the classification error becomes smaller as the class-prior becomes farther from $\frac{1}{2}$. This also agrees with the theoretical results. That is, $C_{\mathscr{G},\ell,\delta}$ in (7.17) has the term $|2\pi_P - 1|$ in the denominator, which will make the upper bound looser when π_P is close to $\frac{1}{2}$.[5]

7.7.3 Comparison of SU Classification with Other Methods

Next, the performance of SU classification was compared with the following methods:

- k-means clustering (KMC): As a simple baseline, we considered KMC (MacQueen, 1967). We ignored pairwise information brought by S data pairs and simply applied KMC with $k = 2$ clusters to U data. Labels of test data were predicted by finding the nearest cluster center.

- Information-theoretic metric learning (ITML): ITML (Davis et al., 2007) is a metric learning method that regularizes the Mahalanobis metric matrix based on prior knowledge and pairwise constraints. We used the identity matrix as prior knowledge, and the slack parameter was fixed to $\gamma = 1$, since cross-validation cannot be performed without any class label information. Using the obtained metric, KMC was applied on test data.

- Semi-supervised metric learning paradigm with hyper-sparsity (SERAPH): SERAPH (Niu et al., 2012) is another metric learning method based on entropy regularization. Hyper-parameters were chosen following the setting SERAPH$_{\text{hyper}}$ in the original paper. Using the obtained metric, KMC was applied on test data.

- Semi-supervised SMI-based clustering (3SMIC): 3SMIC (Calandriello et al., 2014) models the class-posterior probabilities and maximizes mutual information between unlabeled data at hand and the cluster labels. The penalty parameter γ and the kernel parameter t were chosen from $\{10^{-2}, 10^0, 10^2\}$ and $\{4, 7, 10\}$, respectively, using fivefold cross-validation.

- DirtyIMC (Chiang et al., 2015) is an inductive matrix completion method to recover the similarity matrix from a low-rank feature matrix. Similarity matrix S is assumed to be expressed as $UU^\top$, where U is low-rank feature representations of input data. After obtaining U, KMC was applied on U. The two hyper-parameters were set to $\lambda_M = \lambda_N = 10^{-2}$. DIMC can only handle in-sample prediction, so it was trained by using both training and test data.

For this experiment, benchmark datasets with 500 similar data pairs, 500 unlabeled data, and 100 test data were used. The class-prior π_P was fixed to 0.7. As can be seen from table 7.2, SU classification outperforms other methods for most of the datasets. In addition, SU classification often performed better with the double-hinge loss than with the squared loss. We speculate that this is partly because the squared loss penalizes the predictions with very large margins even if they are correct.[6] In contrast, the double-hinge loss does not penalize the correct prediction with large enough margins.

Table 7.2
The mean classification accuracy and standard error on benchmark datasets over 20 trials. Sq/Dh means the squared/double-hinge loss, respectively. For all experiments, class-prior π_P was set to 0.7. In the SU classification method, π_P was estimated. Performances were measured by the clustering accuracy $1 - \min\{r, 1 - r\}$, where r is the misclassification rate. Boldface indicates outperforming methods, determined by the one-sided t-test with significance level 5%. The result of SERAPH with "w8a" was unavailable due to high-dimensionality and memory constraints.

| Dataset | d | SU (Proposed) | | Baselines | | | | |
		Sq	Dh	KMC	ITML	SERAPH	3SMIC	DIMC
adult	123	64.5 (1.2)	**84.5 (0.8)**	58.1 (1.1)	57.9 (1.1)	66.5 (1.7)	58.5 (1.3)	63.7 (1.2)
banana	2	**67.5 (1.2)**	**68.2 (1.2)**	54.3 (0.7)	54.8 (0.7)	55.0 (1.1)	61.9 (1.2)	64.3 (1.0)
cod-rna	8	**82.8 (1.3)**	71.0 (0.9)	63.1 (1.1)	62.8 (1.0)	62.5 (1.4)	56.6 (1.2)	63.8 (1.1)
higgs	28	55.1 (1.1)	**69.3 (0.9)**	**66.4 (1.6)**	**66.6 (1.3)**	63.4 (1.1)	57.0 (0.9)	65.0 (1.1)
ijcnn1	22	65.5 (1.3)	**73.6 (0.9)**	54.6 (0.9)	55.8 (0.7)	59.8 (1.2)	58.9 (1.3)	66.2 (2.2)
magic	10	66.0 (2.0)	**69.0 (1.3)**	53.9 (0.6)	54.5 (0.7)	55.0 (0.9)	59.1 (1.4)	63.1 (1.1)
phishing	68	75.0 (1.4)	**91.3 (0.6)**	64.4 (1.0)	61.9 (1.1)	62.4 (1.1)	60.1 (1.3)	64.8 (1.4)
phoneme	5	**67.8 (1.5)**	**70.8 (1.0)**	65.2 (0.9)	66.7 (1.4)	**69.1 (1.4)**	61.3 (1.1)	64.5 (1.2)
spambase	57	69.7 (1.4)	**85.5 (0.5)**	60.1 (1.8)	54.4 (1.1)	65.4 (1.8)	61.5 (1.3)	63.6 (1.3)
susy	18	59.8 (1.3)	**74.8 (1.2)**	55.6 (0.7)	55.4 (0.9)	58.0 (1.0)	57.1 (1.2)	65.2 (1.0)
w8a	300	62.1 (1.5)	**86.5 (0.6)**	71.0 (0.8)	69.5 (1.5)	0.0 (0.0)	60.5 (1.5)	65.0 (2.0)
waveform	21	77.8 (1.3)	**87.0 (0.5)**	56.1 (0.8)	54.8 (0.7)	56.5 (0.9)	56.5 (0.9)	65.0 (0.9)

7.7.4 Comparison of SDU Classification with Other Methods

Finally, we report the performances of the SDU/SD+DU classification methods. The following methods were used as baselines:

- k-means clustering (KMC): KMC with $k = 2$ was applied to training data by ignoring all pairwise information. Labels of test data were predicted by finding the nearest cluster center.

- Constrained k-means clustering (CKM): CKM (Wagstaff et al., 2001) is a semi-supervised variant of KMC, where pairwise similarities/dissimilarities were treated as must-/cannot-links. k was set to 2.

- Semi-supervised spectral clustering (SSP): Proposed in Chen and Feng (2012), SSP is a method where similar and dissimilar labels are propagated through the affinity matrix. We set $k = 5$ for constructing the affinity matrix with the k-nearest-neighbor graph and $\sigma^2 = 1$ for the precision parameter used in similarity measurement. Both train and test samples were used for clustering to give labels for test samples.

- ITML: Similar and dissimilar pairs were used for regularizing the Mahalanobis metric matrix. For test samples prediction, KMC with $k = 2$ was applied with the obtained metric. We used the identity matrix as prior information, and a slack parameter γ was set to 1.

Table 7.3
The mean classification accuracy and standard error on benchmark datasets over 50 trials. For all experiments, class-prior π_P was set to 0.7. For the proposed methods, π_P was estimated with SD data. Performances were measured by the clustering accuracy $1 - \min\{r, 1 - r\}$, where r is the misclassification rate. Boldface indicates outperforming methods, determined by the one-sided t-test with significance level 5%.

Dataset	n_{SD}	Proposed		Baselines				
		SDU	SD+DU	KMC	CKM	SSP	ITML	OVPC
adult	50	73.6 (1.1)	**76.4 (0.8)**	65.0 (0.8)	66.6 (1.1)	69.4 (0.3)	61.8 (0.6)	58.6 (0.9)
$d = 123$	200	**82.0 (0.5)**	**82.2 (0.3)**	63.3 (0.8)	71.9 (0.9)	69.3 (0.3)	60.9 (0.6)	57.3 (0.8)
banana	50	**66.1 (0.7)**	**66.5 (0.7)**	52.9 (0.4)	52.7 (0.3)	58.7 (0.7)	52.9 (0.4)	62.0 (1.0)
$d = 2$	200	**69.5 (0.4)**	**69.0 (0.4)**	52.5 (0.2)	52.5 (0.2)	66.5 (1.3)	52.5 (0.2)	64.8 (0.8)
cod-rna	50	**79.9 (1.1)**	**81.1 (1.2)**	62.6 (0.5)	61.5 (0.4)	54.6 (1.0)	62.6 (0.5)	62.9 (1.0)
$d = 8$	200	89.2 (0.7)	**90.6 (0.4)**	62.8 (0.5)	59.5 (0.5)	53.2 (0.7)	62.6 (0.5)	72.4 (1.3)
ijcnn1	50	**67.0 (0.7)**	**67.8 (0.6)**	55.5 (0.6)	54.7 (0.5)	60.0 (0.8)	55.1 (0.7)	57.9 (0.9)
$d = 22$	200	**76.7 (0.4)**	**77.4 (0.4)**	54.2 (0.3)	53.1 (0.3)	59.5 (0.8)	53.6 (0.3)	56.9 (0.8)
magic	50	**65.5 (0.8)**	**65.8 (0.9)**	52.4 (0.2)	51.8 (0.2)	52.7 (0.3)	52.4 (0.2)	56.8 (0.7)
$d = 10$	200	**74.1 (0.7)**	**74.4 (0.6)**	52.0 (0.2)	51.7 (0.2)	52.6 (0.3)	52.0 (0.2)	56.6 (0.8)
phishing	50	**77.9 (1.2)**	**78.6 (1.1)**	62.6 (0.3)	62.6 (0.3)	68.1 (0.3)	62.6 (0.3)	60.3 (0.8)
$d = 68$	200	**88.0 (0.4)**	**88.0 (0.4)**	62.6 (0.3)	62.8 (0.3)	68.4 (0.3)	62.6 (0.3)	57.1 (0.7)
phoneme	50	**70.4 (0.8)**	**70.8 (0.8)**	67.8 (0.3)	68.9 (0.5)	66.5 (1.0)	67.8 (0.3)	62.2 (0.9)
$d = 5$	200	**74.5 (0.5)**	**74.5 (0.5)**	67.8 (0.3)	71.0 (0.6)	72.0 (0.7)	68.0 (0.4)	62.9 (0.9)
spambase	50	**79.0 (1.4)**	**79.6 (1.3)**	63.7 (1.1)	64.2 (1.1)	70.4 (0.3)	60.9 (1.1)	60.2 (0.9)
$d = 57$	200	**87.2 (0.3)**	**87.7 (0.2)**	61.8 (1.2)	70.4 (0.6)	70.6 (0.3)	59.6 (1.1)	60.4 (1.1)
w8a	50	63.8 (1.3)	67.0 (1.2)	**69.3 (0.3)**	66.0 (0.7)	64.0 (0.7)	**68.9 (0.3)**	56.6 (0.8)
$d = 300$	200	79.6 (0.8)	**81.2 (0.7)**	69.3 (0.3)	56.3 (0.6)	67.4 (0.6)	68.3 (0.3)	54.9 (0.6)
waveform	50	**82.7 (1.0)**	**84.2 (0.9)**	51.5 (0.2)	51.6 (0.2)	53.3 (0.3)	51.5 (0.2)	63.6 (1.4)
$d = 21$	200	**87.2 (0.3)**	**87.4 (0.3)**	51.5 (0.2)	51.5 (0.1)	53.1 (0.3)	51.6 (0.2)	64.7 (1.3)

- On the value of pairwise constraints (OVPC): A classification-based approach was proposed in Zhang and Yan (2007), where an auxiliary classifier is trained on the feature vectors obtained from pairwise examples. The trained classifier was converted into a function that can be applied to pointwise prediction. The weight of the ridge regularization was chosen from $\{10^{-1}, 10^{-4}, 10^{-7}\}$ by fivefold cross-validation.

We used the same benchmark datasets as those used in section 7.7.3, with $n_U = 500$ and $n_{SD} \in \{50, 200\}$. Here, the class-prior π_P was fixed to 0.7. In each trial, the misclassification rate was measured with 500 test examples. Since there is no explicit positive/negative assignment in clustering methods, their performances were evaluated by the clustering accuracy $1 - \min\{r, 1 - r\}$, where r is the misclassification rate.

The results are reported in table 7.3, showing that the SDU and SD+DU classification methods outperform other methods for almost all datasets. The SD+DU classification method performs slightly better than the SDU classification method. This agrees with our theoretical analysis in corollary 7.4, claiming that DU and SD classification are always better than SU classification asymptotically. In contrast to the baselines, the SDU and

SD+DU classification performances can be improved with more SD data. This is observed in theorem 7.5, where the estimation error of the classification risk decreases as n_S and n_D increase. As a result, the SD+DU classification methods outperformed all the baselines with $n_\mathrm{SD} = 200$.

7.8 Ongoing Research

Finally, we want to briefly summarize extensions and related researches on pairwise-constraint classification.

• Weakly supervised classification from *noisy* pairwise constraints: We assumed that the pairwise constraints are generated from (7.3) and (7.4), where no noise exists in obtaining x once we observe similarity/dissimilarity. However, observation noise is inevitable in human annotations. To handle such observation noise, Dan et al. (2020) have considered two types of observation noise caused by the difficulty in label discrimination and pairwise comparison.

• Weakly supervised classification from *similarity-confidence* (Sconf): Another approach to alleviate the observation noise in pairwise constraints is to average over multiple similarity/dissimilarity labels obtained by crowd-workers, for example. Such averaged labels can naturally deliver the confidence of similarity. Cao et al. (2021) formulated Sconf classification in a similar spirit to Pconf classification (chapter 6). In Sconf classification classifiers can be trained only with Sconf data without U data, unlike SU classification.

• Weakly supervised classification from *triplet comparison data*: Pairwise constraints are collected by comparing two data points. In contrast, relative comparison of *three* data points can often be easier to obtain, in the form that X is closer to Y than Z. Cui et al. (2020) provided a classification risk estimator that can be computed from such triplet comparison data.

• *Uncoupled regression* (UR) from pairwise comparison data: UR is a regression problem where we are not given training data that have correspondence between input and output. Such a situation is ubiquitous in real-world problems such as census data analysis. Xu et al. (2019) introduced pairwise comparison in UR and gave an algorithm to perform regression from training data that consists of pairs of input points and a comparison label of which one takes a higher value.

• Relaxing the assumption $\pi_\mathrm{P} \neq \frac{1}{2}$ in SD classification: In theorem 7.3, $\pi_\mathrm{P} \neq \frac{1}{2}$ was assumed to make R_SD well-defined. On the other hand, Bao et al. (2020) revealed that SD classification is possible even when $\pi_\mathrm{P} = \frac{1}{2}$, by predicting similar/dissimilar labels directly. This direct approach successfully avoids the regularity condition $\pi_\mathrm{P} = \frac{1}{2}$ in classification risk estimation, and the classification performance becomes less sensitive to true π_P.

8 Unlabeled-Unlabeled (UU) Classification

In chapter 7, we discussed pairwise-constraint classification. Compared with other weakly supervised classification settings discussed in earlier chapters, pairwise-constraint classification is close to unsupervised classification in the sense that it does not use explicit class labels for training a classifier. So, a natural question is: What is the minimum degree of supervision to train a classifier in a valid way? In this chapter, we answer this question by showing that a binary classifier can be trained from two sets of unlabeled (U) data with different class-priors but not from one set. We call this problem setup *unlabeled-unlabeled (UU) classification* and present its empirical risk minimization (ERM) based algorithm.

8.1 Introduction

A naive approach to train a classifier only with U data, without any labeled data, is to use *discriminative clustering* (Xu et al., 2005; Valizadegan and Jin, 2007; Li et al., 2009; Gomes et al., 2010; Sugiyama et al., 2014; Hu et al., 2017), which is also known as *unsupervised classification*. However, this solution is usually suboptimal for two major reasons. First, successful translation of clusters into meaningful classes relies on the critical assumption that *one cluster exactly corresponds to one class* (Chapelle et al., 2003), which is rarely satisfied in practice. Second, clustering must introduce additional geometric or information-theoretic assumptions on which the learning objectives of clustering are built.

In this chapter, we show that if one more U set with a different class-prior is additionally available, one can successfully train a binary classifier without such clustering assumptions. We call this problem setup *unlabeled-unlabeled (UU) classification* (see figure 8.1). Here, we introduce two solutions for UU classification from different perspectives:

- The risk estimation approach (Lu et al., 2019) (section 8.3): We prove that it is impossible to estimate the risk of an arbitrary binary classifier in an *unbiased* manner given a single set of U data, but it becomes possible given two sets of U data with *different class-priors*. These results answer a fundamental question—what the minimal supervision is for training any

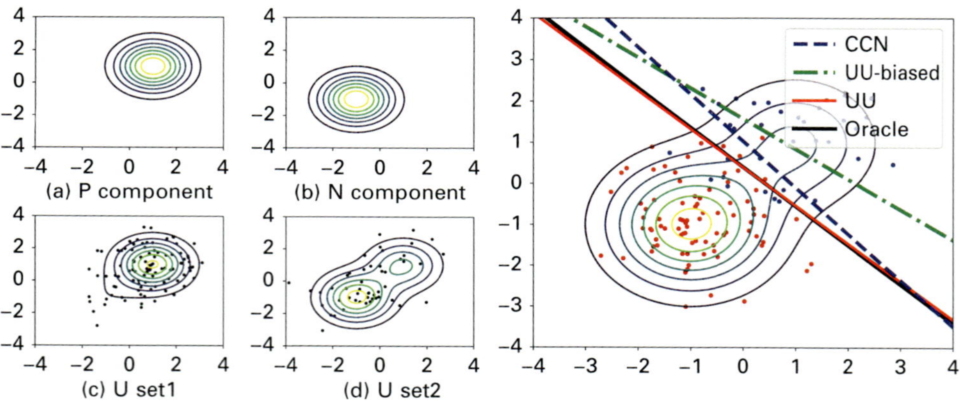

Figure 8.1

Illustrative example of classification from a Gaussian mixture dataset. Black points are unlabeled (U) data, blue points are positive (P) data, and red points are negative (N) data. In the left panel, (a) and (b) show P and N components of the Gaussian mixture; (c) and (d) show two distributions (with class-priors 0.9 and 0.4) from which U data are drawn. The right panel shows the test distribution (with class-prior 0.3), as well as four learned classifiers. In the legend, "CCN" refers to Natarajan et al. (2013), "UU-biased" means supervised classification taking U data with larger/smaller class-prior as P/N data, "UU" is the UU classification method (Lu et al., 2019), and "Oracle" means supervised classification from the same amount of labeled data. We can see that UU performs almost comparable to Oracle and much better than the other two methods.

binary classifier. Following these findings, we obtain an ERM-based classification method from two sets of U data, and then prove its *consistency*.

- The generative approach (du Plessis et al., 2013) (section 8.4): We show that UU classification can be solved by estimating the sign of the difference between the probability densities of the two U sets. We then introduce several estimation methods of such density difference.

The UU classification problem is actually conceivable in many real-world applications. For example, two sets of U data with different class-priors can be naturally collected from temporal or spatial difference. Considering morbidity rates, they can be potential patient datasets collected from urban and rural areas. Likewise, considering approval rates, they can be unlabeled voter datasets collected in different years. In these examples, we may not be able to collect explicit labels for privacy reasons. However, the morbidity or approval rates can be obtained from released health reports or anonymous surveys. These are the class-priors in the context of classification and are very weakly supervised. Nevertheless, such class-priors are all we need for UU classification, as we show in this chapter.

8.2   Problem Formulation

Now, let us formulate the problem of UU classification.

8.2.1   Data Generation Process

Let θ and θ' be two valid class-priors such that $\theta \neq \theta'$,[1] and the UU training data are drawn from the marginal densities:

$$p_{\mathrm{tr}}(\boldsymbol{x}) := \theta p_{\mathrm{P}}(\boldsymbol{x}) + (1-\theta)p_{\mathrm{N}}(\boldsymbol{x}), \tag{8.1}$$

$$p'_{\mathrm{tr}}(\boldsymbol{x}) := \theta' p_{\mathrm{P}}(\boldsymbol{x}) + (1-\theta')p_{\mathrm{N}}(\boldsymbol{x}), \tag{8.2}$$

where $p_{\mathrm{P}}(\boldsymbol{x}) := p(\boldsymbol{x}\,|\,y=+1)$ and $p_{\mathrm{N}}(\boldsymbol{x}) := p(\boldsymbol{x}\,|\,y=-1)$ are the *class-conditional densities*. Equations (8.1) and (8.2) imply that there are $p_{\mathrm{tr}}(\boldsymbol{x},y)$ and $p'_{\mathrm{tr}}(\boldsymbol{x},y)$, whose class-conditional densities are the same and equal to those of $p(\boldsymbol{x},y)$, but class-priors are different:

$$p_{\mathrm{tr}}(\boldsymbol{x}\,|\,y) = p'_{\mathrm{tr}}(\boldsymbol{x}\,|\,y) = p(\boldsymbol{x}\,|\,y),$$

$$p_{\mathrm{tr}}(y=+1) = \theta \neq \theta' = p'_{\mathrm{tr}}(y=+1).$$

If we could sample labeled data from $p_{\mathrm{tr}}(\boldsymbol{x},y)$ or $p'_{\mathrm{tr}}(\boldsymbol{x},y)$, the above setting would reduce to that of supervised classification under *class-prior change* (Quiñonero-Candela et al., 2009). Nonetheless, the problem of interest belongs to weakly supervised classification—U training (and validation) data are supposed to be drawn independently according to (8.1) and (8.2). More specifically, we have

$$\mathscr{X}_{\mathrm{tr}} := \{\boldsymbol{x}_1, \ldots, \boldsymbol{x}_n\} \overset{\mathrm{i.i.d.}}{\sim} p_{\mathrm{tr}}(\boldsymbol{x}),$$

$$\mathscr{X}'_{\mathrm{tr}} := \{\boldsymbol{x}'_1, \ldots, \boldsymbol{x}'_{n'}\} \overset{\mathrm{i.i.d.}}{\sim} p'_{\mathrm{tr}}(\boldsymbol{x}),$$

where n and n' are two natural numbers that represent the sample sizes of $\mathscr{X}_{\mathrm{tr}}$ and $\mathscr{X}'_{\mathrm{tr}}$.

8.2.2   Performance Measures

Let $g : \mathscr{X} \to \mathbb{R}$ be an arbitrary *decision function*, i.e., g may literally be any binary classifier, and let $\ell : \mathbb{R} \times \mathscr{Y} \to \mathbb{R}_+$ be a *loss function*. The *classification risk* of g is given by

$$R(g) := \mathbb{E}_{p(\boldsymbol{x},y)}[\ell(g(\boldsymbol{x}),y)]$$

$$= \pi_{\mathrm{P}}\,\mathbb{E}_{\mathrm{P}}[\ell(g(\boldsymbol{x}),+1)] + (1-\pi_{\mathrm{P}})\,\mathbb{E}_{\mathrm{N}}[\ell(g(\boldsymbol{x}),-1)], \tag{8.3}$$

where $\pi_{\mathrm{P}} := p(y=+1)$, $\mathbb{E}_{\mathrm{P}}[\cdot]$ denotes the expectation over $p_{\mathrm{P}}(\boldsymbol{x})$, and $\mathbb{E}_{\mathrm{N}}[\cdot]$ denotes the expectation over $p_{\mathrm{N}}(\boldsymbol{x})$. If ℓ is chosen to be the *zero-one loss* defined by

$$\ell_{01}(\hat{y},y) := \frac{1 - \mathrm{sign}(\hat{y}y)}{2},$$

the risk is also known as the *classification error* and it is the standard performance measure in classification. A *balanced* version of (8.3) is

$$B(g) := \frac{1}{2}\mathbb{E}_{\mathrm{P}}[\ell(g(\boldsymbol{x}), +1] + \frac{1}{2}\mathbb{E}_{\mathrm{N}}[\ell(g(\boldsymbol{x}), -1], \tag{8.4}$$

and if ℓ is ℓ_{01}, (8.4) is named the *balanced error* (BER) (Brodersen et al., 2010). BER is a popular performance measure in *imbalanced* classification problems where π_{P} is away from $1/2$,[2] since it attempts to balance the errors on the two classes. For reasonably balanced binary classification (i.e, $\pi_{\mathrm{P}} \approx 1/2$), (8.3) would be more natural as the performance measure than (8.4) since the class balance is reflected.

8.2.3　Relation to Classification with Noisy Labels

At first glance, the UU data generation process described above, using the names from Menon et al. (2015), looks quite similar to *class-conditional noise* (CCN, Angluin and Laird, 1987) in *classification with noisy labels* (cf. Natarajan et al., 2013). In fact, Menon et al. (2015) made use of the *mutually contaminated distributions* (MCD, Scott et al., 2013) formulation that is more general than CCN. Denote by $\tilde{y}$ and $\tilde{p}(\cdot)$ the corrupted label and density. Then, CCN and MCD are defined by

$$\begin{pmatrix} \tilde{p}(\tilde{y} = +1 \mid \boldsymbol{x}) \\ \tilde{p}(\tilde{y} = -1 \mid \boldsymbol{x}) \end{pmatrix} = \boldsymbol{T}_{\mathrm{CCN}} \begin{pmatrix} p(y = +1 \mid \boldsymbol{x}) \\ p(y = -1 \mid \boldsymbol{x}) \end{pmatrix},$$

$$\begin{pmatrix} \tilde{p}(\boldsymbol{x} \mid \tilde{y} = +1) \\ \tilde{p}(\boldsymbol{x} \mid \tilde{y} = -1) \end{pmatrix} = \boldsymbol{T}_{\mathrm{MCD}} \begin{pmatrix} p_{\mathrm{P}}(\boldsymbol{x}) \\ p_{\mathrm{N}}(\boldsymbol{x}) \end{pmatrix},$$

where both of $\boldsymbol{T}_{\mathrm{CCN}}$ and $\boldsymbol{T}_{\mathrm{MCD}}$ are two by two matrices but $\boldsymbol{T}_{\mathrm{CCN}}$ is column normalized and $\boldsymbol{T}_{\mathrm{MCD}}$ is row normalized. Menon et al. (2015) proved that CCN is a strict special case of MCD. To be clear, $\tilde{p}(\tilde{y})$ is fixed in CCN once $\tilde{p}(\tilde{y} \mid \boldsymbol{x})$ is specified, while $\tilde{p}(\tilde{y})$ is free in MCD even after $\tilde{p}(\boldsymbol{x} \mid \tilde{y})$ is specified. Furthermore, $\tilde{p}(\boldsymbol{x}) = p(\boldsymbol{x})$ in CCN but $\tilde{p}(\boldsymbol{x}) \neq p(\boldsymbol{x})$ in MCD. Because of this *covariate shift* (Sugiyama and Kawanabe, 2012), CCN methods do not fit the MCD problem setting, though MCD methods can fit the CCN problem setting.

On the other hand, *complementary-label classification* that we will explore in chapter 9 is a special case of classification with noisy labels based on the CCN model (see section 9.2.4 for details).

8.3　Risk Estimation from UU Data

As explained in chapter 3, in the PN classification scenario, $R(g)$ can be empirically approximated by the P data and N data:

$$\widehat{R}_{\mathrm{PN}}(g) = \frac{\pi_{\mathrm{P}}}{n_{\mathrm{P}}} \sum_{i=1}^{n_{\mathrm{P}}} \ell(g(\boldsymbol{x}_i^{\mathrm{P}}), +1) + \frac{1 - \pi_{\mathrm{P}}}{n_{\mathrm{N}}} \sum_{j=1}^{n_{\mathrm{N}}} \ell(g(\boldsymbol{x}_j^{\mathrm{N}}), -1).$$

However, this naive approach does not work in UU classification since we are given only unlabeled data.

In this section, we show how the risk $R(g)$ can be empirically estimated from U sets with different class-priors. Here, a fundamental question is how many sets of U data with *different class-priors* are necessary for estimating the risk? We show that it is impossible given only a single set of U data in section 8.3.1, and it becomes possible given *two* sets of U data in section 8.3.2 (Lu et al., 2019). Based on these findings, we can explicitly design an unbiased risk estimator and derive an ERM-based classification method from two sets of U data. Note that given only unlabeled data, by no means could we learn the class-priors (Menon et al., 2015). Therefore, throughout this section, we assume that all necessary class-priors are given, which is the unique type of supervision we will leverage here.

8.3.1 Risk Estimation from One Set of U Data

We start with classification from a single set of U data, $\mathcal{X}_{\text{tr}}$. The difficulty now is how to estimate the risk with only U data.

Similarly to the previous chapters, let us employ the ERM-enabling risk rewrite approach: First, the risk is rewritten into an equivalent expression such that it could be approximated given $\mathcal{X}_{\text{tr}}$ and/or $\mathcal{X}'_{\text{tr}}$. This step leads to certain risk estimators. Second, the risk is estimated from the given U data, and the resulted empirical training risk is minimized.

Definition 8.1   *We say that $R(g)$ in (8.3) is rewritable given p_{tr}, if and only if[3] there exist constants C_{a} and C_{b}, such that for any g, it holds that*

$$R(g) = \mathbb{E}_{p_{\text{tr}}}[\bar{\ell}(g(\boldsymbol{x}))], \tag{8.5}$$

where $\mathbb{E}_{p_{\text{tr}}}[\cdot]$ denotes the expectation over $p_{\text{tr}}(\boldsymbol{x})$ and

$$\bar{\ell}(g(\boldsymbol{x})) := C_{\text{a}}\ell(g(\boldsymbol{x}), +1) + C_{\text{b}}\ell(g(\boldsymbol{x}), -1)$$

is the corrected loss function.

In what follows, we prove that knowing class-priors π_{P} and θ is insufficient to rewrite $R(g)$ in (8.3).

Theorem 8.2   *(Lu et al., 2019) Let ℓ be ℓ_{01}, or any bounded surrogate loss satisfying*

$$0 \leq \lim_{yg(\boldsymbol{x}) \to +\infty} \ell(g(\boldsymbol{x}), y) < \lim_{yg(\boldsymbol{x}) \to -\infty} \ell(g(\boldsymbol{x}), y) < +\infty. \tag{8.6}$$

Assume that p_{P} and p_{N} are almost surely separable. Then, $R(g)$ is not rewritable though θ is free.

Proof   We prove the theorem by contradiction—namely, for any such $p(\boldsymbol{x}, y)$ (with almost surely separable p_{P} and p_{N}), for all C_{a}, C_{b}, and all θ, we are able to find some g for which (8.5) fails. Our argument goes from the special case of ℓ_{01} to the general case of ℓ satisfying (8.6).

First, let $g(x) = +\infty$ identically so that $\ell(g(x), +1) = 0$ and $\ell(g(x), -1) = 1$. Plugging them into equations (8.3) and (8.5), we obtain

$$C_b = 1 - \pi_P.$$

Second, let $g(x) = -\infty$ identically; this time $\ell(g(x), +1) = 1$ and $\ell(g(x), -1) = 0$, and thus we obtain

$$C_a = \pi_P.$$

Third, let $g(x) = +\infty$ over p_P and $g(x) = -\infty$ over p_N. To be precise, define

$$g(x) = \begin{cases} +\infty & (p_P(x) > 0 \text{ and } p_N(x) = 0), \\ -\infty & (p_P(x) = 0 \text{ and } p_N(x) > 0), \\ 0 & (p_P(x) > 0 \text{ and } p_N(x) > 0). \end{cases}$$

This is possible because g is arbitrary. The last case $g(x) = 0$ should have a zero probability, since p_P and p_N are almost surely separable. Hence, we have $\ell(g(x), +1) = 0$ and $\ell(g(x), -1) = 1$ over p_P and $\ell(g(x), +1) = 1$ and $\ell(g(x), -1) = 0$ over p_N, resulting in

$$0 = \pi_P \mathbb{E}_P[\ell(g(x), +1)] + (1 - \pi_P) \mathbb{E}_N[\ell(g(x), -1)]$$

$$= \theta \mathbb{E}_P[\bar{\ell}(g(x))] + (1 - \theta) \mathbb{E}_N[\bar{\ell}(g(x))]$$

$$= \theta C_b + (1 - \theta) C_a.$$

By solving this equation, we have

$$\theta = \frac{C_a}{C_a - C_b} = \frac{\pi_P}{2\pi_P - 1}. \tag{8.7}$$

By definition, θ must satisfy $0 \leq \theta \leq 1$, whereas

- $\pi_P/(2\pi_P - 1) < 0$ if $0 < \pi_P < 1/2$;

- $\pi_P/(2\pi_P - 1) > 1$ if $1/2 < \pi_P < 1$;

- $\pi_P/(2\pi_P - 1)$ is undefined if $\pi_P = 1/2$.

Therefore, (8.7) must be a contradiction, unless $\pi_P = 0$ or $\pi_P = 1$. This implies that there is just a single class and the problem under consideration is not binary classification anymore.

Finally, given any ℓ satisfying (8.6), it is not difficult to verify that the three g above lead to the same contradiction with exactly the same C_a, C_b, and θ by solving a bit more complicated equations. $\qquad\square$

This theorem shows that under the separability assumption of p_P and p_N, $R(g)$ is not rewritable given a single set of U data. As a consequence, we lack a learning objective—that is, the empirical training risk. It is even worse—we cannot access the empirical validation risk of g after it is trained by other classification methods such as discriminative clustering. In particular, ℓ_{01} satisfies (8.6), which implies that the common practice of hyper-parameter tuning is disabled by theorem 8.2, since U validation data are also drawn from p_{tr}.

8.3.2   Risk Estimation from Two Sets of U Data

Next, let us consider classification from two sets of U data, $\mathscr{X}_{\text{tr}}$ and $\mathscr{X}'_{\text{tr}}$.

8.3.2.1   Risk estimation
Similar to section 8.3.1, we define the notion of risk rewritability for two sets of U data.

Definition 8.3   *We say that $R(g)$ is rewritable given p_{tr} and p'_{tr}, if and only if there exist constants C_a, C_b, C_c, and C_d, such that for any g, it holds that*

$$R(g) = \mathbb{E}_{p_{\text{tr}}}[\bar{\ell}_+(g(\boldsymbol{x}))] + \mathbb{E}_{p'_{\text{tr}}}[\bar{\ell}_-(g(\boldsymbol{x}))],$$

where

$$\bar{\ell}_+(g(\boldsymbol{x})) := C_a\ell(g(\boldsymbol{x}), +1) + C_b\ell(g(\boldsymbol{x}), -1),$$

$$\bar{\ell}_-(g(\boldsymbol{x})) := C_c\ell(g(\boldsymbol{x}), -1) + C_d\ell(g(\boldsymbol{x}), +1)$$

are the corrected loss functions.

In what follows, we prove that knowing π_P, θ, and θ' is sufficient for rewriting $R(g)$.

Theorem 8.4   *Fix θ and θ'. Assume $\theta > \theta'$; otherwise, swap p_{tr} and p'_{tr} to make sure $\theta > \theta'$. Then, $R(g)$ is rewritable, by letting*

$$C_a = \frac{(1-\theta')\pi_P}{\theta - \theta'}, \quad C_b = -\frac{\theta'(1-\pi_P)}{\theta - \theta'}, \quad C_c = \frac{\theta(1-\pi_P)}{\theta - \theta'}, \quad C_d = -\frac{(1-\theta)\pi_P}{\theta - \theta'}. \tag{8.8}$$

Proof   Let $J(g)$ be an alias of $R(g)$ in definition 8.3, serving as the learning objective:

$$J(g) = \mathbb{E}_{p_{\text{tr}}}[\bar{\ell}_+(g(\boldsymbol{x}))] + \mathbb{E}_{p'_{\text{tr}}}[\bar{\ell}_-(g(\boldsymbol{x}))]. \tag{8.9}$$

Then we have

$$J(g) = \mathbb{E}_{p_{\text{tr}}}[C_a\ell(g(\boldsymbol{x}), +1) + C_b\ell(g(\boldsymbol{x}), -1)]$$

$$+ \mathbb{E}_{p'_{\text{tr}}}[C_c\ell(g(\boldsymbol{x}), -1) + C_d\ell(g(\boldsymbol{x}), +1)]$$

$$= \theta \, \mathbb{E}_P[C_a\ell(g(\boldsymbol{x}), +1) + C_b\ell(g(\boldsymbol{x}), -1)]$$

$$+ (1-\theta) \, \mathbb{E}_N[C_a\ell(g(\boldsymbol{x}), +1) + C_b\ell(g(\boldsymbol{x}), -1)]$$

$$+ \theta' \, \mathbb{E}_{\mathrm{P}}[C_{\mathrm{c}}\ell(g(\boldsymbol{x}), -1) + C_{\mathrm{d}}\ell(g(\boldsymbol{x}), +1)]$$

$$+ (1 - \theta') \, \mathbb{E}_{\mathrm{N}}[C_{\mathrm{c}}\ell(g(\boldsymbol{x}), -1) + C_{\mathrm{d}}\ell(g(\boldsymbol{x}), +1)]$$

$$= (C_{\mathrm{a}}\theta + C_{\mathrm{d}}\theta') \, \mathbb{E}_{\mathrm{P}}[\ell(g(\boldsymbol{x}), +1)] + (C_{\mathrm{b}}\theta + C_{\mathrm{c}}\theta') \, \mathbb{E}_{\mathrm{P}}[\ell(g(\boldsymbol{x}), -1)]$$

$$+ [C_{\mathrm{a}}(1 - \theta) + C_{\mathrm{d}}(1 - \theta')] \, \mathbb{E}_{\mathrm{N}}[\ell(g(\boldsymbol{x}), +1)]$$

$$+ [C_{\mathrm{b}}(1 - \theta) + C_{\mathrm{c}}(1 - \theta')] \, \mathbb{E}_{\mathrm{N}}[\ell(g(\boldsymbol{x}), -1)].$$

On the other hand, we have

$$J(g) = \pi_{\mathrm{P}} \, \mathbb{E}_{\mathrm{P}}[\ell(g(\boldsymbol{x}), +1)] + (1 - \pi_{\mathrm{P}}) \, \mathbb{E}_{\mathrm{N}}[\ell(g(\boldsymbol{x}), -1)],$$

since $J(g)$ is an alias of $R(g)$. As a result, in order to minimize $R(g)$ in (8.3), it suffices to minimize $J(g)$ in (8.9), if we can make

$$C_{\mathrm{a}}\theta + C_{\mathrm{d}}\theta' = \pi_{\mathrm{P}},$$

$$C_{\mathrm{b}}\theta + C_{\mathrm{c}}\theta' = 0,$$

$$C_{\mathrm{a}}(1 - \theta) + C_{\mathrm{d}}(1 - \theta') = 0,$$

$$C_{\mathrm{b}}(1 - \theta) + C_{\mathrm{c}}(1 - \theta') = 1 - \pi_{\mathrm{P}}.$$

Solving these equations gives us (8.8), which concludes the proof. $\square$

Theorem 8.4 suggests that *three class-priors* (i.e., two class-prior probabilities are of the training distributions and one is of the test distribution) are all we need to train any binary classifier from only U data, while any two (i.e., one of the training distribution and one of the test distribution, or two of the training distributions) should not be enough.

The above result immediately leads to the following unbiased risk estimator:

$$\widehat{R}_{\mathrm{UU}}(g) = \frac{1}{n} \sum_{i=1}^{n} \left(\frac{(1 - \theta')\pi_{\mathrm{P}}}{\theta - \theta'} \ell(g(\boldsymbol{x}_i), +1) - \frac{\theta'(1 - \pi_{\mathrm{P}})}{\theta - \theta'} \ell(g(\boldsymbol{x}_i), -1) \right)$$

$$\tag{8.10}$$

$$+ \frac{1}{n'} \sum_{j=1}^{n'} \left(-\frac{(1 - \theta)\pi_{\mathrm{P}}}{\theta - \theta'} \ell(g(\boldsymbol{x}'_j), +1) + \frac{\theta(1 - \pi_{\mathrm{P}})}{\theta - \theta'} \ell(g(\boldsymbol{x}'_j), -1) \right).$$

Equation (8.10) is useful for both training a classifier (by plugging U training data into it) and hyper-parameter tuning (by plugging U validation data into it). The process of obtaining the empirical risk minimizer of (8.10) over the function class $\mathcal{G}$, i.e.,

$$\widehat{g}_{\mathrm{UU}} := \operatorname*{argmin}_{g \in \mathcal{G}} \widehat{R}_{\mathrm{UU}},$$

is referred to as *unlabeled-unlabeled (UU) classification*. Since it is ERM-based, $\widehat{g}_{UU}$ can be obtained by powerful *stochastic optimization* algorithms (e.g., Duchi et al., 2011; Kingma and Ba, 2015).

8.3.2.2 Simplification

Equation (8.10) can be simplified by employing ℓ that satisfies the following symmetric condition:

$$\ell(g(\boldsymbol{x}), +1) + \ell(g(\boldsymbol{x}), -1) = 1, \tag{8.11}$$

for any $\boldsymbol{x}$ and g. Equation (8.11) covers the zero-one loss, ramp loss, sigmoid loss, and linear loss (see section 4.3.2). With the help of (8.11), (8.10) can be simplified as

$$\widehat{R}_{UU}^{Sym}(g) = \frac{\alpha}{n} \sum_{i=1}^{n} \ell(g(\boldsymbol{x}_i), +1) + \frac{\alpha'}{n'} \sum_{j=1}^{n'} \ell(g(\boldsymbol{x}_j'), -1) + C, \tag{8.12}$$

where

$$\alpha := \frac{\theta' + \pi_P - 2\theta'\pi_P}{\theta - \theta'},$$

$$\alpha' := \frac{\theta + \pi_P - 2\theta\pi_P}{\theta - \theta'},$$

$$C := \frac{\theta'\pi_P + \theta\pi_P - \theta' - \pi_P}{\theta - \theta'}.$$

In this simplified expression, one set of U data is treated as P data and the other set as N data, and their losses are weighted by α and α'. This form is known as *cost-sensitive classification* (Elkan, 2001).

8.3.2.3 Special cases

UU classification is a very general framework in weakly supervised classification. We consider some special cases of (8.10) by specifying θ and θ':

- If $\theta = 1$ and $\theta' = 0$, (8.10) reduces to (3.2) for supervised classification (see chapter 3).

- If $\theta = 1$ and $\theta' = \pi_P$, (8.10) reduces to

$$\widehat{R}_{PU}(g) = \frac{\pi_P}{n_P} \sum_{i=1}^{n_P} \left(\ell(g(\boldsymbol{x}_i^P), +1) - \ell(g(\boldsymbol{x}_i^P), -1)\right) + \frac{1}{n_U} \sum_{j=1}^{n_U} \ell(g(\boldsymbol{x}_i^U), -1).$$

Thus, we can recover the unbiased risk estimator in PU classification (see chapter 4).

- If $\theta = \pi_P$ and $\theta' = \pi_P^2/(2\pi_P^2 - 2\pi_P + 1)$ or vice versa, (8.10) reduces to the unbiased risk estimator in SU classification (see chapter 7).

8.3.3   Theoretical Analysis

The consistency of UU classification is guaranteed because of the unbiasedness of (8.10). In what follows, we analyze the estimation error $R(\widehat{g}_{\mathrm{UU}}) - R(g^*)$. To this end, assume there exist $C_g > 0$ and $C_\ell > 0$ such that $\sup_{g \in \mathcal{G}} \|g\|_\infty \le C_g$ and $\sup_{|z| \le C_g} \ell(z) \le C_\ell$, where we are using the margin-based loss $\ell(z) = \ell(g(\boldsymbol{x}), y)$ with $z = yg(\boldsymbol{x})$. Assume also that $\ell(z)$ is Lipschitz continuous for all $|z| \le C_g$ with Lipschitz constant L_ℓ. Let $\mathfrak{R}_n(\mathcal{G})$ and $\mathfrak{R}'_{n'}(\mathcal{G})$ be the *Rademacher complexity* of $\mathcal{G}$ over $p_{\mathrm{tr}}(\boldsymbol{x})$ and $p'_{\mathrm{tr}}(\boldsymbol{x})$ (see section 3.1.2.3). For convenience, we define $\chi_{n,n'} = \alpha / \sqrt{n} + \alpha' / \sqrt{n'}$. Then the following lemma holds.

Lemma 8.5   *(Lu et al., 2019) For any $\delta > 0$, let $C_\delta = \sqrt{(\ln 2/\delta)/2}$. Then we have with probability at least $1 - \delta$,*

$$\sup_{g \in \mathcal{G}} |\widehat{R}_{\mathrm{UU}}(g) - R(g)| \le 2 L_\ell \alpha \mathfrak{R}_n(\mathcal{G}) + 2 L_\ell \alpha' \mathfrak{R}'_{n'}(\mathcal{G}) + C_\ell C_\delta \chi_{n,n'},$$

where the probability is over repeated sampling of data for evaluating $\widehat{R}_{\mathrm{UU}}$.

Proof   Consider the one-side uniform deviation $\sup_{g \in \mathcal{G}} \widehat{R}_{\mathrm{UU}}(g) - R(g)$. Since $0 \le \ell(z) \le C_\ell$, its change will be no more than $C_\ell \alpha / n$ if some x_i is replaced, or no more than $C_\ell \alpha' / n'$ if some x'_j is replaced. Subsequently, *McDiarmid's inequality* (McDiarmid, 1989) tells us that

$$\Pr\{\sup_{g \in \mathcal{G}} \widehat{R}_{\mathrm{UU}}(g) - R(g) - \mathbb{E}[\sup_{g \in \mathcal{G}} \widehat{R}_{\mathrm{UU}}(g) - R(g)] \ge \epsilon\}$$

$$\le \exp\left(-\frac{2\epsilon^2}{C_\ell^2 (\alpha^2/n + \alpha'^2/n')}\right),$$

or equivalently, with probability at least $1 - \delta/2$,

$$\sup_{g \in \mathcal{G}} \widehat{R}_{\mathrm{UU}}(g) - R(g)$$

$$\le \mathbb{E}[\sup_{g \in \mathcal{G}} \widehat{R}_{\mathrm{UU}}(g) - R(g)] + C_\ell (\alpha/\sqrt{n} + \alpha'/\sqrt{n'}) \sqrt{(\ln 2/\delta)/2}$$

$$= \mathbb{E}[\sup_{g \in \mathcal{G}} \widehat{R}_{\mathrm{UU}}(g) - R(g)] + C_\ell C_\delta \chi_{n,n'}.$$

By *symmetrization* (see section 3.1.2.5), we can show that

$$\mathbb{E}[\sup_{g \in \mathcal{G}} \widehat{R}_{\mathrm{UU}}(g) - R(g)] \le 2\alpha \mathfrak{R}_n(\ell \circ \mathcal{G}) + 2\alpha' \mathfrak{R}'_{n'}(\ell \circ \mathcal{G}),$$

and according to *Talagrand's contraction lemma* (see section 3.1.2.3),

$$\mathfrak{R}_n(\ell \circ \mathcal{G}) \le L_\ell \mathfrak{R}_n(\mathcal{G}), \quad \mathfrak{R}'_{n'}(\ell \circ \mathcal{G}) \le L_\ell \mathfrak{R}'_{n'}(\mathcal{G}).$$

The one-side uniform deviation $\sup_{g \in \mathcal{G}} R(g) - \widehat{R}_{\mathrm{UU}}(g)$ can be bounded similarly. $\square$

Lemma 8.5 guarantees that with high probability, $\widehat{R}_{\mathrm{UU}}(g)$ concentrates around $R(g)$ for all $g \in \mathcal{G}$, and the degree of such concentration is controlled by $\mathfrak{R}_n(\mathcal{G})$ and $\mathfrak{R}'_{n'}(\mathcal{G})$.

Based on this uniform deviation bound, we can derive an estimation error bound as follows.

Theorem 8.6 *(Lu et al., 2019) For any $\delta > 0$, let $C_\delta = \sqrt{(\ln 2/\delta)/2}$. Then we have with probability at least $1 - \delta$,*

$$R(\widehat{g}_{UU}) - R(g^*) \leq 4L_\ell \alpha \mathfrak{R}_n(\mathscr{G}) + 4L_\ell \alpha' \mathfrak{R}'_{n'}(\mathscr{G}) + 2C_\ell C_\delta \chi_{n,n'}, \tag{8.13}$$

where the probability is over repeated sampling of $\mathscr{X}_{\mathrm{tr}}$ and $\mathscr{X}'_{\mathrm{tr}}$ for training $\widehat{g}_{UU}$.

Proof Based on lemma 8.5, the estimation error bound (8.13) is proved through

$$R(\widehat{g}_{UU}) - R(g^*) = \left(\widehat{R}_{UU}(\widehat{g}_{UU}) - \widehat{R}_{UU}(g^*)\right) + \left(R(\widehat{g}_{UU}) - \widehat{R}_{UU}(\widehat{g}_{UU})\right)$$
$$+ \left(\widehat{R}_{UU}(g^*) - R(g^*)\right)$$
$$\leq 0 + 2 \sup_{g \in \mathscr{G}} |\widehat{R}_{UU}(g) - R(g)|$$
$$\leq 4L_\ell \alpha \mathfrak{R}_n(\mathscr{G}) + 4L_\ell \alpha' \mathfrak{R}'_{n'}(\mathscr{G}) + 2C_\ell C_\delta \chi_{n,n'},$$

where $\widehat{R}_{UU}(\widehat{g}_{UU}) \leq \widehat{R}_{UU}(g^*)$ by the definition of $\widehat{g}_{UU}$. $\qquad\square$

Theorem 8.6 ensures that classification with (8.12) is consistent (and so are all the special cases described in section 8.3.2.3). More specifically, as $n, n' \to \infty$, we have $R(\widehat{g}_{UU}) \to R(g^*)$, since $\mathfrak{R}_n(\mathscr{G}), \mathfrak{R}'_{n'}(\mathscr{G}) \to 0$ for all parametric models with a bounded norm, such as deep networks trained with weight decay. Moreover, we also have $R(\widehat{g}_{UU}) \to R(g^*)$ in $\mathscr{O}_p(\chi_{n,n'})$, where $\mathscr{O}_p$ denotes the order in probability, for all linear-in-parameter models with a bounded norm, including non-parametric kernel models in *reproducing kernel Hilbert spaces* (Schölkopf and Smola, 2002).

8.3.4 Experiments

Here, we empirically demonstrate the effectiveness of the risk estimation method for UU classification and compare it with other methods for classification from two sets of U data.

8.3.4.1 Setup

Table 8.1 summarizes the benchmark datasets. *MNIST*[4] (LeCun et al., 1998), *Fashion-MNIST*[5] (Xiao et al., 2017), *SVHN*[6] (Netzer et al., 2011), and *CIFAR-10*[7] (Krizhevsky,

Table 8.1
Specification of benchmark datasets.

Dataset	# Train	# Test	# Feature	π_P
MNIST	60,000	10,000	784	0.49
Fashion-MNIST	60,000	10,000	784	0.50
SVHN	100,000	26,032	3,072	0.27
CIFAR-10	50,000	10,000	3,072	0.60

2009) have 10 classes originally, and we constructed the P and N classes from them as follows: MNIST was preprocessed in such a way that even digits constitute the P class and odd digits constitute the N class. For Fashion-MNIST, the P class was formed by "T-shirt," "Pullover," "Coat," "Shirt," and "Bag," and the N class was formed by "Trouser," "Dress," "Sandal," "Sneaker," and "Ankle boot." For SVHN, "0," "6," "8," and "9" made up the P class and "1," "2," "3," "4," "5," and "7" made up the N class. For CIFAR-10, the P class was composed of "bird," "cat," "deer," "dog," "frog," and "horse," and the N class was composed of "airplane," "automobile," "ship," and "truck." $\mathcal{X}_{\mathrm{tr}}$ and $\mathcal{X}_{\mathrm{tr}}'$ of the same sample size were drawn according to (8.1) and (8.2), where (θ, θ') were chosen as $(0.9, 0.1)$ or $(0.8, 0.2)$. The test data are drawn from $p(\boldsymbol{x}, y)$.

The following model $g(\boldsymbol{x})$ and optimization algorithm are chosen:

MNIST: We used a *fully connected neural network* (depth 5) d-300-300-300-300-1 with rectified linear units (ReLUs) (Nair and Hinton, 2010) and the stochastic gradient descent (SGD) algorithm (Robbins and Munro, 1951).

Fashion-MNIST: The model and optimization algorithm were the same as MNIST.

SVHN: The model was an *all convolutional net* (Springenberg et al., 2015): (32*32*3)-[C(3*3, 96)]*2-C(3*3, 96, 2)-[C(3*3, 192)]*2-C(3*3, 192, 2)-C(3*3, 192)-C(1*1, 192)-C(1*1, 10)-1000-1000-1, where C(3*3, 96) means 96 channels of 3*3 convolutions followed by ReLU, [·]*2 means 2 such layers, C(3*3, 96, 2) means a similar layer but with stride 2, etc. Adam (Kingma and Ba, 2015) was used as an optimization algorithm.

CIFAR-10: The model was a 32-layer *residual net* (ResNet) (Springenberg et al., 2015): (32*32*3)-C(3*3, 16)-[C(3*3, 16), C(3*3, 16)]*5-[C(3*3, 32), C(3*3, 32)]*5-[C(3*3, 64), C(3*3, 64)]*5-Global Average Pooling-1, where [·, ·] means a building block (He et al., 2016). The optimization setup was the same as SVHN. Furthermore, the sigmoid loss was used as the surrogate loss. Batch normalization and ℓ_2-regularization were also applied.

A Keras[8] implementation of the UU classification algorithm is available from https://github.com/lunanbit/UUlearning.

8.3.4.2   Benchmark experiments with neural network models

In order to analyze the risk estimation method for UU classification, we compare it with three supervised baseline methods:

- *Small PN*: supervised classification from 10% labeled data;

- *PN oracle*: supervised classification from 100% labeled data;

- *Small PN prior-shift*: supervised classification from 10% labeled data under class-prior change.

Notice that the first two baselines have labeled data identically distributed as the test data, which is very advantageous, and thus the experiments here are not really for the purpose of performance comparison but merely for a proof of concept.

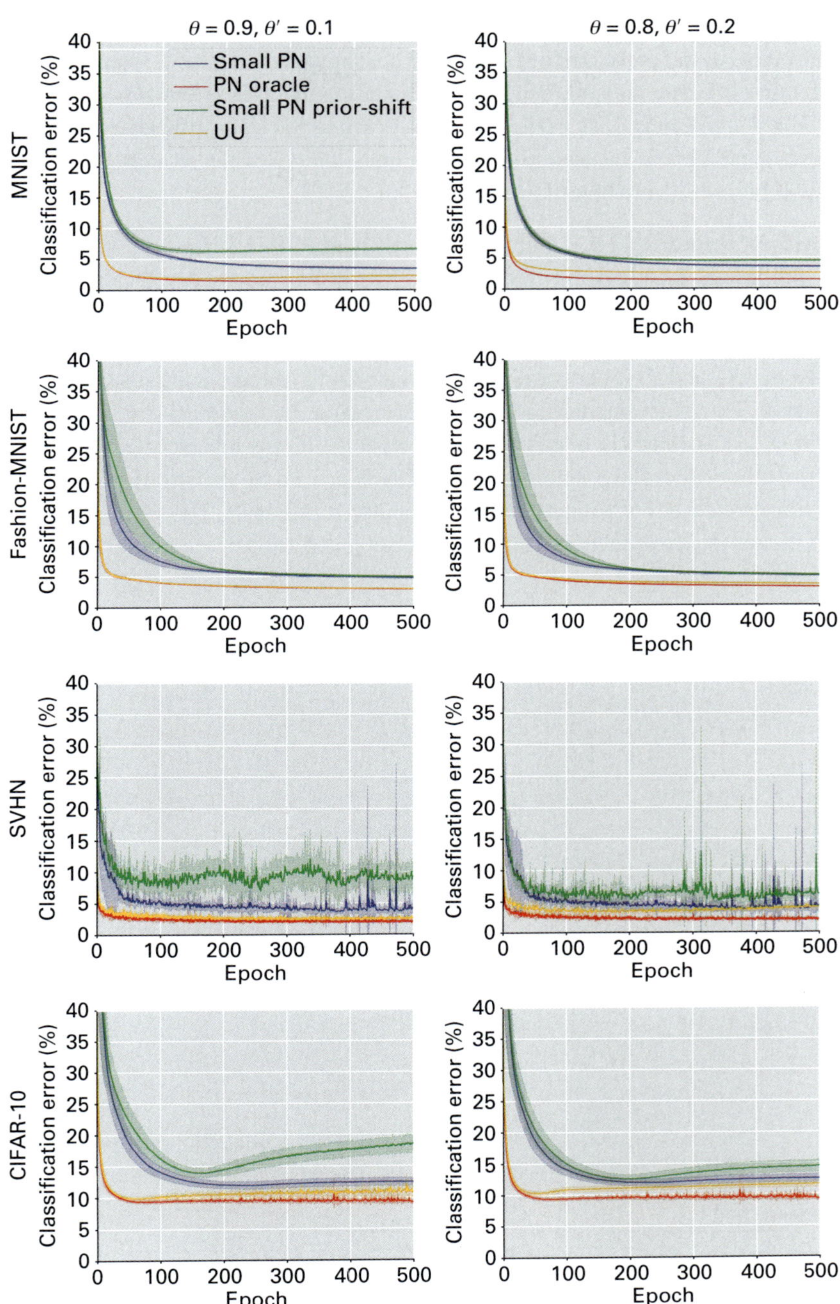

Figure 8.2
Experimental results of training deep neural networks.

The experimental results are reported in figure 8.2, where means and standard deviations of classification errors based on 10 random samplings are shown. When $\theta = 0.9$ and $\theta' = 0.1$ (cf. the left column), UU is comparable to the PN oracle in most cases. When $\theta = 0.8$ and $\theta' = 0.2$ (cf. the right column), UU performs slightly worse but it is still better than the "small PN" baselines. This is because the task becomes harder when θ and θ' become closer, which will be intensively investigated next.

On the closeness of θ and θ':   It is intuitive that if θ and θ' move closer, $\mathcal{X}_{\mathrm{tr}}$ and $\mathcal{X}'_{\mathrm{tr}}$ will be more similar and thus less informative. To investigate this, we test UU and CCN (Natarajan et al., 2013) on MNIST by fixing θ to 0.9 or 0.8 and gradually moving θ' from 0.1 to 0.5. The experimental results are reported in figure 8.3. We can see that when θ' moves closer to θ, UU and CCN become worse, while UU is affected slightly and CCN is affected severely. This phenomenon of UU can be explained by theorem 8.6, where the upper bound in (8.13) is linear in α and α' which, as $\theta' \to \theta$, are inversely proportional to $\theta - \theta'$. On the other hand, the phenomenon of CCN is caused by the stronger covariate shift when θ' moves closer to θ rather than the difficulty of the task. This illustrates that CCN methods do not fit our problem setting, so that we called for some new classification method (i.e., UU).

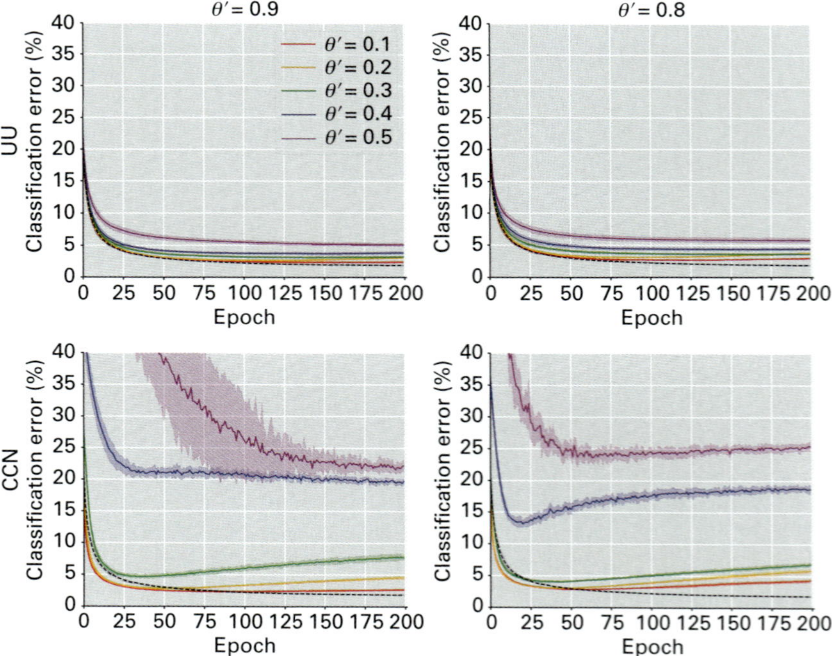

Figure 8.3
Experimental results of moving θ' closer to θ (black dashed lines are the PN oracle).

Table 8.2
Mean errors (standard deviations) in percentage given inaccurate training class-priors.

Dataset	θ,θ'	$\epsilon=0.8$ $\epsilon'=0.8$	$\epsilon=0.9$ $\epsilon'=0.9$	$\epsilon=1.0$ $\epsilon'=1.0$	$\epsilon=1.1$ $\epsilon'=1.1$	$\epsilon=1.2$ $\epsilon'=1.2$
MNIST	0.9, 0.1	2.31 (0.16)	2.31 (0.14)	2.31 (0.14)	2.32 (0.14)	2.35 (0.14)
	0.8, 0.2	3.00 (0.12)	3.00 (0.11)	3.01 (0.10)	3.02 (0.10)	3.01 (0.10)
	0.7, 0.3	4.24 (0.23)	4.24 (0.24)	4.24 (0.26)	4.25 (0.24)	4.25 (0.25)
CIFAR-10	0.9, 0.1	10.19 (0.37)	10.14 (0.29)	10.14 (0.30)	10.11 (0.34)	10.09 (0.35)
	0.8, 0.2	10.84 (0.38)	10.84 (0.40)	10.77 (0.40)	10.73 (0.40)	10.73 (0.40)
	0.7, 0.3	12.04 (0.61)	12.00 (0.54)	11.92 (0.54)	11.91 (0.53)	11.88 (0.53)

Dataset	θ,θ'	$\epsilon=0.8$ $\epsilon'=1.2$	$\epsilon=0.9$ $\epsilon'=1.1$	$\epsilon=1.0$ $\epsilon'=1.0$	$\epsilon=1.1$ $\epsilon'=0.9$	$\epsilon=1.2$ $\epsilon'=0.8$
MNIST	0.9, 0.1	2.30 (0.15)	2.31 (0.16)	2.31 (0.14)	2.30 (0.13)	2.30 (0.14)
	0.8, 0.2	3.00 (0.10)	3.00 (0.12)	3.01 (0.10)	3.02 (0.12)	3.01 (0.11)
	0.7, 0.3	4.19 (0.22)	4.22 (0.23)	4.24 (0.26)	4.24 (0.25)	4.25 (0.23)
CIFAR-10	0.9, 0.1	10.20 (0.33)	10.15 (0.34)	10.14 (0.30)	10.12 (0.35)	10.08 (0.37)
	0.8, 0.2	10.94 (0.46)	10.83 (0.39)	10.77 (0.40)	10.75 (0.37)	10.71 (0.43)
	0.7, 0.3	12.24 (0.71)	12.05 (0.59)	11.92 (0.54)	11.88 (0.53)	11.95 (0.49)

Robustness against inaccurate training class-priors: Hitherto, we have assumed that the values of θ and θ' are accessible, which is rarely satisfied in practice. Fortunately, UU is a robust classification method against inaccurate training class-priors. To show this, let ϵ and ϵ' be real numbers around 1, and let $\vartheta = \epsilon\theta$ and $\vartheta' = \epsilon'\theta'$ be perturbed θ and θ'. We test UU on MNIST and CIFAR-10 by drawing data using θ and θ' but training models using ϑ and ϑ' instead. The experimental results shown in table 8.2 imply that UU is fairly robust to inaccurate ϑ and ϑ' and can be safely applied in the wild.

8.3.4.3   Comparison with other methods
Finally, we compare UU with two methods for dealing with two sets of U data:

- *Proportion-SVM* (pSVM, Yu et al., 2013) that *learns from label proportions* by discriminative clustering together with expectation regularization (Mann and McCallum, 2007);

- *Balanced error minimization* (BEM, Menon et al., 2015) that learns the classifier by taking data from $p_{\mathrm{tr}}(\boldsymbol{x})$ and $p'_{\mathrm{tr}}(\boldsymbol{x})$ as labeled data from $p_{\mathrm{P}}(\boldsymbol{x})$ and $p_{\mathrm{N}}(\boldsymbol{x})$ and minimizing the balanced error.

The information on datasets can be found in table 8.3. In UU and BEM-FC, a fully connected (FC) neural network (depth 5) was adopted and stochastic gradient descent was used as the optimization algorithm. This sampling-and-training process was repeated 10 times for all classification methods on all datasets.

The experimental results are reported in table 8.3. We can see that UU is always either the best method (7 out of 10 cases) or comparable to the best method (3 out of 10 cases). Moreover, the closer π_{P} is to 0.5, the better BEM and BEM-FC are; however, the closer

Table 8.3
Mean errors (standard deviations) in percentage obtained by UU and other methods. Best and comparable methods by the paired t-test at significance level 1% are highlighted in boldface.

Dataset	π_P	$\theta - \theta'$	pSVM	BEM	BEM-FC	UU
pendigits	0.1	0.57	4.03 (0.27)	5.51 (1.35)	5.46 (1.23)	**1.97 (0.78)**
covertype	0.3	0.80	14.63 (1.00)	11.33 (0.26)	**5.17 (0.57)**	**4.97 (0.48)**
MNIST	0.5	0.77	N/A	**3.66 (0.20)**	3.03 (0.25)	2.87 (0.28)
spambase	0.7	0.80	29.18 (1.29)	**11.28 (1.73)**	13.98 (1.63)	**12.53 (1.00)**
letter	0.9	0.60	15.65 (4.18)	15.45 (6.99)	8.45 (2.92)	**3.15 (0.84)**
USPS	0.1	0.57	5.91 (1.52)	12.69 (4.09)	8.57 (2.40)	**3.74 (1.24)**
	0.3	0.80	5.55 (0.46)	5.36 (0.41)	**2.75 (0.28)**	**2.63 (0.18)**
	0.5	0.60	9.27 (0.61)	**7.27 (1.09)**	**5.48 (1.33)**	**5.52 (1.02)**
	0.7	0.80	8.20 (0.73)	7.48 (0.65)	**4.23 (0.50)**	**4.43 (0.94)**
	0.9	0.44	9.80 (2.07)	14.13 (2.02)	18.27 (5.17)	**6.20 (1.33)**

The codes of pSVM and BEM are downloadable from https://github.com/felixyu/pSVM and https://akmenon. github.io/papers/corrupted-labels/index.html, respectively. Since the original codes of BEM train single-hidden-layer neural networks by the L-BFGS quasi-Newton method (which belongs to second-order optimization) in MATLAB, we also added BEM-FC by fixing π_P to 0.5 in UU so that UU and BEM-FC only differ in the performance measure. The first five datasets come from https://cs.nyu.edu/~roweis/data.html. The rows are arranged according to π_P. The cell "N/A" (in the MNIST row and pSVM column) is because pSVM is based on maximum margin clustering and is too slow on MNIST. The task would be harder if π_P is closer to 0.5 or the number of training data or $\theta - \theta'$ is smaller.

π_P is to 0 or 1, the worse they are, and sometimes they are much worse than pSVM. This is because their goal is to minimize the balanced error instead of the classification error. In the experiments reported here, pSVM falls behind because it is based on discriminative clustering and is also not designed to minimize the classification error.

8.4   Generative Approach

In this section, we show that when adopting (8.4) as the performance measure (i.e., assuming $\pi_\mathrm{P} = \frac{1}{2}$), UU classification can be solved by estimating the sign of the difference between the probability densities of the U sets $\mathcal{X}_\mathrm{tr}$ and $\mathcal{X}'_\mathrm{tr}$ (du Plessis et al., 2013).

8.4.1   Analysis of Bayes-Optimal Classifier

We can write the class-posterior probability as

$$q(y|\boldsymbol{x}) = \frac{p(\boldsymbol{x}|y)q(y)}{q(\boldsymbol{x})},$$

where $q(y=+1) = q(y=-1) = \frac{1}{2}$ is the equal class-prior. A class label can then be assigned to the most likely class by

$$d(\boldsymbol{x}) = \mathrm{sign}\left(q(y=+1|\boldsymbol{x}) - q(y=-1|\boldsymbol{x})\right),$$

where $d(\boldsymbol{x})$ denotes the criterion for classification. This is called the *Bayes-optimal classifier*. Below we explain how $d(\boldsymbol{x})$ can be estimated only from $\mathscr{X}_{\mathrm{tr}}$ and $\mathscr{X}'_{\mathrm{tr}}$.

By writing the difference between the class-posterior probabilities as

$$q(y=+1|\boldsymbol{x}) - q(y=-1|\boldsymbol{x}) = \frac{p(\boldsymbol{x}|y=+1)\frac{1}{2}}{q(\boldsymbol{x})} - \frac{p(\boldsymbol{x}|y=-1)\frac{1}{2}}{q(\boldsymbol{x})}$$

$$= \frac{1}{2q(\boldsymbol{x})} \left(p(\boldsymbol{x}|y=+1) - p(\boldsymbol{x}|y=-1) \right),$$

we obtain

$$d(\boldsymbol{x}) = \mathrm{sign}\left(p(\boldsymbol{x}|y=+1) - p(\boldsymbol{x}|y=-1) \right),$$

due to the fact that $1/(2q(\boldsymbol{x}))$ is always positive. However, labeled samples are unavailable in the current setup, so the difference between the class-conditional densities above cannot be directly calculated. Surprisingly, we can show that it is proportional to the difference between marginal densities $p_{\mathrm{tr}}(\boldsymbol{x})$ and $p'_{\mathrm{tr}}(\boldsymbol{x})$ (du Plessis et al., 2013).

More specifically, since

$$p_{\mathrm{tr}}(\boldsymbol{x}) - p'_{\mathrm{tr}}(\boldsymbol{x}) = p_{\mathrm{tr}}(y=+1)p(\boldsymbol{x}|y=+1) + (1 - p_{\mathrm{tr}}(y=+1))\,p(\boldsymbol{x}|y=-1)$$

$$- p'_{\mathrm{tr}}(y=+1)p(\boldsymbol{x}|y=+1) - \left(1 - p'_{\mathrm{tr}}(y=+1)\right)p(\boldsymbol{x}|y=-1)$$

$$= \left(p_{\mathrm{tr}}(y=+1) - p'_{\mathrm{tr}}(y=+1)\right)\left(p(\boldsymbol{x}|y=+1) - p(\boldsymbol{x}|y=-1)\right)$$

$$= \left(\theta - \theta'\right)\left(p(\boldsymbol{x}|y=+1) - p(\boldsymbol{x}|y=-1)\right),$$

the criterion can be expressed as

$$d(\boldsymbol{x}) = A\,\mathrm{sign}\left(p_{\mathrm{tr}}(\boldsymbol{x}) - p'_{\mathrm{tr}}(\boldsymbol{x})\right),$$

where $A = \mathrm{sign}\left(\theta - \theta'\right)$. The expression signifies the following:

- If it is known whether $\theta > \theta'$ or $\theta < \theta'$, we can compute A; thus, we can separate the two classes and tell which class is P and which class is N.

- If we do not know whether $\theta > \theta'$ or $\theta < \theta'$, we cannot compute A; thus, we can only separate the two classes but cannot tell the correct class labels.

Now our challenge is to obtain a good estimator of $\mathrm{sign}\left(p_{\mathrm{tr}}(\boldsymbol{x}) - p'_{\mathrm{tr}}(\boldsymbol{x})\right)$.

8.4.2   KDE-Based Algorithm

A naive approach to estimating the sign of the density difference is to use *kernel density estimation* (KDE) (Silverman, 1986). For Gaussian kernels, the KDE solutions are given by

$$\widehat{p}_{\mathrm{tr}}(\boldsymbol{x}) \propto \sum_{i=1}^{n} \exp\left(-\frac{\|\boldsymbol{x} - \boldsymbol{x}_i\|^2}{2\sigma^2}\right), \quad \widehat{p}'_{\mathrm{tr}}(\boldsymbol{x}) \propto \sum_{i'=1}^{n'} \exp\left(-\frac{\|\boldsymbol{x} - \boldsymbol{x}'_{i'}\|^2}{2\sigma'^2}\right).$$

The Gaussian widths σ and σ' may be determined based on least-squares cross-validation (Härdle et al., 2004). Finally, labels are predicted as

$$y = \mathrm{sign}\left(\widehat{p}_{\mathrm{tr}}(\boldsymbol{x}) - \widehat{p}'_{\mathrm{tr}}(\boldsymbol{x})\right).$$

However, the KDE-based density-difference estimator may not be the best approach because of its two-step nature: Small estimation error incurred in each density estimate can cause a big error in the final density-difference estimate. More intuitively, good density estimators tend to be smooth, and thus a density-difference estimator obtained as the difference of such smooth density estimators tends to be over-smoothed (Hall and Wand, 1988; Anderson et al., 1994).

8.4.3   LSDD-Based Algorithm

The density difference can be estimated in a single shot using the *least-squares density difference* (LSDD) approach (Sugiyama et al., 2013). In this approach, density-difference model $g(\boldsymbol{x})$ is directly fitted to the density difference under the squared loss:

$$\widehat{g} = \underset{g}{\mathrm{argmin}} \; \frac{1}{2} \int \left(g(\boldsymbol{x}) - (p_{\mathrm{tr}}(\boldsymbol{x}) - p'_{\mathrm{tr}}(\boldsymbol{x}))\right)^2 \, \mathrm{d}\boldsymbol{x}.$$

From the computational viewpoint, the use of the following linear-in-parameter model as $g(\boldsymbol{x})$ is advantageous (Sugiyama et al., 2013):

$$g(\boldsymbol{x}) = \sum_{\ell=1}^{b} \theta_\ell \psi_\ell(\boldsymbol{x}) = \boldsymbol{\theta}^\top \boldsymbol{\psi}(\boldsymbol{x}), \tag{8.14}$$

where b denotes the number of basis functions, $\boldsymbol{\psi}(\boldsymbol{x}) = (\psi_1(\boldsymbol{x}), \ldots, \psi_b(\boldsymbol{x}))^\top$ is a b-dimensional basis function vector, $\boldsymbol{\theta} = (\theta_1, \ldots, \theta_b)^\top$ is a b-dimensional parameter vector, and $^\top$ denotes the transpose. In practice, Gaussian kernels may be chosen as basis functions:

$$g(\boldsymbol{x}) = \sum_{\ell=1}^{n+n'} \theta_\ell \exp\left(-\frac{\|\boldsymbol{x} - \boldsymbol{c}_\ell\|^2}{2\sigma^2}\right), \tag{8.15}$$

where $(\boldsymbol{c}_1, \ldots, \boldsymbol{c}_n, \boldsymbol{c}_{n+1}, \ldots, \boldsymbol{c}_{n+n'}) := (\boldsymbol{x}_1, \ldots, \boldsymbol{x}_n, \boldsymbol{x}'_1, \ldots, \boldsymbol{x}'_{n'})$ are Gaussian kernel centers. If $n + n'$ is large, only a subset of $\{\boldsymbol{x}_1, \ldots, \boldsymbol{x}_n, \boldsymbol{x}'_1, \ldots, \boldsymbol{x}'_{n'}\}$ may be used as Gaussian kernel centers.

For the model (8.14), the optimal parameter $\boldsymbol{\theta}^*$ is given by

$$\boldsymbol{\theta}^* := \underset{\boldsymbol{\theta}}{\mathrm{argmin}} \; \frac{1}{2} \int \left(g(\boldsymbol{x}) - (p_{\mathrm{tr}}(\boldsymbol{x}) - p'_{\mathrm{tr}}(\boldsymbol{x}))\right)^2 \, \mathrm{d}\boldsymbol{x}$$

$$= \underset{\boldsymbol{\theta}}{\mathrm{argmin}} \left[\int g(\boldsymbol{x})^2 \mathrm{d}\boldsymbol{x} - 2 \int g(\boldsymbol{x})(p_{\mathrm{tr}}(\boldsymbol{x}) - p'_{\mathrm{tr}}(\boldsymbol{x})) \mathrm{d}\boldsymbol{x}\right]$$

$$= \operatorname*{argmin}_{\boldsymbol{\theta}} \left[\boldsymbol{\theta}^\top \boldsymbol{H} \boldsymbol{\theta} - 2\boldsymbol{h}^\top \boldsymbol{\theta} \right]$$

$$= \boldsymbol{H}^{-1} \boldsymbol{h},$$

where $\boldsymbol{H}$ is the $b \times b$ matrix and $\boldsymbol{h}$ is the b-dimensional vector defined as

$$\boldsymbol{H} := \int \boldsymbol{\psi}(\boldsymbol{x}) \boldsymbol{\psi}(\boldsymbol{x})^\top \mathrm{d}\boldsymbol{x},$$

$$\boldsymbol{h} := \int \boldsymbol{\psi}(\boldsymbol{x}) p_{\mathrm{tr}}(\boldsymbol{x}) \mathrm{d}\boldsymbol{x} - \int \boldsymbol{\psi}(\boldsymbol{x}') p'_{\mathrm{tr}}(\boldsymbol{x}') \mathrm{d}\boldsymbol{x}'.$$

Note that for the Gaussian kernel model (8.15), the integral in $\boldsymbol{H}$ can be computed analytically as

$$H_{\ell,\ell'} = \int \exp\left(-\frac{\|\boldsymbol{x} - \boldsymbol{c}_\ell\|^2}{2\sigma^2} \right) \exp\left(-\frac{\|\boldsymbol{x} - \boldsymbol{c}_{\ell'}\|^2}{2\sigma^2} \right) \mathrm{d}\boldsymbol{x}$$

$$= (\pi \sigma^2)^{d/2} \exp\left(-\frac{\|\boldsymbol{c}_\ell - \boldsymbol{c}_{\ell'}\|^2}{4\sigma^2} \right),$$

where d denotes the dimensionality of $\boldsymbol{x}$. This is the reason why the Gaussian kernel model is preferable in practice (Sugiyama et al., 2013).

By replacing the expectations in $\boldsymbol{h}$ with empirical estimators and adding the ℓ_2-regularizer to the objective function, we arrive at the following optimization problem:

$$\widehat{\boldsymbol{\theta}} := \operatorname*{argmin}_{\boldsymbol{\theta}} \left[\boldsymbol{\theta}^\top \boldsymbol{H} \boldsymbol{\theta} - 2\widehat{\boldsymbol{h}}^\top \boldsymbol{\theta} + \lambda \boldsymbol{\theta}^\top \boldsymbol{\theta} \right], \tag{8.16}$$

where $\lambda \ (\geq 0)$ is the regularization parameter and $\widehat{\boldsymbol{h}}$ is the b-dimensional vector defined as

$$\widehat{\boldsymbol{h}} = \frac{1}{n} \sum_{i=1}^n \boldsymbol{\psi}(\boldsymbol{x}_i) - \frac{1}{n'} \sum_{i'=1}^{n'} \boldsymbol{\psi}(\boldsymbol{x}'_{i'}).$$

Taking the derivative of the objective function in (8.16) and equating it to zero, we can obtain the solution $\widehat{\boldsymbol{\theta}}$ analytically as

$$\widehat{\boldsymbol{\theta}} = (\boldsymbol{H} + \lambda \boldsymbol{I}_b)^{-1} \widehat{\boldsymbol{h}},$$

where $\boldsymbol{I}_b$ denotes the b-dimensional identity matrix.

Finally, labels are predicted as

$$y = \operatorname{sign}[\widehat{g}(\boldsymbol{x})] = \operatorname{sign}[\widehat{\boldsymbol{\theta}}^\top \boldsymbol{\psi}(\boldsymbol{x})].$$

8.4.4   DSDD-Based Algorithm

We expect that an improved solution can be obtained by LSDD over KDEs because of the more direct nature of LSDD. However, LSDD is still indirect because the sign of the density difference is inspected after the density difference is estimated. In what follows, we introduce *direct sign density difference (DSDD)* estimation (du Plessis et al., 2013), which allows us to directly estimate the sign of the density difference.

By lower-bounding the L_1-distance between the target probability densities, defined as

$$\int \left| p_{\mathrm{tr}}(x) - p'_{\mathrm{tr}}(x) \right| \, dx, \tag{8.17}$$

we can obtain the sign of the density difference. We begin by considering the following self-evident relation:

$$|t| \geq tz, \ \text{ if } |z| \leq 1.$$

We can apply this relation at each point x, to obtain

$$\left| p_{\mathrm{tr}}(x) - p'_{\mathrm{tr}}(x) \right| \geq g(x) \left[p_{\mathrm{tr}}(x) - p'_{\mathrm{tr}}(x) \right] \ \text{ if } |g(x)| \leq 1, \ \forall x.$$

By applying the above inequality to (8.17) and maximizing with respect to $g(x)$, we can obtain the tightest lower bound as

$$\int \left| p_{\mathrm{tr}}(x) - p'_{\mathrm{tr}}(x) \right| \, dx \geq \sup_{g} \int g(x) \left[p_{\mathrm{tr}}(x) - p'_{\mathrm{tr}}(x) \right] \, dx \tag{8.18}$$

$$\text{s.t. } |g(x)| \leq 1, \ \forall x.$$

It is straightforward to verify that the above relation will be met with equality when

$$g(x) = \mathrm{sign} \left[p_{\mathrm{tr}}(x) - p'_{\mathrm{tr}}(x) \right].$$

The expression on the right-hand side of (8.18) is especially useful since the probability densities occur linearly in the integral. By replacing the integrals with sample averages and searching $g(x)$ from a parametric family (denoted as $g_\alpha(x)$), we can write the above as

$$\widehat{\alpha} = \ \operatorname{argmin}_\alpha \ \frac{1}{n'} \sum_{i=1}^{n'} g_\alpha(x'_i) - \frac{1}{n} \sum_{j=1}^{n} g_\alpha(x_j)$$

$$\text{s.t.} \quad |g_\alpha(x)| \leq 1, \ \forall x.$$

Next we briefly discuss how to solve the above optimization problem. To satisfy the constraint $|g(x)| \leq 1, \ \forall x$, let us consider the following clipped function:

$$\widetilde{g}(x) := T(g(x)),$$

where T is a clipping function defined as

$$T(z) := \min\{1, \max\{-1, z\}\}.$$

We use a linear-in-parameter model,

$$g(\boldsymbol{x}) = \sum_{\ell=1}^{b} \alpha_\ell \varphi_\ell(\boldsymbol{x}),$$

where $\varphi_\ell(\boldsymbol{x})$ are basis functions. Using the above definitions and including a regularizer, we arrive at the following objective function to be minimized:

$$J(\boldsymbol{\alpha}) = \frac{1}{n'} \sum_{i=1}^{n'} T\left(\sum_{\ell=1}^{b} \alpha_\ell \varphi_\ell(\boldsymbol{x}_i')\right) - \frac{1}{n} \sum_{j=1}^{n} T\left(\sum_{\ell=1}^{b} \alpha_\ell \varphi_\ell(\boldsymbol{x}_j)\right) + \frac{\lambda}{2} \sum_{\ell=1}^{b} \alpha_\ell^2.$$

Although the above objective function is non-convex, we can still efficiently find a local minimizer using the *convex-concave procedure* (CCCP) (Yuille and Rangarajan, 2003).

CCCP requires the objective function to be split into convex and concave parts:

$$J(\boldsymbol{\alpha}) = J_{\text{vex}}(\boldsymbol{\alpha}) + J_{\text{cave}}(\boldsymbol{\alpha}).$$

This is done by expressing $T(z)$ as

$$T(z) = C_{-1}(z) - C_1(z) - 1,$$

where $C_\epsilon(z) = \max\{0, z - \epsilon\}$. This results in the following convex and concave functions:

$$J_{\text{vex}}(\boldsymbol{\alpha}) = \frac{1}{n'} \sum_{i=1}^{n'} C_{-1}\left(\sum_{\ell=1}^{b} \alpha_\ell \varphi_\ell(\boldsymbol{x}_i')\right) + \frac{1}{n} \sum_{j=1}^{n} C_1\left(\sum_{\ell=1}^{b} \alpha_\ell \varphi_\ell(\boldsymbol{x}_j)\right) + \frac{\lambda}{2} \sum_{\ell=1}^{b} \alpha_\ell^2,$$

$$J_{\text{cave}}(\boldsymbol{\alpha}) = -\frac{1}{n'} \sum_{i=1}^{n'} C_1\left(\sum_{\ell=1}^{b} \alpha_\ell \varphi_\ell(\boldsymbol{x}_i')\right) - \frac{1}{n} \sum_{j=1}^{n} C_{-1}\left(\sum_{\ell=1}^{b} \alpha_\ell \varphi_\ell(\boldsymbol{x}_j)\right).$$

Using *Fenchel's inequality* (Boyd and Vandenberghe, 2004), we can bound the function $C_\epsilon(z)$ as

$$C_\epsilon(z) \geq zt - C_\epsilon^*(t),$$

where $C_\epsilon^*(t)$ is the *Fenchel dual* of $C_\epsilon(z)$,

$$C_\epsilon^*(t) = \begin{cases} \epsilon t & (0 \leq t \leq 1), \\ \infty & (\text{otherwise}). \end{cases}$$

Applying this to the concave part gives

$$J_{\text{cave}}(\boldsymbol{\alpha}) \leq \bar{J}_{\text{cave}}(\boldsymbol{\alpha}, \boldsymbol{b}, \boldsymbol{c}),$$

where the bound is specified by $\boldsymbol{b}$ and $\boldsymbol{c}$:

$$\bar{J}_{\text{cave}}(\boldsymbol{\alpha}, \boldsymbol{b}, \boldsymbol{c}) = \frac{1}{n'} \sum_{i=1}^{n'} \left(C_1^*(b_i) - b_i \sum_{\ell=1}^{b} \alpha_\ell \varphi_\ell(\boldsymbol{x}_i') \right)$$
$$+ \frac{1}{n} \sum_{j=1}^{n} \left(C_{-1}^*(c_j) - c_j \sum_{\ell=1}^{b} \alpha_\ell \varphi_\ell(\boldsymbol{x}_j) \right).$$

This bound is convex with respect to $\boldsymbol{b}$ and $\boldsymbol{c}$ if $\boldsymbol{\alpha}$ is fixed. Using this bound, we have

$$J(\boldsymbol{\alpha}) \leq J_{\text{vex}}(\boldsymbol{\alpha}) + \bar{J}_{\text{cave}}(\boldsymbol{\alpha}, \boldsymbol{b}, \boldsymbol{c}).$$

The strategy to minimize $J(\boldsymbol{\alpha})$ is then to alternately minimize the right-hand side by minimizing with respect to $\boldsymbol{\alpha}$ (keeping $\boldsymbol{b}$ and $\boldsymbol{c}$ fixed), and minimize with respect to $\boldsymbol{b}$ and $\boldsymbol{c}$ (keeping $\boldsymbol{\alpha}$ fixed). Minimization with respect to $\boldsymbol{\alpha}$ minimizes the current upper bound and minimization with respect to $\boldsymbol{b}$ and $\boldsymbol{c}$ corresponds to tightening the bound at the current point.

The final optimization algorithm is summarized below:

1. *Initialize the starting value:*

$$\boldsymbol{\alpha}^1 \leftarrow \arg\min_{\boldsymbol{\alpha}} J_{\text{vex}}(\boldsymbol{\alpha}).$$

2. *For $t = 1, \ldots T$:*

   i. *Tighten the upper-bound:* Obtain $\boldsymbol{b}$ and $\boldsymbol{c}$ as

$$\boldsymbol{b}^t, \boldsymbol{c}^t \leftarrow \arg\min_{\boldsymbol{b}, \boldsymbol{c}} \bar{J}_{\text{cave}}(\boldsymbol{\alpha}^t, \boldsymbol{b}, \boldsymbol{c}),$$

   which can be analytically performed as

$$b_i^t \leftarrow \begin{cases} 0 & (\sum_{\ell=1}^{b} \alpha_\ell^t \varphi_\ell(\boldsymbol{x}_i') < 1), \\ 1 & (\text{otherwise}), \end{cases}$$

$$c_j^t \leftarrow \begin{cases} 0 & (\sum_{\ell=1}^{b} \alpha_\ell^t \varphi(\boldsymbol{x}_j) < -1), \\ 1 & (\text{otherwise}). \end{cases}$$

   ii. *Minimize the upper bound:* Set

$$\boldsymbol{\alpha}^{t+1} \leftarrow \arg\min_{\boldsymbol{\alpha}} J_{\text{vex}}(\boldsymbol{\alpha}) + \bar{J}_{\text{cave}}(\boldsymbol{\alpha}, \boldsymbol{b}^t, \boldsymbol{c}^t),$$

which can be performed by solving the following convex *quadratic program* (QP):

$$
\min_{\alpha} \quad -\sum_{\ell=1}^{b} \alpha_\ell \left(\frac{1}{n'} \sum_{i=1}^{n'} b_i^t \varphi_\ell(\boldsymbol{x}_i') + \frac{1}{n} \sum_{j=1}^{n} c_j^t \varphi_\ell(\boldsymbol{x}_j) \right)
$$

$$
+ \frac{1}{n'} \sum_{i=1}^{n'} \xi_i' + \frac{1}{n} \sum_{j=1}^{n} \xi_j + \frac{\lambda}{2} \sum_{\ell=1}^{b} \alpha_\ell^2
$$

$$
\text{s.t.} \quad \xi_i' \geq 0,\ \xi_i' \geq \sum_{\ell=1}^{b} \alpha_\ell \varphi_\ell(\boldsymbol{x}_i') + 1,\ \forall i = 1, \ldots, n'
$$

$$
\xi_j \geq 0,\ \xi_j \geq \sum_{\ell=1}^{b} \alpha_\ell \varphi_\ell(\boldsymbol{x}_j) - 1,\ \forall j = 1, \ldots, n.
$$

The above constrained problem can be solved with an off-the-shelf QP solver. In practice, Gaussian kernels centered at the sample points in $\mathcal{X}_{\text{tr}}$ and $\mathcal{X}_{\text{tr}}'$ are chosen as the basis functions. All hyper-parameters are set by cross-validation.

8.4.5   Experiments

Finally, we compare the performance of the clustering-based methods, KDE-based method, LSDD-based method, and DSDD-based method for UU classification:

- *k*-**means clustering (KMC)**: Cluster the data into two clusters using the KMC algorithm (MacQueen, 1967).

- **Spectral clustering (SC)**: Cluster the data into two clusters using the SC algorithm (Shi and Malik, 2000). The affinity matrix was constructed with 7 nearest neighbors.

- **Squared-loss mutual information-based clustering (SMIC)** : Cluster the data according to the SMIC method (Sugiyama et al., 2014). SMIC was chosen since it provides model selection, avoiding the need for subjective parameter tuning.

- **Kernel density estimation (KDE)**: Estimate $\text{sign}\left[p_{\text{tr}}(\boldsymbol{x}) - p_{\text{tr}}'(\boldsymbol{x}) \right]$ by estimating each density $p_{\text{tr}}(\boldsymbol{x})$ and $p_{\text{tr}}'(\boldsymbol{x})$ with the method described in section 8.4.2. Hyper-parameters are selected using least-squares cross validation.

- **Least-squares density difference (LSDD) estimation**: Estimate $\text{sign}\left[p_{\text{tr}}(\boldsymbol{x}) - p_{\text{tr}}'(\boldsymbol{x}) \right]$ by estimating the difference $p_{\text{tr}}(\boldsymbol{x}) - p_{\text{tr}}'(\boldsymbol{x})$ using the method described in section 8.4.3. Hyper-parameters are selected by cross-validation.

- **Direct sign density difference (DSDD) estimation**: Directly estimate $\text{sign}\left(p_{\text{tr}}(\boldsymbol{x}) - p_{\text{tr}}'(\boldsymbol{x}) \right)$ using the method described in section 8.4.4. Hyper-parameters are selected by cross-validation.

Table 8.4
Mean and standard deviation of the classification error for experiments. The size of each dataset was $|\mathcal{X}_{\mathrm{tr}}| = 40$ and $|\mathcal{X}'_{\mathrm{tr}}| = 40$. The best method in terms of the mean error and comparable methods according to the two-sided paired t-test at the significance level 5% are specified by boldface.

(a) $\theta = 0.2$ and $\theta' = 0.8$

Dataset	DSDD	LSDD	KDE	KMC	SC	SMIC
australian	**.142**(.045)	.174(.110)	.211(.126)	.266(.147)	.381(.033)	.303(.103)
banana	.179(.097)	**.170**(.070)	.237(.147)	.431(.068)	.427(.141)	.424(.141)
diabetes	.246 (.122)	**.223** (.079)	**.226** (.051)	.372 (.080)	.380 (.094)	.370 (.131)
german	.268 (.059)	.281 (.127)	**.211** (.051)	.437 (.114)	.448 (.128)	.439 (.052)
heart	**.176** (.051)	**.174** (.047)	.211 (.074)	.261 (.131)	.310 (.032)	.327 (.107)
image	**.198** (.078)	.206 (.047)	**.201** (.049)	.385 (.093)	.351 (.119)	.384 (.135)
ionosphere	**.157** (.059)	.184 (.106)	.194 (.123)	.329 (.145)	.319 (.113)	.311 (.174)
saheart	.310 (.093)	**.205** (.048)	.238 (.113)	.422 (.121)	.395 (.113)	.384 (.072)
thyroid	**.102** (.052)	.121 (.116)	.207 (.074)	.328 (.113)	.326 (.109)	.305 (.074)
twonorm	.044 (.085)	.051 (.072)	.200 (.028)	**.036** (.054)	.043 (.069)	.048 (.071)

(b) $\theta = 0.35$ and $\theta' = 0.65$

Dataset	DSDD	LSDD	KDE	KMC	SC	SMIC
australian	**.244** (.116)	.259 (.088)	.355 (.104)	.265 (.080)	.376 (.065)	.308 (.107)
banana	**.338** (.094)	**.339** (.100)	.365 (.067)	.433 (.049)	.427 (.069)	.424 (.070)
diabetes	**.340** (.075)	.361 (.124)	.345 (.034)	.373 (.063)	.380 (.048)	.371 (.114)
german	.375 (.042)	.380 (.093)	**.354** (.057)	.437 (.024)	.445 (.057)	.438 (.041)
heart	.270 (.133)	**.247** (.084)	.354 (.052)	.264 (.059)	.315 (.081)	.327 (.089)
image	**.331** (.078)	.350 (.067)	.350 (.039)	.384 (.031)	.354 (.049)	.382 (.050)
ionosphere	**.291** (.099)	.356 (.066)	.345 (.048)	.330 (.070)	.322 (.058)	.314 (.107)
saheart	.378 (.093)	**.353** (.057)	.363 (.066)	.419 (.082)	.395 (.022)	.385 (.040)
thyroid	**.227** (.098)	.251 (.087)	.302 (.022)	.326 (.061)	.329 (.047)	.307 (.076)
twonorm	.164 (.188)	.153 (.121)	.352 (.096)	**.036** (.053)	.042 (.122)	.049 (.120)

For each experiment, the datasets $\mathcal{X}_{\mathrm{tr}}$ and $\mathcal{X}'_{\mathrm{tr}}$ were constructed by drawing n and n' samples from the P and N classes of the binary classification datasets according to the class-priors $\theta = p_{\mathrm{tr}}(y = +1)$ and $\theta' = p'_{\mathrm{tr}}(y = +1)$. Then labels were predicted according to the sign of the density difference.

The *UCI Machine Learning Repository* (Lichman, 2013) was used for the experiments. The performance of the methods was evaluated by varying the class-priors on these datasets. Experiments were carried out with the following two class balances:

- Large difference between the classes ($\theta = 0.2$ and $\theta' = 0.8$);

- Small difference between the classes ($\theta = 0.35$ and $\theta' = 0.65$).

The average and standard deviation of the clustering error rate for the above two setups, with $|\mathcal{X}_{\mathrm{tr}}| = |\mathcal{X}'_{\mathrm{tr}}| = 40$, are given in table 8.4.

From the results, we see that methods of estimating the sign of the density difference (i.e., DSDD, LSDD, and KDE) generally work better than methods using the cluster structure of the data (i.e., KMC, SC, and SMIC). The thyroid dataset lends itself to interpretation of why these methods work better. The labels in the thyroid dataset correspond to healthy and diseased. The diseased label is caused by either a hyper-functioning or hypo-functioning thyroid. These two underlying causes result in within-class multi-modality, which may cause clustering-based methods to fail.

Among the methods which estimate the sign of the density difference, we see that DSDD generally performs better than LSDD, and LSDD in turn performs better than KDE. This is as expected since KDE solves a more general problem than LSDD, and LSDD solves a more general problem than DSDD. This tendency is even more pronounced on the more difficult case where the class-priors are close to each other (see table 8.4b).

III WEAKLY SUPERVISED LEARNING FOR MULTI-CLASS CLASSIFICATION

9 Complementary-Label Classification

In this chapter, we introduce a weakly supervised classification approach for multi-class classification called *complementary-label (CL) classification*. We first introduce single CL classification, and then extend the framework to *multi-complementary-label (MCL) classification*.

9.1 Introduction

Conceptually, a CL specifies a class that a pattern does *not* belong to. If the number of classes is huge, choosing the correct class label from many candidate classes is highly laborious. On the other hand, choosing one of the *incorrect* class labels would be much easier and thus less costly. In the binary classification setup, classification with CLs is equivalent to classification with ordinary-labels (OLs) because a positive CL (i.e., not positive) immediately means a negative OL. In contrast, in multi-class classification with the number of classes more than two, CLs are less informative than OLs because a CL only means any of the other classes.

CLs are also handy when each labeler has expertise tied only to a single domain (or class). For example, a person can identify the language of an audio when it is spoken in their mother language; otherwise, that person will only know that it is *not* the language they are familiar with. In this potential application, labelers cannot give the correct class most of the time, but it will be easier for them to provide a CL.

In this chapter, we first show that an unbiased estimator of the multi-class classification risk can be obtained only from CL data. We then show that classification from CL data can be easily combined with classification from OL data (i.e., ordinary supervised classification).

In practice, it would be useful if a labeler can provide *multiple* incorrect labels for each pattern. We refer to such a set of multiple CLs given to a pattern as a multi-complementary-label (MCL). Naively, one can think of decomposing an MCL into multiple *independent* single CLs and then apply a (single) CL classification method. However, the amount of supervision that an MCL contains is actually diluted after decomposition. To fully utilize the information contained in MCLs, we provide an unbiased risk estimator that takes MCLs directly.

9.2   Risk Estimation from CL Data

First, we want to introduce the framework of CL classification. Then we give a CL classification method and its extension to classification from both OLs and CLs.

9.2.1   Formulation

Suppose that d-dimensional pattern $x \in \mathcal{X} \subset \mathbb{R}^d$ and its class label $y \in \mathcal{Y} := \{1, \ldots, c\}$ follow an unknown probability distribution with density $p(x, y)$. As detailed in section 2.2, the goal of ordinary multi-class classification is to learn a classifier $g : \mathcal{X} \to \mathbb{R}^c$ that minimizes the classification risk with multi-class loss $\mathscr{L}(g(x), y)$:

$$R(g) := \mathbb{E}_{p(x,y)}\left[\mathscr{L}(g(x), y)\right], \tag{9.1}$$

where $\mathbb{E}$ denotes the expectation. With classifier $g(x) := (g_1(x), \ldots, g_c(x))^\top$, prediction for test pattern x is obtained as

$$\hat{y} = \underset{y \in \mathcal{Y}}{\mathrm{argmax}}\, g_y(x), \tag{9.2}$$

where $g_y : \mathbb{R}^d \to \mathbb{R}$ corresponds to a binary classifier for class y versus the rest.

Since $p(x, y)$ is unknown, the expectation in (9.1) cannot be computed directly. In an ordinary supervised classification scenario, the expectation is approximated by the average over independent and identically distributed training samples to obtain an empirical objective for training a classifier. However, the situation considered in this section does not follow this standard setup. Specifically, our goal is to still learn a classifier that minimizes the classification risk (9.1), but only with the CL dataset given by

$$\{(x_i, \bar{y}_i)\}_{i=1}^n.$$

Conceptually, a CL $\bar{y}_i$ specifies a class that the pattern x_i does *not* belong to. However, in our stochastic formulation, the *true* class of x_i is not deterministic but is randomly determined following the class-posterior probability $p(y|x_i)$. Therefore, $\bar{y}_i$ cannot also be specified deterministically. Below, we denote the probability mass and density functions for $\bar{y}$ by $\bar{p}$.

To discuss CLs in the stochastic formulation, we assume the pairs in the CL dataset $\{(x_i, \bar{y}_i)\}_{i=1}^n$ are drawn independently from an unknown probability distribution with density

$$\bar{p}(x, \bar{y}) = \frac{1}{c-1} \sum_{y \neq \bar{y}} p(x, y). \tag{9.3}$$

The coefficient $\frac{1}{c-1}$ is for the normalization purpose so that $\bar{p}(x,\bar{y})$ is a valid density: $\sum_{\bar{y}=1}^{c} \int_{\mathcal{X}} \bar{p}(x,\bar{y})\mathrm{d}x = 1$. Note that (9.3) implies

$$\bar{p}(x) = p(x), \tag{9.4}$$

since marginalization of (9.3) over $\bar{y}$ yields

$$\bar{p}(x) = \frac{1}{c-1}\sum_{\bar{y}=1}^{c}\left(\sum_{y=1}^{c}p(x,y) - p(x,\bar{y})\right) = \frac{c}{c-1}p(x) - \frac{1}{c-1}p(x) = p(x).$$

Then (9.3) yields

$$\bar{p}(\bar{y}|x) = \frac{1}{c-1}\sum_{y\neq\bar{y}}p(y|x). \tag{9.5}$$

Thus, (9.3) means that x is drawn from $p(x)$, and the probability of observing $\bar{y}$ given x is the average of $p(y|x)$ for all $y\neq\bar{y}$.

9.2.2   Risk Estimation

For a multi-class loss $\mathscr{L}(g(x),y)$, let us define the *complementary-loss* $\bar{\mathscr{L}}(g(x),\bar{y})$ as

$$\bar{\mathscr{L}}(g(x),\bar{y}) := -(c-1)\cdot\mathscr{L}(g(x),\bar{y}) + \sum_{y=1}^{c}\mathscr{L}(g(x),y). \tag{9.6}$$

Then the following theorem holds:

Theorem 9.1   *(Ishida et al., 2019) The classification risk (9.1) can be equivalently expressed as*

$$R(g) = \mathbb{E}_{\bar{p}(x,\bar{y})}[\bar{\mathscr{L}}(g(x),\bar{y})]. \tag{9.7}$$

Proof   First of all, we have

$$\bar{p}(x,\bar{y}) = \frac{1}{c-1}\sum_{y\neq\bar{y}}p(x,y) = \frac{1}{c-1}\left(\sum_{y=1}^{c}p(x,y) - p(x,\bar{y})\right)$$

$$= \frac{1}{c-1}(p(x) - p(x,\bar{y})).$$

This and (9.4) yield

$$p(x,y) = \bar{p}(x) - (c-1)\bar{p}(x,y).$$

Then, we have

$$R(g) = \mathbb{E}_{p(x,y)}[\mathscr{L}(g(x),y)]$$

$$= \sum_{y=1}^{c} \int \mathscr{L}(g(x),y)\,(\bar{p}(x) - (c-1)\bar{p}(x,y))\,dx$$

$$= \int \sum_{y=1}^{c} \mathscr{L}(g(x),y) \sum_{\bar{y}=1}^{c} \bar{p}(x,\bar{y})dx - (c-1)\int \sum_{y=1}^{c} \mathscr{L}(g(x),y)\bar{p}(x,y)dx$$

$$= \int \sum_{\bar{y}=1}^{c}\sum_{y=1}^{c} \mathscr{L}(g(x),y)\bar{p}(x,\bar{y})dx - (c-1)\int \sum_{\bar{y}=1}^{c} \mathscr{L}(g(x),\bar{y})\bar{p}(x,\bar{y})dx$$

$$= \mathbb{E}_{\bar{p}(x,\bar{y})}\left[\sum_{y=1}^{c} \mathscr{L}(g(x),y) - (c-1)\mathscr{L}(g(x),\bar{y}) \right]$$

$$= \mathbb{E}_{\bar{p}(x,\bar{y})}\left[\bar{\mathscr{L}}(g(x),\bar{y}) \right],$$

which concludes the proof. $\square$

Equation (9.7) allows us to obtain an empirical risk estimator that can be computed only from CL data:

$$\widehat{R}(g) = \frac{1}{n}\sum_{i=1}^{n}[\bar{\mathscr{L}}(g(x_i),\bar{y}_i)]. \tag{9.8}$$

It is worth noting that, in the above derivation, there are no constraints on the loss function and classifier. Thus, we can use any loss (convex and non-convex) and any model (linear and non-linear) for CL classification.[1]

However, the above CL classification method has a practical issue that the empirical version (9.8) can take a negative value since the complementary-loss $\bar{\mathscr{L}}(g(x),\bar{y})$ defined in (9.6) contains a negative term. This negative risk issue can actually cause severe overfitting (Ishida et al., 2019). In chapter 11, we will have an extensive discussion on this issue and give practical countermeasures.

Here, we focus on certain classes of loss functions and investigate the properties of the CL classification method (Ishida et al., 2017).

9.2.3   Case-Study for Symmetric Losses

As discussed in section 2.1.3, the *one-versus-the-rest* (OVR) loss and the *pairwise-comparison* (PC) loss are defined as

$$\mathscr{L}_{\mathrm{OVR}}(g(x),y) := \ell\big(g_y(x),+1\big) + \frac{1}{c-1}\sum_{y'\neq y} \ell\big(g_{y'}(x),-1\big), \tag{9.9}$$

$$\mathscr{L}_{\mathrm{PC}}\big(\boldsymbol{g}(\boldsymbol{x}),y\big) := \sum_{y' \neq y} \ell\big(g_y(\boldsymbol{x}) - g_{y'}(\boldsymbol{x}), +1\big), \tag{9.10}$$

where $\ell : \mathbb{R} \times \{+1, -1\} \to \mathbb{R}$ is a binary loss. These are popular multi-class loss functions that can also be used for CL classification. As the binary loss ℓ included in the OVR and PC losses, we may use the following ones:

$$\text{Zero-one loss: } \ell_{01}(g(\boldsymbol{x}),y) = \begin{cases} 0 & (yg(\boldsymbol{x}) > 0), \\ 1 & (yg(\boldsymbol{x}) \leq 0), \end{cases}$$

$$\text{Sigmoid loss: } \ell_{\mathrm{S}}(g(\boldsymbol{x}),y) = \frac{1}{1 + \exp(yg(\boldsymbol{x}))}, \tag{9.11}$$

$$\text{Ramp loss: } \ell_{\mathrm{R}}(g(\boldsymbol{x}),y) = \frac{1}{2} \max\Big\{0, \min\{2, 1 - yg(\boldsymbol{x})\}\Big\}.$$

Note that these binary losses are margin-based, non-convex, bounded in $[0, 1]$, and symmetric:

$$\ell(g(\boldsymbol{x}), +1) + \ell(g(\boldsymbol{x}), -1) = 1.$$

Such symmetric and bounded losses are expected to provide more robustness in classification from CL data (Feng et al., 2020b) or classification with noisy labels (Ghosh et al., 2017).

For the OVR and PC losses with a binary loss ℓ described above, the classification risk can be expressed as follows (Ishida et al., 2017):

$$R_{\mathrm{OVR}}(\boldsymbol{g}) = \mathbb{E}_{\bar{p}(\boldsymbol{x},\bar{y})}\left[\sum_{y \neq \bar{y}} \ell\big(g_y(\boldsymbol{x}), +1\big) + (c-1)\ell\big(g_{\bar{y}}(\boldsymbol{x}), -1\big) - (c-2) \right],$$

$$R_{\mathrm{PC}}(\boldsymbol{g}) = (c-1)\mathbb{E}_{\bar{p}(\boldsymbol{x},\bar{y})}\left[\sum_{y \neq \bar{y}} \ell\big(g_y(\boldsymbol{x}) - g_{\bar{y}}(\boldsymbol{x}), +1\big) - \frac{(c-2)}{2} \right].$$

Later in section 9.5, we will illustrate the numerical behavior of the CL classification method for PC loss (9.10) with binary sigmoid loss (9.11).

9.2.4 Relation to Classification with Noisy Labels

The formulation for CL classification can be viewed as a *class-conditional noise* (CCN) model in *classification with noisy labels* (Angluin and Laird, 1987; Natarajan et al., 2013). This becomes more clear by rewriting (9.5) as

$$\begin{pmatrix} \bar{p}(\bar{y} = 1|\boldsymbol{x}) \\ \vdots \\ \bar{p}(\bar{y} = c|\boldsymbol{x}) \end{pmatrix} = \boldsymbol{T} \begin{pmatrix} p(y = 1|\boldsymbol{x}) \\ \vdots \\ p(y = c|\boldsymbol{x}) \end{pmatrix},$$

where $T \in \mathbb{R}^{c \times c}$ is a *noise transition matrix* (Patrini et al., 2017). $T_{y,\bar{y}}$ specifies the probability that label y is flipped to label $\bar{y}$. In the case of CL classification, T becomes a specific matrix that takes zero on diagonals and $1/(c-1)$ on non-diagonals. Therefore, CL classification is actually a special case of classification with noisy labels based on the CCN model. Indeed, unbiased risk estimation based on theorem 9.1 can be regarded as an instance of *backward correction* of label noise (Patrini et al., 2017). Backward correction is a general method that modifies the loss function so that the resulting empirical risk estimator is consistent to the true risk. In Patrini et al. (2017), another approach to classification with noisy labels called *forward correction* has also been proposed. Forward correction explicitly corrects the classifier output in the loss function so that the learned classifier is consistent to the Bayes optimal classifier. Forward correction has been explored in Yu et al. (2018) for CL classification and in Feng et al. (2020a) for *partial-label* (PL) classification. We will discuss PL classification in chapter 10 in detail.

As discussed in section 8.2.3, UU classification corresponds to classification with noisy labels based on the noise model called *mutually contaminated distributions* (MCD). While the MCD model corrupts the class-conditional density $p(x|y)$, the CCN model corrupts the class-posterior probability $p(y|x)$. Therefore, CL classification discussed in this chapter is substantially different from the weakly supervised classification methods explored in part II.

Note that CCN is a special case of MCD in the context of classification with noisy labels (Menon et al., 2015). On the other hand, in the context of weakly supervised classification, CL classification discussed in this chapter is defined specifically for multi-class classification—UU classification and other methods explored in part II are basically specialized to binary classification.

9.3   Theoretical Analysis

In this section, we derive estimation error bounds of CL classification.

Following section 3.2, we use c sets of data, where the $\bar{y}$-th set is

$$\mathscr{X}_{\bar{y}} := \{x_i^{(\bar{y})}\}_{i=1}^{n_{\bar{y}}} \overset{\text{i.i.d.}}{\sim} \bar{p}(x|\bar{y}).$$

We derive estimation error bounds based on the *Rademacher complexity* (see section 3.1.2.3). The estimation error is given by $R(\hat{g}) - R(g^*)$, where the empirical risk minimizer is $\hat{g} := \operatorname{argmin}_{g \in \mathscr{G}^c} \widehat{R}_{\text{CL}}(g)$ and the true risk minimizer is $g^* := \operatorname{argmin}_{g \in \mathscr{G}^c} R(g)$. The empirical risk can be written as

$$\widehat{R}_{\text{CL}}(g) = \sum_{\bar{y}=1}^{c} \frac{\bar{\pi}_{\bar{y}}}{n_{\bar{y}}} \sum_{i=1}^{n_{\bar{y}}} \left(-(c-1)\mathscr{L}(g(x_i^{(\bar{y})}),\bar{y}) + \sum_{y=1}^{c} \mathscr{L}(g(x_i^{(\bar{y})}),y) \right),$$

where $\bar{\pi}_{\bar{y}} := \bar{p}(\bar{y})$. Then we have the following theorem.

Theorem 9.2 *Under the same assumptions as theorem 3.6, suppose that there exists $C_{\mathscr{L}} > 0$ such that $\mathscr{L}(\mathbf{g}(\mathbf{x}), y) \leq C_{\mathscr{L}}$ and there exists $L_{\mathscr{L}} > 0$ such that the Rademacher complexity satisfies $\mathfrak{R}_{n_{\bar{y}}, p_{\bar{y}}}(\mathscr{L}(\cdot, \bar{y}) \circ \mathscr{G}^c) \leq L_{\mathscr{L}} \mathfrak{R}_{n_{\bar{y}}, p_{\bar{y}}}(\mathscr{G})$. Then, for any $\delta > 0$, with probability at least $1 - \delta$,*

$$R(\hat{\mathbf{g}}) - R(\mathbf{g}^*) \leq \sum_{\bar{y}=1}^{c} 4(2c-1)\bar{\pi}_{\bar{y}} L_{\mathscr{L}} \mathfrak{R}_{n_{\bar{y}}, p_{\bar{y}}}(\mathscr{G})$$

$$+ (c-1)\sqrt{2 \ln \frac{2}{\delta}} C_{\mathscr{L}} \left(\sum_{\bar{y}=1}^{c} \frac{\bar{\pi}_{\bar{y}}}{\sqrt{n_{\bar{y}}}} \right).$$

Proof We use the proof techniques in section 3.2.2. When $x_i^{(\bar{y})}$ in class $\bar{y}$ is replaced, the change of $\widehat{R}_{\mathrm{CL}}(\mathbf{g})$ will be no more than $\bar{\pi}_{\bar{y}}(c-1)C_{\mathscr{L}}/n_{\bar{y}}$. Hence, for any $\delta > 0$, with probability at least $1 - \delta/2$,

$$\sup_{\mathbf{g} \in \mathscr{G}^c} \widehat{R}_{\mathrm{CL}}(\mathbf{g}) - R(\mathbf{g}) \leq \mathbb{E}\left[\sup_{\mathbf{g} \in \mathscr{G}^c} \widehat{R}_{\mathrm{CL}}(\mathbf{g}) - R(\mathbf{g}) \right]$$

$$+ \sqrt{\frac{1}{2} \ln \frac{2}{\delta}} (c-1) C_{\mathscr{L}} \left(\sum_{\bar{y}=1}^{c} \frac{\bar{\pi}_{\bar{y}}}{\sqrt{n_{\bar{y}}}} \right),$$

where $\mathbb{E}$ is over the repeated sampling of $(\mathscr{X}_1, \ldots, \mathscr{X}_c)$ for evaluating $\widehat{R}_{\mathrm{CL}}(\mathbf{g})$. Supposing that $\mathscr{X}_{\bar{y}}'$ is a ghost sample, we bound the first term of the right-hand side by using symmetrization (see section 3.1.2.5) as follows:

$$\mathbb{E}\left[\sup \widehat{R}_{\mathrm{CL}}(\mathbf{g}) - R(\mathbf{g}) \right]$$

$$\leq \sum_{\bar{y}=1}^{c} \frac{\bar{\pi}_{\bar{y}}}{n_{\bar{y}}} \mathbb{E}_{\mathscr{X}_{\bar{y}}, \mathscr{X}_{\bar{y}}'}\left[\sup_{\mathbf{g} \in \mathscr{G}^c} \sum_{i=1}^{n_{\bar{y}}} \bar{\mathscr{L}}(\mathbf{g}(x_i^{(\bar{y})}), \bar{y}) - \bar{\mathscr{L}}(\mathbf{g}(x_i^{(\bar{y})'}), \bar{y}) \right]$$

$$= \sum_{\bar{y}=1}^{c} \frac{\bar{\pi}_{\bar{y}}}{n_{\bar{y}}} \mathbb{E}_{\sigma^{(\bar{y})}, \mathscr{X}_{\bar{y}}, \mathscr{X}_{\bar{y}}'}\left[\sup_{\mathbf{g} \in \mathscr{G}^c} \sum_{i=1}^{n_{\bar{y}}} \sigma_i^{(\bar{y})} \Big[(1-c)\{\mathscr{L}(\mathbf{g}(x_i^{(\bar{y})}), \bar{y}) - \mathscr{L}(\mathbf{g}(x_i^{(\bar{y})'}), \bar{y})\} \right.$$

$$\left. + \sum_{y=1}^{c} (\mathscr{L}(\mathbf{g}(x_i^{(\bar{y})}), y) - \mathscr{L}(\mathbf{g}(x_i^{(\bar{y})'}), y)) \Big] \right]$$

$$\leq \sum_{\bar{y}=1}^{c} 2(c-1)\frac{\bar{\pi}_{\bar{y}}}{n_{\bar{y}}} \mathbb{E}_{\sigma^{(\bar{y})}, \mathscr{X}_{\bar{y}}}\left[\sup_{\mathbf{g} \in \mathscr{G}^c} \sum_{i=1}^{n_{\bar{y}}} \sigma_i^{(\bar{y})} \mathscr{L}(\mathbf{g}(x_i^{(\bar{y})}), \bar{y}) \right]$$

$$+ \sum_{\bar{y}=1}^{c} \frac{2\bar{\pi}_{\bar{y}}}{n_{\bar{y}}} \mathbb{E}_{\boldsymbol{\sigma}^{(\bar{y})}, \mathcal{X}_{\bar{y}}} \left[\sup_{\boldsymbol{g}\in\mathcal{G}^c} \sum_{i=1}^{n_{\bar{y}}} \sigma_i^{(\bar{y})} \sum_{y=1}^{c} \mathcal{L}(\boldsymbol{g}(\boldsymbol{x}_i^{(\bar{y})}), y) \right]$$

$$= \sum_{\bar{y}=1}^{c} 2(c-1)\bar{\pi}_{\bar{y}} \mathbb{E}_{\boldsymbol{\sigma}^{(\bar{y})}, \mathcal{X}_{\bar{y}}} \left[\sup_{\boldsymbol{g}\in\mathcal{G}^c} \frac{1}{n_{\bar{y}}} \sum_{i=1}^{n_{\bar{y}}} \sigma_i^{(\bar{y})} \mathcal{L}(\boldsymbol{g}(\boldsymbol{x}_i^{(\bar{y})}), \bar{y}) \right]$$

$$+ \sum_{\bar{y}=1}^{c} 2\bar{\pi}_{\bar{y}} \sum_{y=1}^{c} \mathbb{E}_{\boldsymbol{\sigma}^{(\bar{y})}, \mathcal{X}_{\bar{y}}} \left[\sup_{\boldsymbol{g}\in\mathcal{G}^c} \frac{1}{n_{\bar{y}}} \sum_{i=1}^{n_{\bar{y}}} \sigma_i^{(\bar{y})} \mathcal{L}(\boldsymbol{g}(\boldsymbol{x}_i^{(\bar{y})}), y) \right]$$

$$= \sum_{\bar{y}=1}^{c} 2(2c-1)\bar{\pi}_{\bar{y}} \mathfrak{R}_{n_{\bar{y}}, p_{\bar{y}}}(\mathcal{L}(\cdot, \bar{y}) \circ \mathcal{G}^c) \leq \sum_{\bar{y}=1}^{c} 2(2c-1)\bar{\pi}_{\bar{y}} L_{\mathcal{L}} \mathfrak{R}_{n_{\bar{y}}, p_{\bar{y}}}(\mathcal{G}),$$

where $\boldsymbol{\sigma}^{(\bar{y})} = (\sigma_1^{(\bar{y})}, \ldots, \sigma_{n_{\bar{y}}}^{(\bar{y})})$ are Rademacher variables. Then,

$$\sup_{\boldsymbol{g}\in\mathcal{G}^c} \widehat{R}_{\mathrm{CL}}(\boldsymbol{g}) - R(\boldsymbol{g}) \leq \sum_{\bar{y}=1}^{c} 2(2c-1)\bar{\pi}_{\bar{y}} L_{\mathcal{L}} \mathfrak{R}_{n_{\bar{y}}, p_{\bar{y}}}(\mathcal{G})$$

$$+ \sqrt{\frac{1}{2} \ln \frac{2}{\delta}} (c-1) C_{\mathcal{L}} \left(\sum_{\bar{y}=1}^{c} \frac{\bar{\pi}_{\bar{y}}}{\sqrt{n_{\bar{y}}}} \right).$$

We can derive the other direction similarly. Hence, for any $\delta > 0$, with probability at least $1 - \delta$,

$$\sup_{\boldsymbol{g}\in\mathcal{G}^c} |\widehat{R}_{\mathrm{CL}}(\boldsymbol{g}) - R(\boldsymbol{g})| \leq \sum_{\bar{y}=1}^{c} 2(2c-1)\bar{\pi}_{\bar{y}} L_{\mathcal{L}} \mathfrak{R}_{n_{\bar{y}}, p_{\bar{y}}}(\mathcal{G})$$

$$+ \sqrt{\frac{1}{2} \ln \frac{2}{\delta}} (c-1) C_{\mathcal{L}} \left(\sum_{\bar{y}=1}^{c} \frac{\bar{\pi}_{\bar{y}}}{\sqrt{n_{\bar{y}}}} \right).$$

This proves the theorem since the estimation error can be bounded by $2 \sup_{\boldsymbol{g}\in\mathcal{G}^c} |\widehat{R}_{\mathrm{CL}}(\boldsymbol{g}) - R(\boldsymbol{g})|$. $\qquad\square$

As shown in theorem 3.9, we have
$C_{\mathcal{L}} = 2C_\ell$ and $L_{\mathcal{L}} = 2L_\ell$ for the OVR loss;
$C_{\mathcal{L}} = (c-1)C_\ell$ and $L_{\mathcal{L}} = 2(c-1)L_\ell$ for the PC loss;
$C_{\mathcal{L}} = 2C_{\mathrm{g}} + \ln c$ and $L_{\mathcal{L}} = c$ for the *softmax cross-entropy* (CE) loss:

$$\mathcal{L}_{\mathrm{CE}}(\boldsymbol{g}(\boldsymbol{x}), y) := -g_y(\boldsymbol{x}) + \ln \left(\sum_{y'=1}^{c} \exp\left(g_{y'}(\boldsymbol{x})\right) \right).$$

Note that $C_\ell = 1$ can be derived when symmetric binary loss functions are used. This means $C_{\mathscr{L}} = 2$ for the OVR loss and $C_{\mathscr{L}} = c - 1$ for the PC loss.

By using corollary 3.5, we can simplify theorem 9.2. Let $\mathscr{G}$ be a function class with $C_{\mathscr{G}} > 0$, satisfying $\mathfrak{R}_{n_{\bar{y}}, p_{\bar{y}}}(\mathscr{G}) \leq C_{\mathscr{G}} / \sqrt{n_{\bar{y}}}$ for $y = 1, \ldots, c$. For any $\delta > 0$, define

$$C_\delta := 2(2c - 1) L_{\mathscr{L}} C_{\mathscr{G}} + (c - 1)\sqrt{2 \ln(2/\delta)} C_{\mathscr{L}}.$$

Then, with probability at least $1 - \delta$, we have

$$R(\widehat{g}) - R(g^*) \leq C_\delta \sum_{\bar{y}=1}^{c} \frac{\bar{\pi}_{\bar{y}}}{\sqrt{n_{\bar{y}}}},$$

which establishes the consistency of CL classification, with convergence rate $\mathcal{O}_p(\sum_{\bar{y}=1}^{c} 1/\sqrt{n_{\bar{y}}})$.

9.4 Incorporation of Ordinary-Labels

In many practical situations, we may also have OL data in addition to CL data. For example, in *crowd-sourcing* (Howe, 2008), we may choose one of the classes randomly by following the uniform distribution, with probability $\frac{1}{c-1}$ for each class, and ask crowd-workers whether a pattern belongs to the chosen class or not. Then the pattern is equipped with an OL if the answer is yes; otherwise, the pattern is equipped with a CL.

In such cases, we want to leverage both kinds of labeled data to obtain more accurate classifiers. To this end, motivated by PNU classification in chapter 5, let us consider a convex combination of the classification risks derived from OL and CL data:

$$R(g) = \alpha \mathbb{E}_{p(x,y)}[\mathscr{L}(g(x), y)] + (1 - \alpha)\mathbb{E}_{\bar{p}(x,\bar{y})}[\bar{\mathscr{L}}(g(x), \bar{y})], \tag{9.12}$$

where $\alpha \in [0, 1]$ is a hyper-parameter that interpolates between the two risks. The combined risk (9.12) can be naively approximated by the sample averages as

$$\widehat{R}_{\mathrm{CL+OL}}(g) = \frac{\alpha}{m} \sum_{j=1}^{m} \mathscr{L}(g(x_j), y_j) + \frac{1 - \alpha}{n} \sum_{i=1}^{n} \bar{\mathscr{L}}(g(x_i), \bar{y}_i), \tag{9.13}$$

where $\{(x_j, y_j)\}_{j=1}^{m}$ are OL data and $\{(x_i, \bar{y}_i)\}_{i=1}^{n}$ are CL data. We expect that, with this combined empirical risk estimator (9.13), we can obtain better classifiers since both OL and CL data are incorporated.[2] In section 9.5.2, we experimentally show that this is really true.

9.5 Experiments

Here, the performance of the CL classification method is evaluated experimentally.

9.5.1   Experiments with CL

The following methods are compared:

- CL classification (section 9.2.3): The multi-class PC loss (9.10) with the binary sigmoid loss (9.11) was used.

- Partial-label (PL) classification:[3] The PL method proposed by Cour et al. (2011) with the squared-hinge loss $\ell(g(x), y) = (\max\{0, 1 - yg(x)\})^2$ was used.

- Multi-label (ML) classification:[4] CL $\bar{y}$ was translated into a negative label for class $\bar{y}$ and positive labels for the other $c - 1$ classes. This formulation yields the following multi-class loss:

$$\mathscr{L}_{\mathrm{ML}}(g(x), \bar{y}) = \sum_{y \neq \bar{y}} \ell(g_y(x), +1) + \ell(g_{\bar{y}}(x), -1),$$

where the squared loss $\ell(g(x), y) = (yg(x) - 1)^2$ was used as the binary loss.

Another interesting comparison is between classification with CLs and OLs. As the OL classification method, we used the unnormalized version of (9.9) with the squared loss. For better comparison of the two settings, we gave to the OL classification method only $\frac{1}{c-1}$ times as many samples as the CL classification method, since one OL corresponds to $c - 1$ CLs. We used a one-hidden-layer neural network (d-3-1) with rectified linear units (ReLUs) (Nair and Hinton, 2010) as activation functions, and weight decay candidates were chosen from $\{10^{-7}, 10^{-4}, 10^{-1}\}$.

We evaluated the classification performance of the above methods with the benchmark datasets listed in table 9.1. Among them, MNIST and USPS were downloaded from the website of the late Sam Roweis,[5] and all other datasets were downloaded from the *UCI Machine Learning Repository*.[6] For datasets with 10 or more classes, we chose five classes with an equal number of samples. In table 9.1, the mean and standard deviation of the classification accuracy over 20 trials are reported. This shows that the CL classification method compares favorably with other methods for many of the datasets. It is also note-worthy that the CL classification method is comparable to the OL classification method for many datasets.

9.5.2   Experiments with CL and OL

Next, we evaluate the effect of combining OL and CL data. The training dataset was divided with the $1 : (c - 1)$ ratio, where one subset contains OL data while the other contains CL data.[7] From the training dataset, 25 percent of the data was left out for val-idating hyper-parameters based on the zero-one loss version of (9.13). Other details such as standardization, the model and optimization, and weight-decay candidates followed the previous experiments.

A classifier was trained with (9.13) with hyper-parameter α fixed at $1/2$. This combina-tion (OL+CL) method (with $\alpha = 1/2$) was compared with the OL method (corresponding

Table 9.1
Mean and standard deviation of the classification accuracy over 20 trials in percentage. "CL" is complementary-label classification, "PL" is partial-label classification, "ML" is multi-label classification, and "OL" is ordinary-label classification. Best and equivalent methods evaluated by the t-test with significance level 5 percent (excluding "OL") are in boldface. "Dim" denotes the input dimensionality, "# train" denotes the total number of training and validation samples in each class, and "# test" denotes the number of test samples in each class.

Dataset	Dim	Class	# train	# test	CL	PL	ML	OL
WAVEFORM1	21	$1 \sim 3$	1230	406	**85.7(0.9)**	84.1(1.5)	84.7(1.6)	85.8(0.9)
WAVEFORM2	40	$1 \sim 3$	1221	400	**84.4(1.3)**	**83.1(2.7)**	81.8(2.3)	86.7(1.8)
SATIMAGE	36	$1 \sim 7$	415	211	**67.2(7.0)**	54.8(6.8)	51.6(6.0)	67.9(4.2)
SHUTTLE	9	$1, 4, 5$	2458	809	94.9(9.7)	**97.5(0.7)**	90.4(11.8)	97.5(0.8)
SEGMENTATION	19	$1 \sim 7$	29	299	**36.1(6.8)**	31.7(5.8)	26.6(5.4)	58.6(4.5)
PENDIGITS	16	$1 \sim 5$	719	336	**79.4(9.5)**	73.2(6.4)	**75.9(7.7)**	78.8(2.9)
		$6 \sim 10$	719	335	**77.7(3.8)**	65.5(6.4)	72.0(8.6)	74.7(4.6)
MNIST	784	$1 \sim 5$	5842	980	**88.4(4.2)**	71.5(7.4)	56.6(12.4)	77.9(0.4)
		$6 \sim 10$	5421	892	**83.4(2.6)**	67.4(8.1)	50.5(13.7)	77.0(4.5)
DRIVE	48	$1 \sim 5$	3931	1280	**87.6(5.9)**	72.7(7.0)	64.2(12.6)	79.3(5.1)
		$6 \sim 10$	3958	1318	**84.9(5.7)**	73.1(5.8)	69.7(9.3)	81.6(2.9)
LETTER	16	$1 \sim 5$	565	171	**79.6(5.5)**	67.6(6.0)	71.0(9.3)	82.2(4.3)
		$6 \sim 10$	550	178	**73.2(6.3)**	63.9(6.1)	61.2(10.6)	75.9(5.6)
		$11 \sim 15$	556	177	**73.3(5.9)**	66.6(3.4)	59.0(10.1)	75.4(5.0)
		$16 \sim 20$	550	184	**71.5(5.9)**	64.9(5.2)	63.5(7.0)	73.9(5.3)
		$21 \sim 25$	585	167	**76.2(6.0)**	68.3(8.1)	63.1(11.2)	77.1(5.1)
VOWEL	12	$1 \sim 5$	48	42	**35.6(9.0)**	**37.0(9.3)**	31.5(6.7)	54.9(6.7)
		$6 \sim 10$	48	42	**32.6(7.5)**	**34.1(7.7)**	**30.0(9.8)**	53.0(4.4)
USPS	256	$1 \sim 5$	652	166	**70.1(5.2)**	62.8(7.2)	45.8(5.9)	76.2(2.3)
		$6 \sim 10$	542	147	**64.3(4.7)**	**61.4(5.9)**	41.7(5.3)	76.9(5.1)

to $\alpha = 1$) and the CL method (corresponding to $\alpha = 0$), where the multi-class PC loss with the binary sigmoid loss was commonly used for all methods.

The means and standard deviations of the classification accuracy over 10 trials are reported in table 9.2. From the results, we can see that OL+CL tends to outperform OL and CL, demonstrating the usefulness of combining OL and CL data.

9.6 Incorporation of Multi-Complementary-Labels

In this section, we extend CL classification so that a set of multiple CLs given to each pattern, called a *multi-complementary-label (MCL)*, can be directly handled (Feng et al., 2020b).

9.6.1 Formulation

Suppose that we are given an MCL dataset, which is represented as

$$\bar{\mathscr{D}} = \{(\boldsymbol{x}_i, \bar{Y}_i)\}_{i=1}^n.$$

Table 9.2
Means and standard deviations of classification accuracy over 10 trials in percentage. "OL" is the ordinary-label classification method, "CL" is the complementary-label classification method, and "OL+CL" is a combination method that uses both OL and CL data. Best and equivalent methods are highlighted in boldface. "Class" denotes the class labels used for the experiment and "Dim" denotes the dimensionality d of patterns to be classified. # train denotes the number of OL/CL data for training and validation in each class, and # test denotes the number of test data in each class.

Dataset	Class	Dim	# train	# test	OL $(\alpha = 1)$	CL $(\alpha = 0)$	OL+CL $(\alpha = \frac{1}{2})$
WAVEFORM1	$1 \sim 3$	21	413/826	408	85.3(0.8)	86.0(0.4)	**86.9(0.5)**
WAVEFORM2	$1 \sim 3$	40	411/821	411	82.7(1.3)	82.0(1.7)	**84.7(0.6)**
SATIMAGE	$1 \sim 7$	36	69/346	211	74.9(4.9)	70.1(5.6)	**81.2(1.1)**
PENDIGITS	$1 \sim 5$	16	144/575	336	**91.3(2.1)**	84.7(3.2)	**93.1(2.0)**
	$6 \sim 10$		144/575	335	**86.3(3.5)**	78.3(6.2)	**87.8(2.8)**
	$1 \sim 10$		72/647	335	61.7(4.3)	41.1(5.7)	**66.9(2.0)**
DRIVE	$1 \sim 5$		780/3121	1305	92.1(2.6)	89.0(2.1)	**94.2(1.0)**
	$6 \sim 10$	48	795/3180	1290	**87.0(3.0)**	86.5(3.1)	**89.5(2.1)**
	$1 \sim 10$		397/3570	1292	**75.2(2.8)**	40.5(7.2)	**77.6(2.2)**
LETTER	$1 \sim 5$	16	113/452	171	85.2(1.3)	77.2(6.1)	**89.5(1.6)**
	$6 \sim 10$		110/440	178	81.0(1.7)	77.6(3.7)	**84.6(1.0)**
	$11 \sim 15$		111/445	177	81.1(2.7)	76.0(3.2)	**87.3(1.6)**
	$16 \sim 20$		110/440	184	81.3(1.8)	77.9(3.1)	**84.7(2.0)**
	$21 \sim 25$		117/468	167	86.8(2.7)	81.2(3.4)	**91.1(1.0)**
	$1 \sim 25$		22/528	167	11.9(1.7)	6.5(1.7)	**31.0(1.7)**
USPS	$1 \sim 5$	256	130/522	166	83.8(1.7)	76.5(5.3)	**89.5(1.3)**
	$6 \sim 10$		108/434	147	79.2(2.1)	67.6(4.3)	**85.5(2.4)**
	$1 \sim 10$		54/488	147	43.7(2.6)	28.5(3.6)	**59.3(2.2)**

Here, $\bar{Y}_i$ is an MCL, which is a set of CLs given to pattern $x_i \in \mathcal{X} \subset \mathbb{R}^d$. We assume that $\bar{Y}_i$ is neither empty nor the full label set:

$$\bar{Y}_i \in \bar{\mathcal{Y}}, \text{ where } \bar{\mathcal{Y}} = \{2^{\mathcal{Y}} \setminus \emptyset \setminus \mathcal{Y}\}.$$

Here, $2^{\mathcal{Y}}$ denotes the *power set*, and thus $|\bar{\mathcal{Y}}| = 2^c - 2$. Let us represent the size of an MCL by a random variable s defined on $\{1, \ldots, c - 1\}$. If $s = 1$ (i.e., $\bar{Y}_i$ contains only one CL), the above formulation reduces to the original (single) CL classification problem. If $s = c - 1$ (i.e., $\bar{Y}_i$ contains $c - 1$ CLs), the above formulation reduces to the ordinary multi-class classification problem.

Since directly choosing the correct label is often hard for labelers, it would be easier if a labeling system can randomly choose a label set and ask labelers whether the correct label is included in the proposed label set or not. Given a pattern x, suppose that the labeling system first randomly samples the size s of the label set from $p_s(s)$ and then randomly chooses a specific label set with size s from $\bar{\mathcal{Y}}$. Then, an MCL $(x_i, \bar{Y}_i)$ follows the probability distribution with density

$$\bar{p}(\boldsymbol{x}, \bar{Y}) = \sum_{s=1}^{c-1} p_s(s)\bar{p}(\boldsymbol{x}, \bar{Y}|s), \tag{9.14}$$

where

$$\bar{p}(\boldsymbol{x}, \bar{Y}|s) := \begin{cases} \dfrac{1}{\binom{c-1}{s}} \displaystyle\sum_{y \notin \bar{Y}} p(\boldsymbol{x}, y) & (|\bar{Y}| = s), \\[2ex] 0 & (\text{otherwise}). \end{cases}$$

Note that $\bar{p}(\boldsymbol{x}, \bar{Y})$ is a valid probability density such that $\sum_{\bar{Y} \in \bar{\mathcal{Y}}} \int_{\mathcal{X}} \bar{p}(\boldsymbol{x}, \bar{Y}) d\boldsymbol{x} = 1$ (Feng et al., 2020b).

9.6.2 Comparison with Multiple Single CLs

Given an MCL, we may want to decompose it into multiple single CLs and apply a single CL classification method. However, the decomposed CL data may not match the originally assumed data distribution (9.3) in the single CL classification framework. Therefore, this decomposition approach may not be theoretically justified.

Moreover, since CL classification can only handle a single CL for each pattern $\boldsymbol{x}$ at a time and treat each sample independently, the supervision information contained in an MCL would be conceptually diluted by such decomposition. To illustrate this, in table 9.3, we compare two settings according to whether an MCL is decomposed into multiple single CLs or not. Since any non-CL (a label that is different from the CL) can be an OL, we count how many times the OL is included in a non-CL (denoted by #TP); we also count how many times the other labels except the OL is included in a non-CL (denoted by #FP). Then we also calculate the *purity* by #TP/(#TP+#FP).

Table 9.3
Comparison between direct handling of an MCL (with size $s \geq 2$) and its decomposition into multiple single CLs. "#TP" indicates how many times an OL is included in a non-CL, while "#FP" means how many times the other labels except the OL is included in a non-CL. "Purity" is calculated by #TP/(#TP+#FP), which decreases after decomposition.

Setting	#TP	#FP	Purity
Direct	1	$c - s - 1$	$\dfrac{1}{c - s}$
Decomposition	s	$(c - 2)s$	$\dfrac{1}{c - 1}$

Let us consider an MCL with size s. Then, for the decomposition scheme, #TP equals s since there are s single CLs after decomposition, and the OL is included in a non-CL for each of the s single CLs. Similarly, the other $c-2$ labels (all except the OL and the single CL) are included in a non-CL s times each, resulting in $\#FP = (c-2)s$. Therefore, the purity is $1/(c-1)$. In contrast, if an MCL is directly handled without decomposition, we can easily see that the OL is included in a non-CL only once, and the other labels except the OL are included in a non-CL for $c-s-1$ times. Therefore, the purity is $1/(c-s)$. These observations show that the supervision information is diluted (in the purity sense) after decomposing MCLs (with $s \geq 2$).

Motivated by this discussion, we solve the MCL classification problem directly without decomposition (Feng et al., 2020b).

9.6.3   Unbiased Risk Estimator

Here, we introduce an unbiased estimator of the classification risk that can be computed directly from MCL data without decomposition.

As shown in Feng et al. (2020b), the probability density for OLs is related to that for MCLs as

$$p(\boldsymbol{x}, y) = 1 - \sum_{s=1}^{c-1} \frac{c-1}{s} \sum_{\bar{Y} \in \bar{\mathcal{Y}}_s^y} \bar{p}(\boldsymbol{x}, \bar{Y}, s). \tag{9.15}$$

Here, $\bar{\mathcal{Y}}_s^y$ is the set of all possible MCLs with size s that includes a specific label $y \in \mathcal{Y}$:

$$\bar{\mathcal{Y}}_s^y := \{\bar{Y} \in \bar{\mathcal{Y}} \mid y \in \bar{Y}, |\bar{Y}| = s\}.$$

Based on (9.15), the risk expression used in CL classification (9.7) can naturally be extended to the MCL classification scenario.

Theorem 9.3   *(Feng et al., 2020b) The ordinary classification risk (9.1) can be equivalently expressed as*

$$R(\boldsymbol{g}) = \sum_{s=1}^{c-1} p_{\mathrm{S}}(s) \bar{R}_s(\boldsymbol{g}), \tag{9.16}$$

where

$$\bar{R}_s(\boldsymbol{g}) := \mathbb{E}_{\bar{p}(\boldsymbol{x}, \bar{Y}|s)}[\bar{\mathscr{L}}_s(\boldsymbol{g}(\boldsymbol{x}), \bar{Y})], \tag{9.17}$$

$$\bar{\mathscr{L}}_s(\boldsymbol{g}(\boldsymbol{x}), \bar{Y}) := \sum_{y \notin \bar{Y}} \mathscr{L}(\boldsymbol{g}(\boldsymbol{x}), y) - \frac{c-1-s}{s} \sum_{y' \in \bar{Y}} \mathscr{L}(\boldsymbol{g}(\boldsymbol{x}), y'). \tag{9.18}$$

Since the above theorem is a direct extension of theorem 9.1, we omit its proof here. We can easily verify that (9.16) reduces to (9.7) when $p_{\mathrm{S}}(s=1) = 1$ (and $p_{\mathrm{S}}(s \neq 1) = 0$).

Given an MCL dataset $\bar{\mathscr{D}} = \{(\boldsymbol{x}_i, \bar{Y}_i)\}_{i=1}^n$, we can empirically approximate $p_{\mathrm{s}}(s)$ by n_s/n where n_s denotes the number of MCL samples with size s. By further taking into account equations (9.16) through (9.18), we can obtain the following unbiased risk estimator:

$$\widehat{R}(\boldsymbol{g}) = \frac{1}{n}\sum_{i=1}^n\left(\sum_{y\notin\bar{Y}_i}\mathscr{L}\big(\boldsymbol{g}(\boldsymbol{x}_i),y\big) - \frac{c-1-|\bar{Y}_i|}{|\bar{Y}_i|}\sum_{y'\in\bar{Y}_i}\mathscr{L}\big(\boldsymbol{g}(\boldsymbol{x}_i),y'\big)\right). \tag{9.19}$$

9.6.4 Estimation Error Bound

Finally, an estimation error bound for the above MCL classification method is given based on the *Rademacher complexity* (see section 3.1.2.3).

Let $\mathscr{G}\subset\{\boldsymbol{g}:\mathbb{R}^d\to\mathbb{R}^c\}$ be the hypothesis class, $\widehat{\boldsymbol{g}}:=\operatorname{argmin}_{\boldsymbol{g}\in\mathscr{G}}\widehat{R}(\boldsymbol{g})$ be the empirical risk minimizer, and $\boldsymbol{g}^* = \operatorname{argmin}_{\boldsymbol{g}\in\mathscr{G}} R(\boldsymbol{g})$ be the true risk minimizer. Besides, we define the function space $\mathscr{G}_y$ for label $y\in\mathscr{Y}$ as $\mathscr{G}_y := \{f:\boldsymbol{x}\to g_y(\boldsymbol{x})\,|\,\boldsymbol{g}\in\mathscr{G}\}$. Then, the following theorem holds.

Theorem 9.4 *(Feng et al., 2020b) Assume that the loss function $\mathscr{L}(\boldsymbol{g}(\boldsymbol{x}),y)$ is ρ-Lipschitz with respect to $\boldsymbol{g}(\boldsymbol{x})$ $(0<\rho<\infty)$ for all $y\in\mathscr{Y}$. Let $C_{\mathscr{L}}:=\sup_{\boldsymbol{x}\in\mathscr{X},\boldsymbol{g}\in\mathscr{G},y\in\mathscr{Y}}\mathscr{L}(\boldsymbol{g}(\boldsymbol{x}),y)$ and $\mathfrak{R}_n(\mathscr{G}_y)$ be the Rademacher complexity of $\mathscr{G}_y$ for the sample size n. Then, for any $\delta>0$, with probability at least $1-\delta$,*

$$R(\widehat{\boldsymbol{g}}) - R(\boldsymbol{g}^*) \le \sum_{s=1}^{c-1}p_{\mathrm{s}}(s)\left(\frac{4\sqrt{2}\rho c(c-1)}{s}\sum_{y=1}^c\mathfrak{R}_{n_s}(\mathscr{G}_y) + \frac{C_s}{\sqrt{n_s}}\right),$$

where $C_s := (4c-4s-2)C_{\mathscr{L}}\sqrt{\dfrac{\log\frac{2(c-1)}{\delta}}{2}}$ for all $s\in\{1,\dots,c-1\}$ and n_s denotes the number of samples whose MCL size is s.

Since the above theorem is a direct extension of theorem 9.2, we omit its proof here. When $p_{\mathrm{s}}(s=1)=1$ (and $p_{\mathrm{s}}(s\neq1)=0$), we can easily verify that theorem 9.4 reduces to theorem 9.2.

Generally, $\mathfrak{R}_{n_s}(\mathscr{G}_y)$ can be upper-bounded by $C_{\mathscr{G}}/\sqrt{n_s}$ for a positive constant $C_{\mathscr{G}}$ (Golowich et al., 2018). Hence, theorem 9.4 establishes the consistency, i.e., the empirical risk minimizer $\widehat{\boldsymbol{g}}$ converges to the true risk minimizer $\boldsymbol{g}^\star$ as $n\to\infty$ in terms of the classification risk R. It is also worth noting that the above bound is not only related to the Rademacher complexity of the function class but also to s and c. This observation accords with our intuition that the classification task will be harder if the number of classes, c, increases or the size of the MCL, s, decreases.

10   Partial-Label Classification

Another weakly supervised classification setting for multi-class classification is called *partial-label (PL) classification*. We give a PL classification method and investigate its behavior both theoretically and experimentally.

10.1   Introduction

In multi-class classification, giving labels to a large number of patterns is often time-consuming. The conceptual idea behind PL classification is that instead of exactly specifying which class is correct, labelers are asked only to select a set of candidate labels that contains the correct one. Classification with such PL data has been investigated mainly from the theoretical viewpoint—for example, from the perspective of *statistical consistency* (Cour et al., 2011) and *learnability* (Liu and Dietterich, 2014). On the other hand, in chapter 9, we have explored classification with *multi-complementary-labels (MCLs)*, which is a set of complementary-labels (CLs) that do not contain ordinary-labels. From the illustration in figure 10.1, we can see that a set of labels that are not included in a PL is an MCL. Therefore, by systematically converting PLs to MCLs by taking their complements, it is possible to solve the PL classification problem with an MCL classification method.

However, is the use of an MCL classification method for solving PL classification the best way to go? In this chapter, we show that *directly* solving the PL classification problem is more promising. Specifically, we give an unbiased estimator of the classification risk for PL data and its estimation error bound (Feng et al., 2020a). Finally, we introduce a general formulation called *proper PL (PPL) classification* (Wu and Sugiyama, 2021), which includes (single) CL classification, MCL classification, and PL classification as special cases.

10.2   Formulation and Assumptions

To begin, we formulate the PL classification problem and discuss the details of data generation.

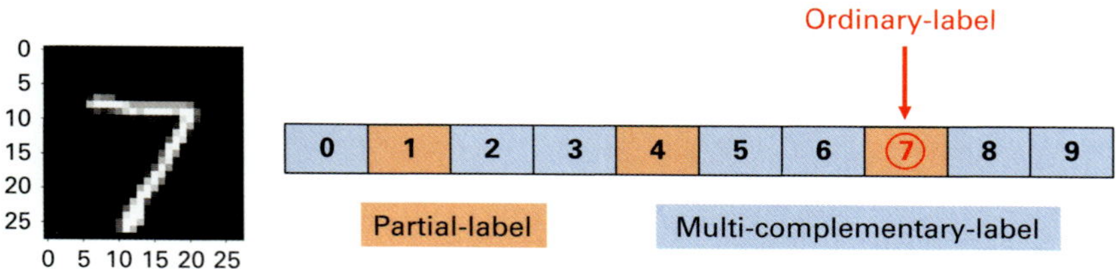

Figure 10.1
Relation among an ordinary-label, a multi-complementary-label (MCL), and a partial-label (PL). The PL and MCL are complementary to each other in the illustration, but they can be overlapped and they do not have to cover all classes in general. The image is taken from the MNIST dataset (LeCun et al., 1998).

10.2.1   Formulation

First, we review multi-class classification with ordinary-labels (see section 3.2). Let the feature space be $\mathcal{X} \in \mathbb{R}^d$ and the label space be $\mathcal{Y} = \{1, \ldots, c\}$, where c denotes the number of classes. Assume that each sample $(\boldsymbol{x}, y) \in \mathcal{X} \times \mathcal{Y}$ is independently drawn from an unknown probability distribution with density $p(\boldsymbol{x}, y)$. The goal of multi-class classification to obtain a multi-class classifier $\boldsymbol{g} : \mathcal{X} \to \mathbb{R}^c$ that minimizes the following classification risk:

$$R(\boldsymbol{g}) := \mathbb{E}_{p(\boldsymbol{x}, y)}[\mathcal{L}(\boldsymbol{g}(\boldsymbol{x}), y)], \tag{10.1}$$

where $\mathbb{E}_{p(\boldsymbol{x}, y)}[\cdot]$ denotes the expectation over $p(\boldsymbol{x}, y)$ and $\mathcal{L} : \mathbb{R}^c \times \mathcal{Y} \to \mathbb{R}$ is a multi-class loss function.

On the other hand, in PL classification, each input $\boldsymbol{x}$ is provided with a set of candidate labels, Y, called a PL. We denote such a PL dataset as

$$\widetilde{\mathcal{D}} = \{(\boldsymbol{x}_i, Y_i)\}_{i=1}^n,$$

where Y_i is a PL for $\boldsymbol{x}_i$. We require each PL not to be empty nor the whole label set, i.e.,

$$Y_i \in \mathscr{C}, \text{ where } \mathscr{C} = \{2^{\mathcal{Y}} \setminus \emptyset \setminus \mathcal{Y}\},$$

where $2^{\mathcal{Y}}$ denotes the *power set* and thus $|\mathscr{C}| = 2^c - 2$. From such PL data, the goal of PL classification is to learn a multi-class classifier $\boldsymbol{g} : \mathcal{X} \to \mathbb{R}^c$ that minimizes the classification risk (10.1).

10.2.2   Data Generation Assumption

A conceptual assumption in PL classification is that the *correct* label is included in the candidate label set. However, in our stochastic formulation, the correct class of pattern $\boldsymbol{x}_i$ is not deterministic but is randomly determined following the class-posterior probability $p(y|\boldsymbol{x}_i)$. Therefore, Y_i cannot also be specified deterministically. Below, we denote the probability mass and density functions for Y by $\widetilde{p}$.

To discuss PLs in the stochastic formulation, we assume

$$\widetilde{p}(Y|\boldsymbol{x},y)=\widetilde{p}(Y|y)=\begin{cases}\dfrac{1}{2^{c-1}-1} & (y\in Y),\\ 0 & (y\notin Y).\end{cases}$$

This means that, independent of pattern $\boldsymbol{x}$, a PL Y is generated uniformly from all Y that contains y, and its probability is zero otherwise. On top of this assumption, we further assume that each PL sample $(\boldsymbol{x}, Y)$ is independently drawn from a probability distribution with density

$$\widetilde{p}(\boldsymbol{x}, Y) = \sum_{y=1}^{c}\widetilde{p}(Y|\boldsymbol{x},y)p(\boldsymbol{x},y) = \frac{1}{2^{c-1}-1}\sum_{y\in Y}p(\boldsymbol{x},y). \tag{10.2}$$

Note that $\widetilde{p}(\boldsymbol{x}, Y)$ is a valid probability density such that $\sum_{Y\in\mathcal{C}}\int_{\mathcal{X}}\widetilde{p}(\boldsymbol{x}, Y)\mathrm{d}\boldsymbol{x}=1$ (Feng et al., 2020a). Equation (10.2) implies

$$\widetilde{p}(\boldsymbol{x}) = p(\boldsymbol{x}),$$

since marginalization of (10.2) over Y yields

$$\bar{p}(\boldsymbol{x}) = \frac{1}{2^{c-1}-1}\sum_{Y\in\mathcal{C}}\sum_{y\in Y}p(\boldsymbol{x},y) = \frac{2^{c-1}-1}{2^{c-1}-1}\sum_{y=1}^{c}p(\boldsymbol{x},y) = p(\boldsymbol{x}).$$

Then (10.2) yields

$$\widetilde{p}(Y|\boldsymbol{x}) = \frac{1}{2^{c-1}-1}\sum_{y\in Y}p(y|\boldsymbol{x}).$$

Thus, (10.2) means that $\boldsymbol{x}$ is drawn from $p(\boldsymbol{x})$, and the probability of observing Y given $\boldsymbol{x}$ is proportional to the sum of $p(y|\boldsymbol{x})$ over $y\in Y$.

10.3 Risk Estimation

Based on the data generating density given in (10.2), we have the following theorem.

Theorem 10.1 *The classification risk (10.1) can be equivalently expressed as*

$$R_{\mathrm{PL}}(\boldsymbol{g}) = \mathbb{E}_{\widetilde{p}(\boldsymbol{x},Y)}\left[\sum_{y\in Y}\frac{p(y|\boldsymbol{x})}{\sum_{y'\in Y}p(y'|\boldsymbol{x})}\mathcal{L}(\boldsymbol{g}(\boldsymbol{x}),y)\right]. \tag{10.3}$$

We omit the proof of the above theorem since a risk expression for a more general setup will be derived in section 10.5.2. Note that in Feng et al. (2020a), an equivalent expression of $R_{\mathrm{PL}}(\boldsymbol{g})$ was derived where (10.3) is multiplied by $1/2$ and the summation over $y\in Y$ is replaced with the summation over all $y\in\{1,\ldots,c\}$.

The expectation over unknown $\widetilde{p}(\boldsymbol{x}, Y)$ in (10.3) can be approximated by the average over its independent samples $\{(\boldsymbol{x}_i, Y_i)\}_{i=1}^{n}$:[1]

$$\widehat{R}_{\mathrm{PL}}(\boldsymbol{g}) = \frac{1}{n} \sum_{i=1}^{n} \sum_{y \in Y} \frac{p(y|\boldsymbol{x}_i)}{\sum_{y' \in Y_i} p(y'|\boldsymbol{x}_i)} \mathscr{L}\big(\boldsymbol{g}(\boldsymbol{x}_i), y\big). \tag{10.4}$$

For the minimizer of the above empirical risk, we can establish an estimation error bound as follows. For a hypothesis class $\mathscr{G} \subset \{\boldsymbol{g} : \mathbb{R}^d \to \mathbb{R}^c\}$, let $\widehat{\boldsymbol{g}}_{\mathrm{PL}} := \operatorname{argmin}_{\boldsymbol{g} \in \mathscr{G}} \widehat{R}_{\mathrm{PL}}(\boldsymbol{g})$ be the empirical risk minimizer and $\boldsymbol{g}^{\star} := \operatorname{argmin}_{\boldsymbol{g} \in \mathscr{G}} R(\boldsymbol{g})$ be the true risk minimizer. For label $y \in \mathscr{Y}$, we define function space $\mathscr{H}_y := \{h : \boldsymbol{x} \mapsto g_y(\boldsymbol{x}) | \boldsymbol{g} \in \mathscr{G}\}$. Let $\mathfrak{R}_n(\mathscr{H}_y)$ be the expected Rademacher complexity of $\mathscr{H}_y$ with sample size n. Then the following theorem holds.

Theorem 10.2　*(Feng et al., 2020a) Assume that the loss function $\mathscr{L}(\boldsymbol{g}(\boldsymbol{x}), y)$ is ρ-Lipschitz $(0 < \rho < \infty)$ with respect to $\boldsymbol{g}(\boldsymbol{x})$ for all $y \in \mathscr{Y}$ and upper-bounded by M, i.e., $M = \sup_{\boldsymbol{x} \in \mathscr{X}, \boldsymbol{g} \in \mathscr{G}, y \in \mathscr{Y}} \mathscr{L}(\boldsymbol{g}(\boldsymbol{x}), y)$. Then, for any $\delta > 0$, with probability at least $1 - \delta$,*

$$R(\widehat{\boldsymbol{g}}_{\mathrm{PL}}) - R(\boldsymbol{g}^{\star}) \leq 4\sqrt{2}\rho \sum_{y=1}^{c} \mathfrak{R}_n(\mathscr{H}_y) + M\sqrt{\frac{\log \frac{2}{\delta}}{2n}}.$$

We omit the proof of the above theorem here since a similar bound in a more general setup will be proved in section 10.5.3.[2]

Generally, $\mathfrak{R}_n(\mathscr{H}_y)$ can be upper-bounded by $C_{\mathscr{H}}/\sqrt{n}$ for a positive constant $C_{\mathscr{H}}$ (Golowich et al., 2018). Hence, theorem 10.2 establishes the consistency, i.e., the empirical risk minimizer $\widehat{\boldsymbol{g}}_{\mathrm{PL}}$ converges to the true risk minimizer $\boldsymbol{g}^{\star}$ as $n \to \infty$ in terms of the classification risk R.

10.4　Experiments

In this section, we numerically investigate the empirical behavior of the PL classification method, which we refer to as the *risk-consistent* (RC) method since the unbiased risk estimator used is consistent to the true classification risk.

We used four benchmark datasets: MNIST (LeCun et al., 1998),Kuzushiji-MNIST (Clanuwat et al., 2018), Fashion-MNIST (Xiao et al., 2017), and CIFAR-10 (Krizhevsky, 2009).For each sample $(\boldsymbol{x}_i, y_i)$ in the dataset, a PL Y_i was uniformly sampled from all possible PLs that contains y_i. In addition, we also used five real-world PL datasets: Lost (Cour et al., 2011), BirdSong (Briggs et al., 2012), MSRCv2 (Liu and Dietterich, 2012), Soccer Player (Zeng et al., 2013), and Yahoo! News (Guillaumin et al., 2010). Each dataset was split into 90 percent and 10 percent for training and testing, respectively.

Since the RC method does not rely on specific classification models, we tested several different models: the linear model, the three-layer (d-500-k) multi-layer perceptron

Table 10.1
Test performance (mean±std over five trials) on benchmark datasets using ResNet for CIFAR-10 and MLP
for the other three datasets. The paired *t*-test at 5 percent significance level was conducted, and •/∘ represents whether the *best* of RC and CC is significantly better/worse than other methods. The best results are
highlighted in bold.

	MNIST	Kuzushiji-MNIST	Fashion-MNIST	CIFAR-10
RC	**98.00±0.11%**	**89.38±0.28%**	**88.38±0.16%**	**77.93±0.59%**
CC	97.87±0.10%•	88.83±0.40%•	87.88±0.25%•	75.78±0.27%•
GA	96.37±0.13%•	84.23±0.19%•	85.57±0.16%•	72.22±0.19%•
NN	96.75±0.08%•	82.36±0.41%•	86.25±0.14%•	68.09±0.31%•
Free	88.48±0.37%•	70.31±0.68%•	81.34±0.47%•	17.74±1.20%•
PC	92.47±0.13%•	73.45±0.20%•	83.37±0.31%•	46.53±2.01%•
Forward	97.64±0.11%•	87.64±0.13%•	86.73±0.15%•	71.18±0.92%•
EXP	97.81±0.04%•	88.48±0.29%•	87.96±0.06%•	73.22±0.66%•
LOG	97.86±0.11%•	88.24±0.08%•	88.31±0.26%	75.38±0.34%•
MAE	97.82±0.11%•	88.43±0.32%•	87.83±0.22%•	66.91±3.08%•
MSE	96.95±0.14%•	85.16±0.44%•	85.72±0.26%•	66.15±2.13%•
GCE	96.71±0.08%•	85.19±0.39%•	86.88±0.16%•	68.39±0.71%•
Phuber-CE	95.10±0.34%•	80.66±0.41%•	85.33±0.23%•	58.60±0.95%•

(MLP), and the 34-layer ResNet (He et al., 2016). RC is compared with SURE (Feng
and An, 2019), CLPL (Cour et al., 2011), IPAL (Zhang and Yu, 2015), PLSVM (Elkan
and Noto, 2008), PLECOC (Zhang et al., 2017), PLKNN (Hüllermeier and Beringer,
2006), and CC (Feng et al., 2020a). In addition, CL classification methods were also
included in the comparison as baselines: *gradient ascent* (GA) (see chapter 11); *non-negative correction* (NN) (see chapter 11); Free (eqn. 9.8) with the CE loss and the PC
loss, Forward (Yu et al., 2018); the MCL classification method (section 9.6.3) with bounded
losses (MAE, MSE, GCE, Phuber-CE) and surrogate losses (EXP and LOG). All hyper-parameters were specified or searched according to the suggested parameter settings in each
paper.

In the RC method (and also the CC method), hyper-parameters to be optimized are the
learning rate and weight decay (Feng et al., 2020a). We chose them so that the accuracy is
maximized on PL data in a validation set (10 percent of the training set). We implemented
them using PyTorch (Paszke et al., 2019), employing the Adam (Kingma and Ba, 2015)
optimizer with the mini-batch size set to 256 and the number of epochs set to 250. For all
the parametric methods, the same base model was adopted for fair comparisons.

Table 10.1 reports the test performance on the benchmark datasets for neural network
models. From the results, we can observe that RC always achieves the best performance
and significantly outperforms other compared methods in most cases. Table 10.2 reports
the test performance on the real-world PL datasets for linear models, showing that RC
generally achieves superior performance against other compared methods.

Table 10.2
Test performance (mean±std over 10 runs) on real-world PL datasets for linear models. The paired t-test at 5 percent significance level was conducted, and •/∘ represents whether the *best* of RC and CC is significantly better/worse than other methods. The best results are highlighted in bold.

	Lost	MSRCv2	BirdSong	Soccer Player	Yahoo! News
RC	**79.43±3.26%**	46.56±2.71%	71.94±1.72%	**57.00±0.97%**	**68.23±0.83%**
CC	79.29±3.19%	47.22±3.02%	**72.22±1.71%**	56.32±0.64%	68.14±0.81%
SURE	71.33±3.57%•	46.88±4.67%	58.92±1.28%•	49.41±086%•	45.49±1.15%•
CLPL	74.87±4.30%•	36.53±4.59%•	63.56±1.40%•	36.82±1.04%•	46.21±0.90%•
PLECOC	49.03±8.36%•	41.53±3.25%•	71.58±1.81%	53.70±2.02%•	66.22±1.01%•
PLSVM	75.31±3.81%•	35.85±4.41%•	49.90±2.07%•	46.29±0.96%•	56.85±0.91%•
PLKNN	36.73±2.99%•	41.36±2.89%•	64.94±1.42%•	49.62±0.67%•	41.07±1.02%•
IPAL	72.12±4.48%•	**50.80±4.46%∘**	72.06±1.55%	55.03±0.77%•	66.79±1.22%•

10.5   Proper Partial-Label (PPL) Classification

Next, we introduce an extension of PL classification called *PPL classification* (Wu and Sugiyama, 2021).

10.5.1   Data Generation Assumption

In section 10.2, we assumed

$$\widetilde{p}(Y|\boldsymbol{x},y) = \begin{cases} \frac{1}{2^{c-1}-1} & (y \in Y), \\ 0 & (y \notin Y). \end{cases} \tag{10.5}$$

Here we generalize it as

$$\widetilde{p}(Y|\boldsymbol{x},y) = C(\boldsymbol{x},Y)\mathbf{1}_{\{y \in Y\}} = \begin{cases} C(\boldsymbol{x},Y) & (y \in Y), \\ 0 & (y \notin Y), \end{cases} \tag{10.6}$$

where C is a function of $\boldsymbol{x}$ and Y. Equation (10.6) means that the conditional probability $\widetilde{p}(Y|\boldsymbol{x},y)$ depends only on $\boldsymbol{x}$ and Y if $y \in Y$, and it is zero otherwise. PL classification that satisfies the above assumption is called *proper* (Wu and Sugiyama, 2021).

This PPL assumption accommodates various settings as special cases (see table 10.3). For example, the *conditional multinomial mixture* (CMM) model (Liu and Dietterich, 2012) adopted a stronger assumption:

$$\widetilde{p}(Y|\boldsymbol{x},y) = \widetilde{p}(Y|y) = C(Y)\mathbf{1}_{\{y \in Y\}},$$

where C only depends on Y. The PL assumption (10.5) is even stronger as

$$\widetilde{p}(Y|\boldsymbol{x},y) = \frac{1}{2^{c-1}-1}\mathbf{1}_{\{y \in Y\}},$$

Table 10.3
Special cases of PPL classification. CMM denotes the *conditional multinomial mixture model* (Liu and Dietterich, 2012), CL denotes single CL classification (section 9.2), MCL denotes MCL classification (section 9.6), and PL denotes PL classification (section 10.2) The expression of $C(x, Y)$ for PPL comes from theorem 10.3.

	CMM	CL	MCL	PL	PPL
$C(x, Y)$	$C(Y)$	$\frac{1}{c-1}$	$\frac{p_{s}(c-|Y|)}{\binom{c-1}{|Y|-1}}$	$\frac{1}{2^{c-1}-1}$	$\frac{p(Y|x)}{\sum_{y' \in Y} p(y'|x)}$

where C is a constant. Furthermore, single CL classification and MCL classification explored in chapter 9 are also accommodated in this PPL assumption as shown below.

Recall the assumption (9.3) in single CL classification (section 9.2):

$$\bar{p}(x, \bar{y}) = \frac{1}{c-1} \sum_{y \neq \bar{y}} p(x, y).$$

If we treat $Y = \{1, \ldots, c\} \setminus \{\bar{y}\}$ as a PL, we have

$$\bar{p}(x, \bar{y}) = \tilde{p}(x, Y) = \sum_{y=1}^{c} \tilde{p}(Y|x, y) p(x, y).$$

Therefore, the implicit assumption here is

$$\tilde{p}(Y|x, y) = \frac{1}{c-1} \mathbf{1}_{\{y \in Y\}}.$$

Similarly, recall the assumption (9.14) in MCL classification (section 9.6):

$$\bar{p}(x, \bar{Y}) = \frac{p_{s}(|\bar{Y}|)}{\binom{c-1}{|\bar{Y}|}} \sum_{y \notin \bar{Y}} p(x, y),$$

where $p_{s}(s)$ is the probability distribution that the set size $s = |\bar{Y}|$ follows. If we treat $Y = \{1, \ldots, c\} \setminus \bar{Y}$ as a PL, we have

$$\bar{p}(x, \bar{Y}) = \tilde{p}(x, Y) = \sum_{y=1}^{c} \tilde{p}(Y|x, y) p(x, y).$$

Therefore, the implicit assumption here is

$$\tilde{p}(Y|x, y) = \frac{p_{s}(c - |Y|)}{\binom{c-1}{|Y|-1}} \mathbf{1}_{\{y \in Y\}},$$

where $|\bar{Y}| = c - |Y|$ and $\binom{a}{b} = \binom{a}{a-b}$ were used.

10.5.2　Risk Estimation

To derive an unbiased risk estimator for PPL classification, we first characterize the PPL assumption (10.6) with the following theorem.

Theorem 10.3　*(Wu and Sugiyama, 2021) The PPL assumption* (10.6) *holds if and only if*

$$p(y|\boldsymbol{x}, Y) = \frac{p(y|\boldsymbol{x})}{\sum_{y' \in Y} p(y'|\boldsymbol{x})} \mathbf{1}_{\{y \in Y\}}. \tag{10.7}$$

Furthermore, in this case, $C(\boldsymbol{x}, Y)$ *in* (10.6) *is given by*

$$C(\boldsymbol{x}, Y) = \frac{p(Y|\boldsymbol{x})}{\sum_{y' \in Y} p(y'|\boldsymbol{x})}. \tag{10.8}$$

Proof　If (10.6) holds, i.e., $p(Y|\boldsymbol{x}, y) = C(\boldsymbol{x}, Y)\mathbf{1}_{\{y \in Y\}}$, then we have

$$p(y|\boldsymbol{x}, Y) = \frac{p(y, Y|\boldsymbol{x})}{p(Y|\boldsymbol{x})} = \frac{p(y, Y|\boldsymbol{x})}{\sum_{y' \in Y} p(y', Y|\boldsymbol{x})} = \frac{p(Y|\boldsymbol{x}, y)p(y|\boldsymbol{x})}{\sum_{y' \in Y} p(Y|\boldsymbol{x}, y')p(y'|\boldsymbol{x})}$$

$$= \frac{C(\boldsymbol{x}, Y)\mathbf{1}_{\{y \in Y\}}p(y|\boldsymbol{x})}{\sum_{y' \in Y} C(\boldsymbol{x}, Y)\mathbf{1}_{\{y' \in Y\}}p(y'|\boldsymbol{x})} = \frac{p(y|\boldsymbol{x})}{\sum_{y' \in Y} p(y'|\boldsymbol{x})}\mathbf{1}_{\{y \in Y\}},$$

which agrees with (10.7). Conversely, if (10.7) holds, then we have

$$p(Y|\boldsymbol{x}, y) = \frac{p(y, Y|\boldsymbol{x})}{p(y|\boldsymbol{x})} = \frac{p(y|\boldsymbol{x}, Y)p(Y|\boldsymbol{x})}{p(y|\boldsymbol{x})} = \frac{p(Y|\boldsymbol{x})}{\sum_{y' \in Y} p(y'|\boldsymbol{x})}\mathbf{1}_{\{y \in Y\}},$$

which is (10.6) with (10.8). $\qquad\square$

At a glance, the PPL assumption (10.6) allows us to choose $C(\boldsymbol{x}, Y)$ arbitrarily. However, the above theorem shows that $C(\boldsymbol{x}, Y)$ is actually always given by (10.8) (see table 10.3) and $p(y|\boldsymbol{x}, Y)$ does not depend on $C(\boldsymbol{x}, Y)$.

Based on the above theorem, the classification risk can be written as follows:

Theorem 10.4　*(Wu and Sugiyama, 2021) The classification risk* (10.1) *can be equivalently expressed as*

$$R_{\mathrm{PPL}}(\boldsymbol{g}) = \mathbb{E}_{\widetilde{p}(\boldsymbol{x}, Y)}\left[\sum_{y \in Y} \frac{p(y|\boldsymbol{x})}{\sum_{y' \in Y} p(y'|\boldsymbol{x})} \mathcal{L}(\boldsymbol{g}(\boldsymbol{x}), y) \right].$$

Proof　$R(\boldsymbol{g}) = \mathbb{E}_{p(\boldsymbol{x}, y)}\left[\mathcal{L}(\boldsymbol{g}(\boldsymbol{x}), y) \right] = \int \sum_{y=1}^{c} \mathcal{L}(\boldsymbol{g}(\boldsymbol{x}), y)p(\boldsymbol{x}, y)\mathrm{d}\boldsymbol{x}$

$$= \int \sum_{y=1}^{c} \sum_{Y \in \mathscr{C}} \mathscr{L}(g(x), y) \widetilde{p}(x, y, Y) \mathrm{d}x$$

$$= \int \sum_{y=1}^{c} \sum_{Y \in \mathscr{C}} \mathscr{L}(g(x), y) p(y|x, Y) \widetilde{p}(x, Y) \mathrm{d}x$$

$$= \mathbb{E}_{\widetilde{p}(x,Y)} \left[\sum_{y=1}^{c} p(y|x, Y) \mathscr{L}(g(x), y) \right].$$

Substituting (10.7) into the above equation concludes the proof. $\qquad\square$

The above theorem shows that the classification risk under the PPL assumption does not depend on $C(x, Y)$, surprisingly. Then, from samples $\{(x_i, Y_i)\}_{i=1}^{n}$ independently drawn from $\widetilde{p}(x, Y)$, an unbiased risk estimator can be immediately obtained as

$$\widehat{R}_{\mathrm{PPL}}(g) = \frac{1}{n} \sum_{i=1}^{n} \sum_{y \in Y_i} \frac{p(y|x_i)}{\sum_{y' \in Y_i} p(y'|x_i)} \mathscr{L}(g(x_i), y),$$

which is exactly the same as $\widehat{R}_{\mathrm{PL}}$ given by (10.4).

10.5.3   Theoretical Analysis

Here, we establish an estimation error bound for PPL classification.

Let $\widehat{g}_{\mathrm{PPL}} := \min_{g \in \mathscr{G}} \widehat{R}_{\mathrm{PPL}}(g)$ be the empirical risk minimizer and $g^{\star} := \min_{g \in \mathscr{G}} R(g)$ be the true risk minimizer. We first introduce the following lemma.

Lemma 10.5   *The following inequality holds:*

$$R(\widehat{g}_{\mathrm{PPL}}) - R(g^{\star}) \leq 2 \sup_{g \in \mathscr{G}} |\widehat{R}_{\mathrm{PPL}}(g) - R(g)|.$$

Proof   As we saw in theorem 3.6,

$$R(\widehat{g}_{\mathrm{PPL}}) - R(g^{\star})$$

$$\leq \left(R(\widehat{g}_{\mathrm{PPL}}) - \widehat{R}_{\mathrm{PPL}}(\widehat{g}_{\mathrm{PPL}})\right) + \left(\widehat{R}_{\mathrm{PPL}}(\widehat{g}_{\mathrm{PPL}}) - \widehat{R}_{\mathrm{PPL}}(g^{\star})\right) + \left(\widehat{R}_{\mathrm{PPL}}(g^{\star}) - R(g^{\star})\right)$$

$$\leq \left(R(\widehat{g}_{\mathrm{PPL}}) - \widehat{R}_{\mathrm{PPL}}(\widehat{g}_{\mathrm{PPL}})\right) + \left(\widehat{R}_{\mathrm{PPL}}(g^{\star}) - R(g^{\star})\right)$$

$$\leq 2 \sup_{g \in \mathscr{G}} |\widehat{R}_{\mathrm{PPL}}(g) - R(g)|,$$

which completes the proof. $\qquad\square$

Next, we define a function space for PPL classification as

$$
\mathscr{F}_{\text{PPL}} := \left\{ (\boldsymbol{x}, Y) \mapsto \sum_{y \in Y} \frac{p(y|\boldsymbol{x})}{\sum_{y' \in Y} p(y'|\boldsymbol{x})} \mathscr{L}(g(\boldsymbol{x}), y) \,\middle|\, \boldsymbol{g} \in \mathscr{G} \right\},
$$

where $(\boldsymbol{x}, Y)$ is randomly sampled from $\widetilde{p}(\boldsymbol{x}, Y)$. Let $\widetilde{\mathfrak{R}}_n(\mathscr{F}_{\text{PPL}})$ be the expected Rademacher complexity of $\mathscr{F}_{\text{PPL}}$, i.e.,

$$
\widetilde{\mathfrak{R}}_n(\mathscr{F}_{\text{PPL}}) := \mathbb{E}_{\widetilde{p}(\boldsymbol{x},Y)} \mathbb{E}_{\sigma} \left[\sup_{f \in \mathscr{F}_{\text{PPL}}} \frac{1}{n} \sum_{i=1}^{n} \sigma_i g(\boldsymbol{x}_i, Y_i) \right].
$$

Then we have the following lemma.

Lemma 10.6　*Suppose the loss function $\mathscr{L}$ is bounded by M, i.e., $M = \sup_{\boldsymbol{x} \in \mathscr{X}, \boldsymbol{g} \in \mathscr{G}, y \in \mathscr{Y}} \mathscr{L}(g(\boldsymbol{x}), y)$. Then, for any $\delta > 0$, with probability at least $1 - \delta$,*

$$
\sup_{\boldsymbol{g} \in \mathscr{G}} \left| R_{\text{PPL}}(\boldsymbol{g}) - \widehat{R}_{\text{PPL}}(\boldsymbol{g}) \right| \leq 2 \widetilde{\mathfrak{R}}_n(\mathscr{F}_{\text{PPL}}) + M \sqrt{\frac{\log \frac{2}{\delta}}{2n}}.
$$

Proof　To prove this lemma, we first show that the one direction $\sup_{\boldsymbol{g} \in \mathscr{G}} R_{\text{PPL}}(\boldsymbol{g}) - \widehat{R}_{\text{PPL}}(\boldsymbol{g})$ is bounded with probability at least $1 - \delta/2$, and the other direction can be similarly shown. Suppose that a sample $(\boldsymbol{x}_i, Y_i)$ is replaced by another arbitrary sample $(\boldsymbol{x}'_i, Y'_i)$. Then, the change of $\sup_{\boldsymbol{g} \in \mathscr{G}} R_{\text{PPL}}(\boldsymbol{g}) - \widehat{R}_{\text{PPL}}(\boldsymbol{g})$ is no greater than M/n, since $\mathscr{L}$ is bounded by M. By applying McDiarmid's inequality (McDiarmid, 1989), for any $\delta > 0$, with probability at least $1 - \delta/2$,

$$
\sup_{\boldsymbol{g} \in \mathscr{G}} R_{\text{PPL}}(\boldsymbol{g}) - \widehat{R}_{\text{PPL}}(\boldsymbol{g}) \leq \mathbb{E} \left[\sup_{\boldsymbol{g} \in \mathscr{G}} R_{\text{PPL}}(\boldsymbol{g}) - \widehat{R}_{\text{PPL}}(\boldsymbol{g}) \right] + M \sqrt{\frac{\log \frac{2}{\delta}}{2n}}.
$$

Using the same trick as that used in Mohri et al. (2012), we can obtain

$$
\mathbb{E} \left[\sup_{\boldsymbol{g} \in \mathscr{G}} R_{\text{PPL}}(\boldsymbol{g}) - \widehat{R}_{\text{PPL}}(\boldsymbol{g}) \right] \leq 2 \widetilde{\mathfrak{R}}_n(\mathscr{F}_{\text{PPL}}).
$$

By further taking into account the other side $\sup_{\boldsymbol{g} \in \mathscr{G}} \widehat{R}_{\text{PPL}}(\boldsymbol{g}) - R_{\text{PPL}}(\boldsymbol{g})$, we have for any $\delta > 0$, with probability at least $1 - \delta$,

$$
\sup_{\boldsymbol{g} \in \mathscr{G}} \left| R_{\text{PPL}}(\boldsymbol{g}) - \widehat{R}_{\text{PPL}}(\boldsymbol{g}) \right| \leq 2 \widetilde{\mathfrak{R}}_n(\mathscr{F}_{\text{PPL}}) + M \sqrt{\frac{\log \frac{2}{\delta}}{2n}},
$$

which concludes the proof.　　　　　　　　　　　　　　　　　　　　　　　　　　　　□

Next, we will bound the expected Rademacher complexity of $\mathscr{F}_{\mathrm{PPL}}$, i.e., $\widetilde{\mathfrak{R}}_n(\mathscr{F}_{\mathrm{PPL}})$, by using the following lemma.

Lemma 10.7 *Assume that the loss function $\mathscr{L}(\boldsymbol{g}(\boldsymbol{x}), y)$ is ρ-Lipschitz with respect to $\boldsymbol{g}(\boldsymbol{x})$ $(0 < \rho < \infty)$ for all $y \in \mathscr{Y}$. Then, the following inequality holds:*

$$\widetilde{\mathfrak{R}}_n(\mathscr{G}_{\mathrm{PPL}}) \le \sqrt{2}\rho \sum_{y=1}^{c} \mathfrak{R}_n(\mathscr{H}_y),$$

where

$$\mathscr{H}_y := \{h : \boldsymbol{x} \mapsto g_y(\boldsymbol{x}) | \boldsymbol{g} \in \mathscr{G}\},$$

$$\mathfrak{R}_n(\mathscr{H}_y) := \mathbb{E}_{p(\boldsymbol{x})} \mathbb{E}_\sigma \left[\sup_{h \in \mathscr{H}_y} \frac{1}{n} \sum_{i=1}^{n} \sigma_i h(\boldsymbol{x}_i) \right].$$

Proof First of all, we introduce $p_y(\boldsymbol{x}) = \frac{p(y|\boldsymbol{x})}{\sum_{y' \in Y} p(y'|\boldsymbol{x})}$ for each sample $(\boldsymbol{x}, Y)$. Thus we have $0 \le p_y(\boldsymbol{x}) \le 1, \forall y \in Y$ and $\sum_{y \in Y} p_y(\boldsymbol{x}) = 1$. In this way, we can obtain $\widetilde{\mathfrak{R}}_n(\mathscr{F}_{\mathrm{PPL}}) \le \mathfrak{R}_n(\mathscr{L} \circ \mathscr{G})$ where $\mathscr{L} \circ \mathscr{G}$ denotes $\{\mathscr{L} \circ \boldsymbol{g} | \boldsymbol{g} \in \mathscr{G}\}$. Since $\mathscr{H}_y = \{h : \boldsymbol{x} \mapsto g_y(\boldsymbol{x}) | \boldsymbol{g} \in \mathscr{G}\}$ and the loss function $\mathscr{L}(\boldsymbol{g}(\boldsymbol{x}), y)$ is ρ-Lipschitz with respect to $\boldsymbol{g}(\boldsymbol{x})$ $(0 < \rho < \infty)$ for all $y \in \mathscr{Y}$, by the Rademacher vector contraction inequality (Maurer, 2016), we have $\mathfrak{R}_n(\mathscr{L} \circ \mathscr{G}) \le \sqrt{2}\rho \sum_{y=1}^{c} \mathfrak{R}_n(\mathscr{H}_y)$, which concludes the proof. $\qquad\square$

Finally, we derive the following theorem.

Theorem 10.8 *Assume that the loss function $\mathscr{L}(\boldsymbol{g}(\boldsymbol{x}), y)$ is ρ-Lipschitz with respect to $\boldsymbol{g}(\boldsymbol{x})$ $(0 < \rho < \infty)$ for all $y \in \mathscr{Y}$ and upper-bounded by M, i.e., $M = \sup_{\boldsymbol{x} \in \mathscr{X}, \boldsymbol{g} \in \mathscr{G}, y \in \mathscr{Y}} \mathscr{L}(\boldsymbol{g}(\boldsymbol{x}), y)$. Then, for any $\delta > 0$, with probability at least $1 - \delta$,*

$$R(\widehat{\boldsymbol{g}}_{\mathrm{PPL}}) - R(\boldsymbol{g}^\star) \le 4\sqrt{2}\rho \sum_{y=1}^{c} \mathfrak{R}_n(\mathscr{H}_y) + 2M\sqrt{\frac{\log \frac{2}{\delta}}{2n}}.$$

Proof Combining lemma 10.5, lemma 10.6, and lemma 10.7 will conclude the proof. $\quad\square$

Generally, $\mathfrak{R}_n(\mathscr{H}_y)$ can be bounded by $C_{\mathscr{H}}/\sqrt{n}$ for a positive constant $C_{\mathscr{H}}$ (Golowich et al., 2018). Therefore, theorem 10.8 establishes the consistency, i.e., the empirical risk minimizer $\widehat{\boldsymbol{g}}_{\mathrm{PPL}}$ converges to the true risk minimizer $\boldsymbol{g}^\star$ as $n \to \infty$.

IV ADVANCED TOPICS AND PERSPECTIVES

11 Non-Negative Correction for Weakly Supervised Classification

So far, we have shown weakly supervised classification methods for various scenarios based on *unbiased* risk estimators. In this chapter, we discuss how their classification performance can be further improved by the technique called *non-negative correction*.

11.1 Introduction

In part II and part III, we have considered various weakly supervised classification settings in binary and multi-class classification. Our systematic "recipe" to solve these problems was *risk-rewrite* empirical risk minimization (ERM), which consists of the following steps:

- First, the classification risk is rewritten into an equivalent expression, such that it only involves the expectations of available weakly supervised data.

- Then, an empirical approximation is applied to the rewritten risk, resulting in a learning objective that is *unbiased*.

- Finally, the obtained learning objective is minimized by any off-the-shelf optimizers, such as stochastic optimization algorithms (Robbins and Munro, 1951; Kingma and Ba, 2015).

A notable advantage of the above ERM-based approaches is their high flexibility. Indeed, they are compatible with *any* losses, models, and optimizers. Also, the risk estimators can be used for *hyper-parameter tuning* only from weakly supervised data, which is practically a highly useful property since fully supervised validation data are not usually available in weakly supervised classification. Furthermore, thanks to the unbiasedness of the obtained risk estimators, the consistency of the learned classifier can be systematically guaranteed through the *Rademacher complexity* analysis.

However, unbiasedness does not always mean a good estimator, since an unbiased estimator can have a large variance; then the obtained empirical risk minimizer can be unreliable. Indeed, if the model being trained is very flexible (such as a deep neural network), the above unbiased learning scheme can produce *negative empirical risks*, resulting in severe overfitting (Kiryo et al., 2017; Ishida et al., 2019; Lu et al., 2020). This chapter is devoted to explaining why this phenomenon occurs and introducing techniques for mitigating the overfitting.[1]

11.2　Overfitting of Unbiased Learning Objectives

In this section, we discuss how overfitting of the unbiased learning objectives occurs.

11.2.1　Binary Classification

In section 3.1.1, the classification risk for PN classification was decomposed as

$$R(g) = \pi_P \, \mathbb{E}_P[\ell(g(x), +1)] + \pi_N \, \mathbb{E}_N[\ell(g(x), -1)]. \tag{11.1}$$

On the other hand, in *unbiased* PU (uPU) classification shown in section 4.3.1, the classification risk was decomposed as

$$R(g) = \pi_P \, \mathbb{E}_P[\ell(g(x), +1) - \ell(g(x), -1)] + \mathbb{E}_U[\ell(g(x), -1)]. \tag{11.2}$$

Thus, in uPU classification, the risk for negative data, $\pi_N \, \mathbb{E}_N[\ell(g(x), -1)]$, was replaced with its equivalent expression:

$$\mathbb{E}_U[\ell(g(x), -1)] - \pi_P \, \mathbb{E}_P[\ell(g(x), -1)]. \tag{11.3}$$

Since the risk $\pi_N \, \mathbb{E}_N[\ell(g(x), -1)]$ is non-negative given that loss ℓ is non-negative, (11.3) is also non-negative. However, if (11.3) is approximated by samples as

$$\frac{1}{n_U} \sum_{i=1}^{n_U} \ell(g(x_i^U), -1) - \frac{\pi_P}{n_P} \sum_{i=1}^{n_P} \ell(g(x_i^P), -1), \tag{11.4}$$

non-negativity is no longer guaranteed. Indeed, for complex models such as neural networks, the above empirical approximation tends to yield negative values (Kiryo et al., 2017). Moreover, if the loss function is not upper-bounded, the empirical risk is not lower-bounded, i.e., it may diverge to $-\infty$.

Let us illustrate this phenomenon using a toy example under zero-one loss $\ell_{0\text{-}1}(m)$:

$$\ell_{0\text{-}1}(m) = \begin{cases} 1 & (m > 0), \\ 0 & (m \le 0). \end{cases}$$

Figure 11.1 shows two classifiers (plotted by blue lines) where all positive samples (red circles) are classified correctly and thus $\ell_{0\text{-}1}(g(x_i^P), -1) = 1$ holds for all $i = 1, \ldots, n_P$. Figure 11.1a illustrates a favorable classifier that misclassifies a certain number of unlabeled samples (black squares) corresponding to the positive samples. In this case, $\ell_{0\text{-}1}(g(x_i^U), -1) = 1$ occurs for some samples (approximately fraction π_P) and thus empirical risk estimate (11.4) is close to zero.

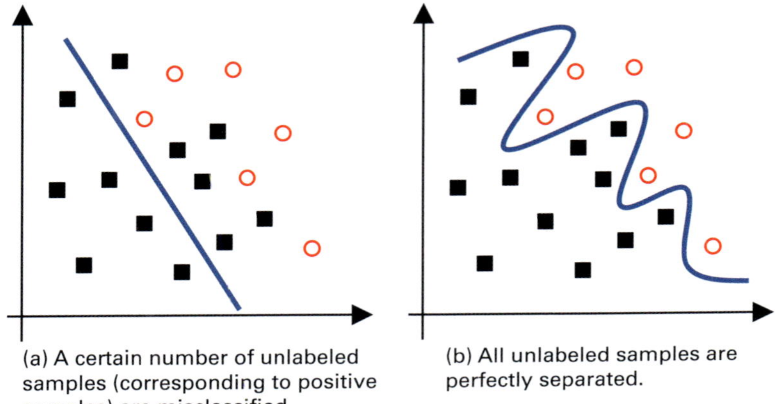

(a) A certain number of unlabeled
samples (corresponding to positive
samples) are misclassified.

(b) All unlabeled samples are
perfectly separated.

Figure 11.1
Illustration of overfitting in PU classification, where the empirical risk goes negative. Red circles are positive
samples, black squares are unlabeled samples, and blue lines are classifiers.

On the other hand, Figure 11.1b shows an overfitted classifier that perfectly sepa-
rates all positive samples and unlabeled samples. In this case, $\ell_{0\text{-}1}(g(x_i^{\mathrm{U}}), -1) = 0$ and
$\ell_{0\text{-}1}(g(x_i^{\mathrm{P}}), -1) = 1$ hold for all samples, and thus empirical risk estimate (11.4) yields
$-\pi_{\mathrm{P}}$. This illustrative example indicates that, to avoid overfitting, it is essential to keep
empirical risk estimate (11.4) non-negative.

Such an overfitting problem also occurs in PNU classification shown in section 5.5,
whose empirical risk approximator is a combination of the PU and PN risk approxi-
mators.

In a similar fashion to uPU classification, *unbiased* UU (uUU) classification shown in
section 8.3 also suffers from overfitting. In uUU classification, the classification risk was
decomposed as

$$R(g) = C_{\mathrm{a}}\mathbb{E}_{p_{\mathrm{tr}}}[\ell(g(x), +1)] - C_{\mathrm{b}}\mathbb{E}_{p_{\mathrm{tr}}}[\ell(g(x), -1)]$$
$$- C_{\mathrm{c}}\mathbb{E}_{p'_{\mathrm{tr}}}[\ell(g(x'), +1)] + C_{\mathrm{d}}\mathbb{E}_{p'_{\mathrm{tr}}}[\ell(g(x'), -1)], \tag{11.5}$$

where $C_{\mathrm{a}} = \frac{(1-\theta')\pi_{\mathrm{P}}}{\theta-\theta'}$, $C_{\mathrm{b}} = \frac{\theta'(1-\pi_{\mathrm{P}})}{\theta-\theta'}$, $C_{\mathrm{c}} = \frac{(1-\theta)\pi_{\mathrm{P}}}{\theta-\theta'}$, and $C_{\mathrm{d}} = \frac{\theta(1-\pi_{\mathrm{P}})}{\theta-\theta'}$. In (11.5), the risk for
positive data, $\pi_{\mathrm{P}}\,\mathbb{E}_{\mathrm{P}}[\ell(g(x), +1)]$, and the risk for negative data, $\pi_{\mathrm{N}}\,\mathbb{E}_{\mathrm{N}}[\ell(g(x), -1)]$, were
replaced with their equivalent expressions:

$$C_{\mathrm{a}}\mathbb{E}_{p_{\mathrm{tr}}}[\ell(g(x), +1)] - C_{\mathrm{c}}\mathbb{E}_{p'_{\mathrm{tr}}}[\ell(g(x'), +1)], \tag{11.6}$$

$$C_{\mathrm{d}}\mathbb{E}_{p'_{\mathrm{tr}}}[\ell(g(x'), -1)] - C_{\mathrm{b}}\mathbb{E}_{p_{\mathrm{tr}}}[\ell(g(x), -1)]. \tag{11.7}$$

If (11.6) and (11.7) are approximated by samples as

$$\frac{C_{\mathrm{a}}}{n}\sum_{i=1}^{n}\ell(g(x_i),+1) - \frac{C_{\mathrm{c}}}{n'}\sum_{j=1}^{n'}\ell(g(x'_j),+1), \tag{11.8}$$

$$\frac{C_{\mathrm{d}}}{n'}\sum_{j=1}^{n'}\ell(g(x'_j),-1) - \frac{C_{\mathrm{b}}}{n}\sum_{i=1}^{n}\ell(g(x_i),-1), \tag{11.9}$$

non-negativity is no longer guaranteed in the same way as uPU classification. For flexible models, such empirical approximations tend to yield negative values and lead to overfitted classifiers (Lu et al., 2020).

11.2.2   Multi-Class Classification

Similarly, in section 3.2, the classification risk for multi-class classification was decomposed as

$$R(g) = \sum_{y=1}^{c} \pi_y \mathbb{E}_{p(x|y)}\Big[\mathscr{L}(g(x),y)\Big]. \tag{11.10}$$

On the other hand, in *unbiased* complementary-label (uCL) classification shown in section 9.2, the classification risk was decomposed as

$$R(g) = \sum_{\bar{y}=1}^{c} \bar{\pi}_{\bar{y}} \mathbb{E}_{\bar{p}(x|\bar{y})}\Big[-(c-1)\cdot\mathscr{L}(g(x),\bar{y}) + \sum_{y=1}^{c}\mathscr{L}(g(x),y)\Big], \tag{11.11}$$

where $\bar{\pi}_{\bar{y}} = \bar{p}(\bar{y})$. Thus, in uCL classification, each partial risk $\pi_y \mathbb{E}_{p(x|y)}[\mathscr{L}(g(x),y)]$ in the original expression of the classification risk (11.10) was replaced with its equivalent expression:

$$-(c-1)\bar{\pi}_{\bar{y}}\cdot\mathbb{E}_{\bar{p}(x|\bar{y})}\Big[\mathscr{L}(g(x),\bar{y})\Big] + \sum_{y=1}^{c}\bar{\pi}_{\bar{y}}\cdot\mathbb{E}_{\bar{p}(x|\bar{y})}\Big[\mathscr{L}(g(x),y)\Big]. \tag{11.12}$$

Since each partial risk $\pi_y \mathbb{E}_{p(x|y)}[\mathscr{L}(g(x),y)]$ is non-negative given that the loss $\mathscr{L}$ is non-negative, (11.12) is also non-negative. However, if (11.12) is approximated by samples as

$$-(c-1)\cdot\frac{\bar{\pi}_{\bar{y}}}{|\mathscr{X}_{\bar{y}}|}\sum_{x_i\in\mathscr{X}_{\bar{y}}}\mathscr{L}(g(x_i),\bar{y}) + \sum_{y=1}^{c}\frac{\bar{\pi}_{\bar{y}}}{|\mathscr{X}_{\bar{y}}|}\sum_{x_{i'}\in\mathscr{X}_{\bar{y}}}\mathscr{L}(g(x_{i'}),y), \tag{11.13}$$

non-negativity is no longer guaranteed due to the negative terms, which leads to overfitting.

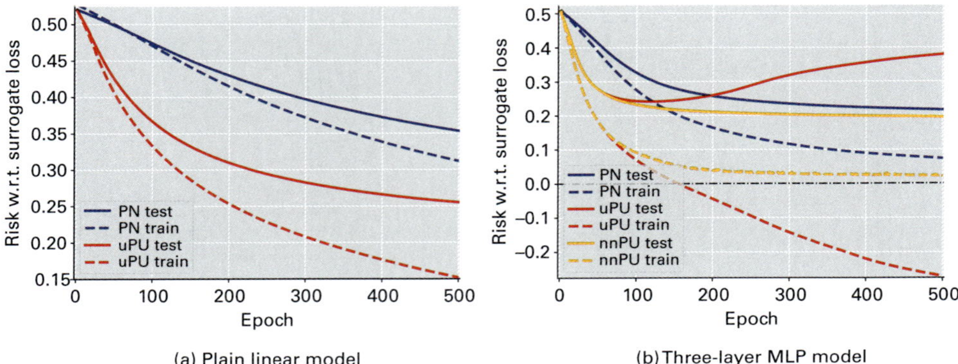

Figure 11.2
Illustrative experimental results for PN, uPU, and nnPU classification. The dataset is MNIST (LeCun et al., 1998); even/odd digits are regarded as the P/N class with $\pi_P \approx 1/2$; $n_P = 100$ and $n_N = 50$ for PN classification, and $n_P = 100$ and $n_U = 59{,}900$ for uPU and nnPU classification. In (a), the model is a plain linear model (784-1), and in (b), the model is a three-layer MLP (784-100-1) with rectified linear units (ReLUs) (Nair and Hinton, 2010). Solid curves denote test risks while dashed curves indicate empirical risks.

11.3   Numerical Illustration

Now, let us numerically illustrate the overfitting of uPU, uUU, and uCL classification.

Overfitting of uPU classification   First, as our analysis in section 11.2 suggested, we compare PN and uPU classification using a simple linear model and a more flexible three-layer *multi-layer perceptron* (MLP) model in figure 11.2.[2]

Since the linear model is so simple, overfitting is not observed in figure 11.2a—in both PN and uPU methods, the training and test risks both monotonically decrease. On the other hand, the three-layer MLP model is so flexible—the number of parameters is 500 times more than the total number of P and N data, which causes overfitting. Indeed, Figure 11.2b shows that the PN and uPU training risks both monotonically decrease. However the uPU training risk decreases faster and goes negative. The PN test risk also monotonically decreases, whereas the uPU risk does not. The uPU test risk is lower than the PN test risk at the beginning, but it becomes higher than the PN test risk at the end.

This illustrative experiment demonstrates that the overfitting problem of uPU classification is not negligible; it has a strong impact on the final classification performance.

Overfitting of uUU classification   Next, in figure 11.3, we show how uUU classification behaves on different models, optimizers, and loss functions. From the results, we observe a strong co-occurrence[3] between severe overfitting and negative uUU training risks regardless of datasets, models, optimizers, and loss functions. Specifically, for all the experiments with the MNIST dataset (LeCun et al., 1998) using the linear model, since the model is

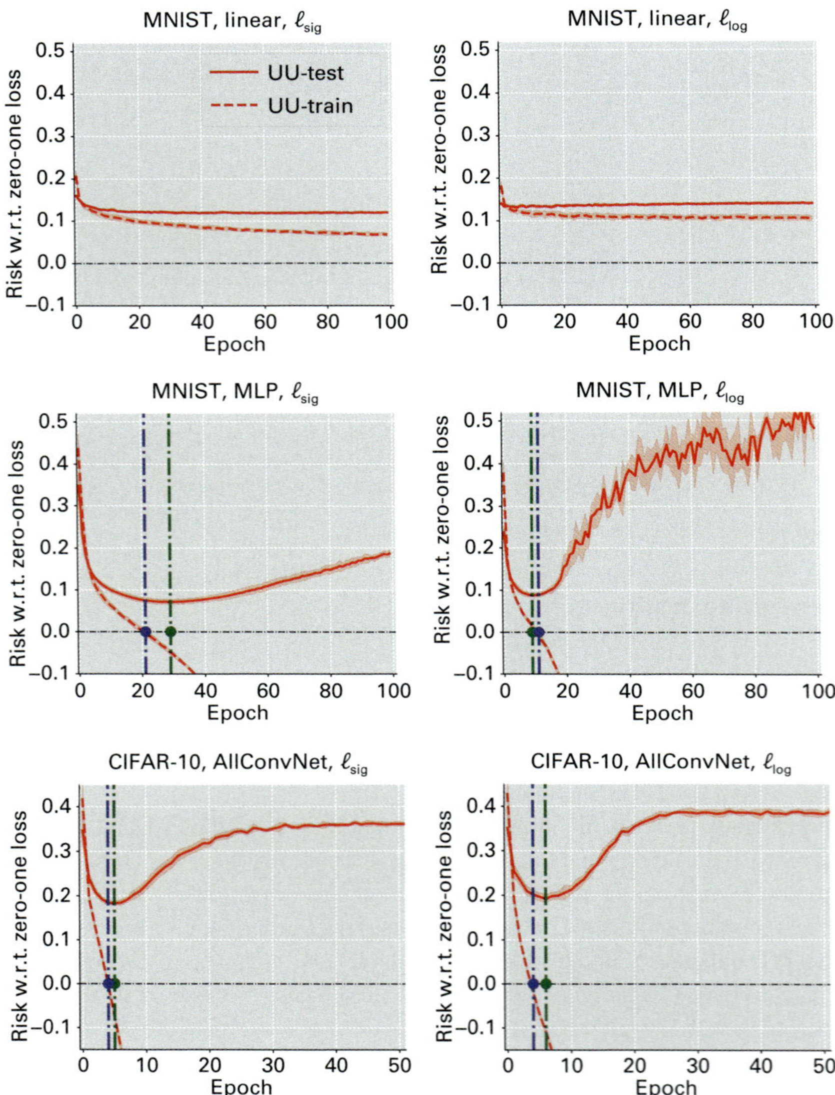

Figure 11.3
Illustrative experimental results for uUU classification. In the top half, the dataset is MNIST (even vs. odd); a linear-in-input model (linear) and a five-layer MLP (d-300-300-300-300-1) were trained by stochastic gradient descent (SGD) using the sigmoid (ℓ_{sig}) and logistic (ℓ_{log}) losses. In the bottom half, the dataset is CIFAR-10 (transportation vs. animal); the *all convolutional net* (AllConvNet) and the 32-layer *residual network* (ResNet-32) were trained by Adam (Kingma and Ba, 2015) using the same losses. The class-priors θ and θ' were set to 0.6 and 0.4. The blue dashed lines indicate when the empirical risk computed from UU training data goes negative, while the green dashed lines indicate when the test error turns around and severe overfitting begins.

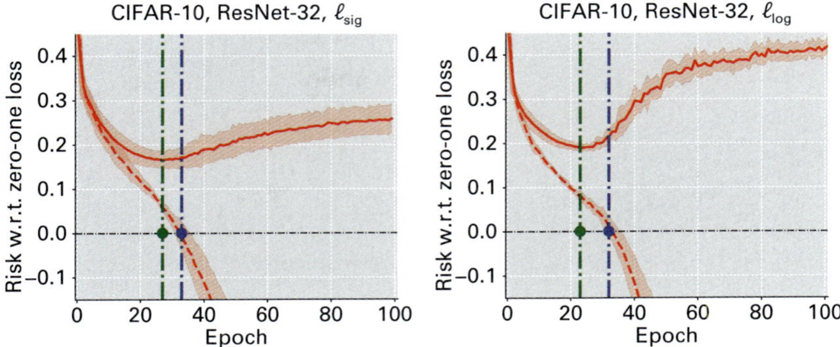

Figure 11.3
(continued)

so simple, overfitting does not happen; for all the experiments with the MNIST dataset and the CIFAR-10 dataset (Krizhevsky, 2009) using flexible deep neural network models, overfitting is observable when the uUU training risk goes negative, regardless of optimizers, models, and loss functions.

Overfitting of uCL classification   Finally, we illustrate how uCL classification behaves on different models in figure 11.4. From the results, we can see that when the linear model is used (top-left graph), the uCL training risk continues decreasing and can go below zero at around epoch 100. The test accuracy hits the peak also at around epoch 100 and then gradually deteriorates. This issue stands out even more significantly when a more flexible three-layer MLP model (top-right graph) is used: The uCL training risk decreases much more quickly and goes negative. Correspondingly, the test accuracy drops significantly after the empirical risk goes negative (bottom graph). To further analyze each partial risk (11.13), we also plot the decomposed risks with respect to each ordinary class for both the linear model and MLP model experiments. We confirmed that the decomposed risks for all classes become negative eventually.

11.4   Non-Negative Correction

As shown above, the overfitting of unbiased learning objectives can be empirically severe. In this section, we introduce the technique called *non-negative correction* to mitigate this overfitting.

11.4.1   nnPU Classification

For uPU classification, let us explicitly enforce its empirical risk estimate (11.4) to be non-negative as

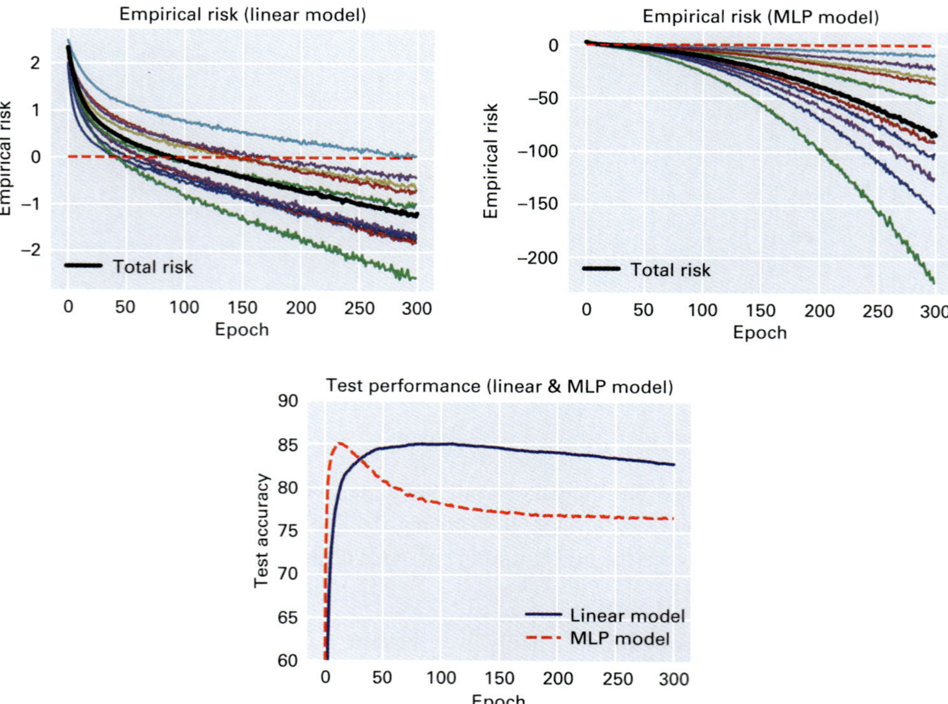

Figure 11.4
Illustrative experimental results for uCL classification. The dataset is MNIST; the model is a plain linear model (784-10) for the top-left graph and a three-layer MLP (784-500-1) with ReLUs for the top-right graph. The top graphs show the total empirical uCL risk in black color and the class-wise empirical risks in other colors. The bottom graph shows the corresponding test accuracy for both models.

$$\widehat{R}_{\mathrm{nnPU}}(g) = \frac{\pi_{\mathrm{P}}}{n_{\mathrm{P}}} \sum_{i=1}^{n_{\mathrm{P}}} \ell(g(\boldsymbol{x}_i^{\mathrm{P}}), +1)$$

$$+ \max\left(0, \frac{1}{n_{\mathrm{U}}} \sum_{i=1}^{n_{\mathrm{U}}} \ell(g(\boldsymbol{x}_i^{\mathrm{U}}), -1) - \frac{\pi_{\mathrm{P}}}{n_{\mathrm{P}}} \sum_{i=1}^{n_{\mathrm{P}}} \ell(g(\boldsymbol{x}_i^{\mathrm{P}}), -1)\right). \qquad (11.14)$$

We refer to the learning method that minimizes this empirical risk as *non-negative PU (nnPU) classification* (Kiryo et al., 2017).

For big data, stochastic optimization is helpful to scale up learning algorithms. In stochastic optimization, training data

$$\mathscr{X}_{\mathrm{P}} := \{\boldsymbol{x}_i^{\mathrm{P}}\}_{i=1}^{n_{\mathrm{P}}} \overset{\text{i.i.d.}}{\sim} p_{\mathrm{P}}(\boldsymbol{x}) := p(\boldsymbol{x}|y = +1),$$

$$\mathscr{X}_{\mathrm{U}} := \{\boldsymbol{x}_i^{\mathrm{U}}\}_{i=1}^{n_{\mathrm{U}}} \overset{\text{i.i.d.}}{\sim} p_{\mathrm{U}}(\boldsymbol{x}) := \pi_{\mathrm{P}} p_{\mathrm{P}}(\boldsymbol{x}) + \pi_{\mathrm{N}} p_{\mathrm{N}}(\boldsymbol{x}),$$

are split into N disjoint mini-batches $\mathcal{X}_P^i$ and $\mathcal{X}_U^i$ for $i = 1, \ldots, N$. Then, in each optimization step, instead of the entire samples $\mathcal{X}_P$ and $\mathcal{X}_U$, only the mini-batches $\mathcal{X}_P^i$ and $\mathcal{X}_U^i$ are used.

We can naively minimize the uPU empirical risk by stochastic optimization since it is *point-wise*. However, stochastically minimizing the nnPU empirical risk is not straightforward since it is not point-wise due to the max operator. Nevertheless, the max term in (11.14) can be upper-bounded as

$$
\max\left(0, \frac{1}{|\mathcal{X}_U|} \sum_{x^U \in \mathcal{X}_U} \ell(g(x^U), -1) - \frac{\pi_P}{|\mathcal{X}_P|} \sum_{x^P \in \mathcal{X}_P} \ell(g(x^P), -1)\right)
$$
$$
\leq \frac{1}{N} \sum_{i=1}^{N} \max\left(0, r_i\right),
$$

where

$$
r_i := \frac{1}{|\mathcal{X}_U^i|} \sum_{x^U \in \mathcal{X}_U^i} \ell(g(x^U), -1) - \frac{\pi_P}{|\mathcal{X}_P^i|} \sum_{x^P \in \mathcal{X}_P^i} \ell(g(x^P), -1).
$$

Since this upper bound is point-wise, it can be minimized by stochastic optimization.

If r_i is negative for a certain mini-batch, the max operator will prevent r_i from further decreasing, and thus overfitting is not worsened. However, the max operator cannot *mitigate* overfitting since negative r_i cannot be increased to be non-negative. From this perspective, it would be useful to increase r_i if it is negative. Thus, when r_i is positive, gradient descent is performed as usual, and when r_i is negative, gradient *ascent* may be performed.

Algorithm 1 describes a practical implementation of large-scale nnPU classification based on stochastic optimization with the gradient ascent trick. Here, a tolerance parameter β was introduced, which satisfies

$$
0 \leq \beta \leq \pi_P \sup_m \max_y \ell(m, y).
$$

With this β, we divide the cases not by $r_i < 0$ but by $r_i < -\beta$. If $\beta = 0$, there is no tolerance, and if $\beta = \pi_P \sup_m \max_y \ell(m, y)$, $r_i < -\beta$ never occurs and thus nnPU is reduced to uPU. In algorithm 1, we have also discounted the amount of gradient ascent with discount factor $0 \leq \gamma \leq 1$, i.e., when $r_i < -\beta$, we go along $-\nabla_\theta r_i$ with a step size discounted by γ.

We will come back to this discounted gradient ascent issue in section 11.4.5.

11.4.2   nnPNU Classification

Given nnPU classification, it is straightforward to develop a non-negative counterpart of *unbiased PNU classification* introduced in section 5.5. Specifically, we define the empirical risk estimator for *non-negative PNU (nnPNU) classification* as

---

Algorithm 1 Implementation of large-scale nnPU classification based on stochastic optimization with discounted gradient ascent.

Input: training data $(\mathscr{X}_{\mathrm{P}}, \mathscr{X}_{\mathrm{U}})$

hyper-parameters $0 \le \beta \le \pi_{\mathrm{P}} \sup_m \max_y \ell(m, y)$ and $0 \le \gamma \le 1$

Output: model parameter θ

1: Let $\mathscr{A}$ be an external stochastic gradient algorithm such as Kingma and Ba (2015) or Duchi et al. (2011)

2: **while** no stopping criterion has been met

3:     Split $(\mathscr{X}_{\mathrm{P}}, \mathscr{X}_{\mathrm{U}})$ into N disjoint mini-batches $\{(\mathscr{X}_{\mathrm{P}}^i, \mathscr{X}_{\mathrm{U}}^i)\}_{i=1}^N$

4:     **for** $i = 1$ **to** N

5:        $r_i = \dfrac{1}{|\mathscr{X}_{\mathrm{U}}^i|} \sum_{x^{\mathrm{U}} \in \mathscr{X}_{\mathrm{U}}^i} \ell(g(x^{\mathrm{U}}), -1) - \dfrac{\pi_{\mathrm{P}}}{|\mathscr{X}_{\mathrm{P}}^i|} \sum_{x^{\mathrm{P}} \in \mathscr{X}_{\mathrm{P}}^i} \ell(g(x^{\mathrm{P}}), -1)$

6:        **if** $r_i \ge -\beta$

7:           Set gradient $\nabla_\theta \left(\dfrac{1}{|\mathscr{X}_{\mathrm{P}}^i|} \sum_{x^{\mathrm{P}} \in \mathscr{X}_{\mathrm{P}}^i} \ell(g(x^{\mathrm{P}})) + r_i \right)$

8:           Update θ by $\mathscr{A}$ with its current step size η

9:        **else**

10:          Set gradient $-\nabla_\theta r_i$

11:          Update θ by $\mathscr{A}$ with a discounted step size $\gamma \eta$

$$\widehat{R}_{\mathrm{nnPNU}}^{\eta_{\mathrm{PNU}}}(g) := \begin{cases} \widehat{R}_{\mathrm{nnPU+PN}}^{\eta_{\mathrm{PNU}}}(g) & (\eta_{\mathrm{PNU}} \ge 0), \\ \widehat{R}_{\mathrm{nnNU+PN}}^{-\eta_{\mathrm{PNU}}}(g) & (\eta_{\mathrm{PNU}} < 0), \end{cases}$$

where $-1 \le \eta_{\mathrm{PNU}} \le 1$ is a hyper-parameter to control the trade-off between nnPU+PN classification and nnNU+PN classification, which are defined as

$$\widehat{R}_{\mathrm{nnPU+PN}}^{\gamma_{\mathrm{PNU}}}(g) := (1 - \gamma_{\mathrm{PNU}})\widehat{R}_{\mathrm{PN}}(g) + \gamma_{\mathrm{PNU}}\widehat{R}_{\mathrm{nnPU}}(g),$$

$$\widehat{R}_{\mathrm{nnNU+PN}}^{\gamma_{\mathrm{PNU}}}(g) := (1 - \gamma_{\mathrm{PNU}})\widehat{R}_{\mathrm{PN}}(g) + \gamma_{\mathrm{PNU}}\widehat{R}_{\mathrm{nnNU}}(g).$$

11.4.3   nnUU Classification

Following the same idea, we can also impose non-negativity on uUU classification to obtain *non-negative UU (nnUU) classification* (Lu et al., 2020). Specifically, empirical estimates (11.8) and (11.9) are required to be non-negative as

$$\widehat{R}_{\mathrm{nnUU}}(g) = \max\left\{ 0, \frac{C_{\mathrm{a}}}{n} \sum_{i=1}^{n} \ell(g(x_i), +1) - \frac{C_{\mathrm{c}}}{n'} \sum_{j=1}^{n'} \ell(g(x'_j), +1) \right\}$$

$$+ \max\left\{0, \frac{C_{\mathrm{d}}}{n'}\sum_{j=1}^{n'}\ell(g(x_j'), -1) - \frac{C_{\mathrm{b}}}{n}\sum_{i=1}^{n}\ell(g(x_i), -1)\right\}. \qquad (11.15)$$

11.4.4 nnCL Classification

The same non-negative technique can be applied to multi-class cases. Indeed, for uCL classification, we can similarly enforce non-negativity on the empirical partial risk estimate (11.13) as

$$\widehat{R}_{\mathrm{nnCL}} = \sum_{\bar{y}=1}^{c} \max\left\{0, \left[-(c-1)\cdot\frac{\overline{\pi}_{\bar{y}}}{|\mathcal{X}_{\bar{y}}|}\sum_{x_i\in\mathcal{X}_{\bar{y}}}\mathcal{L}(g(x_i),\bar{y})\right.\right.$$

$$\left.\left.+\sum_{y=1}^{c}\frac{\overline{\pi}_{\bar{y}}}{|\mathcal{X}_{\bar{y}}|}\sum_{x_{i'}\in\mathcal{X}_{\bar{y}}}\mathcal{L}(g(x_{i'}),y)\right]\right\}. \qquad (11.16)$$

The learning method that minimizes this empirical risk is called *non-negative CL (nnCL) classification* (Ishida et al., 2019).

11.4.5 ccUU Classification

We have used the max operator not to worsen overfitting so far, and we have also introduced a gradient ascent trick as a heuristic to mitigate overfitting in section 11.4.1. Here, we show that the gradient ascent trick can be derived naturally from an extended max operator called the *consistent correction (CC) function* (Lu et al., 2020).

Let us define the CC function as follows.

Definition 11.1 (Consistent correction function) *A function $f : \mathbf{R} \to \mathbf{R}$ is called a CC function if it is Lipschitz continuous, non-negative, and $f(x) = x$ for all $x \geq 0$. Let $\mathscr{F}$ be a class of all consistent correction functions.*

The following functions are members of the CC function class $\mathscr{F}$ (see figure 11.5):

- The rectified linear unit (ReLU) function: $f(r) = \max(0, r)$;

- The absolute value function: $f(r) = |r|$;

- The generalized leaky ReLU function: $f_\gamma(r) = \begin{cases} r & (r \geq 0), \\ -\gamma r & (r < 0), \end{cases}$ for $\gamma \geq 0$.

Based on the notion of CC functions, non-negative risk estimators can be naturally extended to consistently corrected risk estimators. A notable advantage of such consistently corrected risk estimators with the generalized leaky ReLU function is that it naturally yields

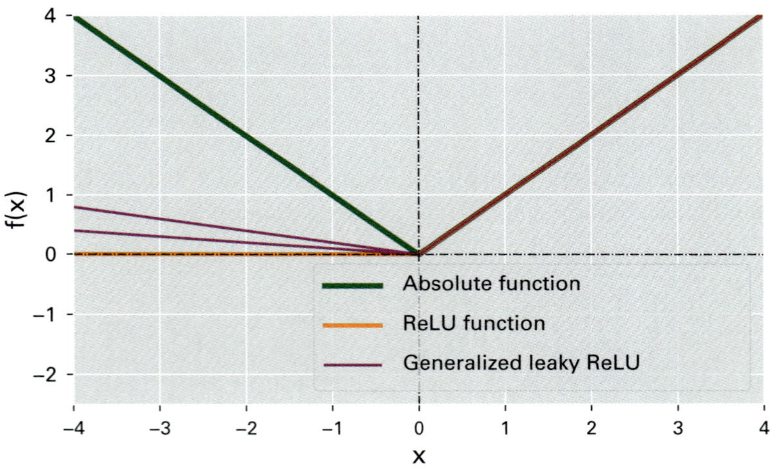

Figure 11.5
Examples of consistent correction functions.

the gradient ascent trick in section 11.4.1:

$$
\nabla_\theta f(r) = \begin{cases} \nabla_\theta r & (r \geq 0), \\ -\gamma \nabla_\theta r & (r < 0). \end{cases}
$$

As an example, a practical implementation of *consistently corrected UU (ccUU) classification* is described in algorithm 2, which uses the following risk estimator (Lu et al., 2020):

$$
\widehat{R}_{\text{ccUU}}(g) = f_1 \left(\frac{C_{\text{a}}}{n} \sum_{i=1}^{n} \ell(g(\boldsymbol{x}_i), +1) - \frac{C_{\text{c}}}{n'} \sum_{j=1}^{n'} \ell(g(\boldsymbol{x}'_j), +1) \right)
$$

$$
+ f_2 \left(\frac{C_{\text{d}}}{n'} \sum_{j=1}^{n'} \ell(g(\boldsymbol{x}'_j), -1) - \frac{C_{\text{b}}}{n} \sum_{i=1}^{n} \ell(g(\boldsymbol{x}_i), -1) \right), \tag{11.17}
$$

where f_1 and f_2 are CC functions. When $f_1(r) = f_2(r) = \max(0, r)$, ccUU classification is reduced to the original nnUU classification.

11.5   Theoretical Analyses

In this section, we theoretically analyze properties of the ccUU empirical risk and its minimizer. Due to the generality of non-negative correction functions used in ccUU classification, theoretical analysis of its special cases can be developed analogously.

Algorithm 2 Implementation of ccUU classification based on stochastic optimization.

Input: two sets of U training data $(\mathscr{X}_{\mathrm{tr}}, \mathscr{X}'_{\mathrm{tr}})$

Output: learned model parameter θ

1: Initialize θ

2: Let $\mathscr{A}$ be an SGD-like optimizer working on θ

3: **for** $t = 1$ **to** number_of_epochs:

4:     Shuffle $(\mathscr{X}_{\mathrm{tr}}, \mathscr{X}'_{\mathrm{tr}})$

5:     **for** $i = 1$ **to** number_of_mini-batches:

6:        Let $(\overline{\mathscr{X}}_{\mathrm{tr}}, \overline{\mathscr{X}}'_{\mathrm{tr}})$ be the current mini-batch

7:        Forward $\overline{\mathscr{X}}_{\mathrm{tr}}$ and $\overline{\mathscr{X}}'_{\mathrm{tr}}$

8:        Compute
$$L^+ = C_{\mathrm{a}}\ell(g(\overline{\mathscr{X}}_{\mathrm{tr}}), +1)/|\overline{\mathscr{X}}_{\mathrm{tr}}| - C_{\mathrm{c}}\ell(g(\overline{\mathscr{X}}'_{\mathrm{tr}}), +1)/|\overline{\mathscr{X}}'_{\mathrm{tr}}|$$
$$L^- = C_{\mathrm{d}}\ell(g(\overline{\mathscr{X}}'_{\mathrm{tr}}), -1)/|\overline{\mathscr{X}}'_{\mathrm{tr}}| - C_{\mathrm{b}}\ell(g(\overline{\mathscr{X}}_{\mathrm{tr}}), -1)/|\overline{\mathscr{X}}_{\mathrm{tr}}|$$

9:        Correct them by $L^+_{\mathrm{ccUU}} = f(L^+),\, L^-_{\mathrm{ccUU}} = f(L^-)$

10:       Backward $L_{\mathrm{ccUU}} = L^+_{\mathrm{ccUU}} + L^-_{\mathrm{ccUU}}$

11:       Update θ by $\mathscr{A}$

For convenience, let us express the PN, uUU, and ccUU empirical risks as

$$\widehat{R}_{\mathrm{PN}}(g) = \pi_{\mathrm{P}}\widehat{R}^+_{\mathrm{P}}(g) + \pi_{\mathrm{N}}\widehat{R}^-_{\mathrm{N}}(g),$$

$$\widehat{R}_{\mathrm{UU}}(g) = C_{\mathrm{a}}\widehat{R}^+_{\mathrm{U}}(g) - C_{\mathrm{b}}\widehat{R}^-_{\mathrm{U}}(g) - C_{\mathrm{c}}\widehat{R}^+_{\mathrm{U}'}(g) + C_{\mathrm{d}}\widehat{R}^-_{\mathrm{U}'}(g),$$

$$\widehat{R}_{\mathrm{ccUU}}(g) = f_1\left(C_{\mathrm{a}}\widehat{R}^+_{\mathrm{U}}(g) - C_{\mathrm{c}}\widehat{R}^+_{\mathrm{U}'}(g)\right) + f_2\left(C_{\mathrm{d}}\widehat{R}^-_{\mathrm{U}'}(g) - C_{\mathrm{b}}\widehat{R}^-_{\mathrm{U}}(g)\right),$$

where

$$\widehat{R}^+_{\mathrm{P}}(g) = \frac{1}{n_{\mathrm{P}}}\sum_{i=1}^{n_{\mathrm{P}}}\ell(g(x^{\mathrm{P}}_i), +1), \quad \widehat{R}^-_{\mathrm{N}}(g) = \frac{1}{n_{\mathrm{N}}}\sum_{i=1}^{n_{\mathrm{N}}}\ell(g(x^{\mathrm{N}}_i), -1),$$

$$\widehat{R}^+_{\mathrm{U}}(g) = \frac{1}{n}\sum_{i=1}^{n}\ell(g(x_i), +1), \quad \widehat{R}^-_{\mathrm{U}}(g) = \frac{1}{n}\sum_{i=1}^{n}\ell(g(x_i), -1),$$

$$\widehat{R}^+_{\mathrm{U}'}(g) = \frac{1}{n'}\sum_{i=1}^{n'}\ell(g(x'_i), +1), \quad \widehat{R}^-_{\mathrm{U}'}(g) = \frac{1}{n'}\sum_{i=1}^{n'}\ell(g(x'_i), -1).$$

11.5.1 Bias and Consistency

$\widehat{R}_{\mathrm{UU}}(g)$ is unbiased, and $\widehat{R}_{\mathrm{ccUU}}(g) \geq \widehat{R}_{\mathrm{UU}}(g)$ holds for any $(\mathscr{X}_{\mathrm{tr}}, \mathscr{X}'_{\mathrm{tr}})$ and fixed g. Therefore, $\widehat{R}_{\mathrm{ccUU}}(g)$ is biased in general. Then a fundamental question is whether $\widehat{R}_{\mathrm{ccUU}}(g)$ is consistent. Below, we establish its consistency.

Let

$$R_a := C_a \widehat{R}_U^+(g), \quad R_b := C_b \widehat{R}_U^-(g), \quad R_c := C_c \widehat{R}_{U'}^+(g), \quad R_d := C_d \widehat{R}_{U'}^-(g),$$

and L_f be the Lipschitz constant of f_1 and f_2. To begin with, partition all possible $(\mathscr{X}_{tr}, \mathscr{X}_{tr}')$ into

$$\mathfrak{D}^+(g) := \{(\mathscr{X}_{tr}, \mathscr{X}_{tr}') \mid R_a - R_c \geq 0, R_d - R_b \geq 0\},$$

$$\mathfrak{D}^-(g) := \{(\mathscr{X}_{tr}, \mathscr{X}_{tr}') \mid R_a - R_c < 0\} \cup \{(\mathscr{X}_{tr}, \mathscr{X}_{tr}') \mid R_d - R_b < 0\}.$$

Assume that there exist $C_g > 0$ and $C_\ell > 0$ such that $\sup_{g \in \mathscr{G}} \|g\|_\infty \leq C_g$ where $\mathscr{G}$ is the model class and $\sup_{|z| \leq C_g} \ell(z) \leq C_\ell$. By *McDiarmid's inequality* (McDiarmid, 1989), we have the following lemma.

Lemma 11.2 *(Lu et al., 2020) The bias of $\widehat{R}_{ccUU}(g)$ is positive if and only if the probability measure of $\mathfrak{D}^-(g)^4$ is non-zero. Further, by assuming that there exist $\alpha_g > 0$ and $\beta_g > 0$ such that*

$$R_P^+(g) = \mathbb{E}_{p(x|y=+1)}[\ell(g(x), +1)] \geq \alpha_g / \pi_P,$$

$$R_N^-(g) = \mathbb{E}_{p(x|y=-1)}[\ell(g(x), -1)] \geq \beta_g / \pi_N,$$

the probability measure of $\mathfrak{D}^-(g)$ can be bounded by

$$\Pr(\mathfrak{D}^-(g)) \leq \exp\left(-\frac{2\alpha_g^2 / C_\ell^2}{C_a^2/n + C_c^2/n'}\right) + \exp\left(-\frac{2\beta_g^2 / C_\ell^2}{C_d^2/n' + C_b^2/n}\right), \tag{11.18}$$

where $\Pr$ denotes the probability.

Proof Let

$$p_{tr}(\mathscr{X}_{tr}) = p_{tr}(x_1) \cdots p_{tr}(x_n),$$

$$p_{tr}'(\mathscr{X}_{tr}') = p_{tr}'(x_1') \cdots p_{tr}'(x_{n'}')$$

be the probability density functions of $\mathscr{X}_{tr}$ and $\mathscr{X}_{tr}'$ (due to the i.i.d. assumption). Then, the measure of $\mathfrak{D}^-(g)$ is defined by

$$\Pr(\mathfrak{D}^-(g)) = \int_{(\mathscr{X}_{tr}, \mathscr{X}_{tr}') \in \mathfrak{D}^-(g)} p_{tr}(\mathscr{X}_{tr}) p_{tr}'(\mathscr{X}_{tr}') \, d\mathscr{X}_{tr} d\mathscr{X}_{tr}',$$

where $d\mathscr{X}_{tr} = dx_1 \cdots dx_n$ and $d\mathscr{X}_{tr}' = dx_1' \cdots dx_{n'}'$. Since $\widehat{R}_{UU}(g)$ is unbiased and $\widehat{R}_{ccUU}(g) - \widehat{R}_{UU}(g) = 0$ on $\mathfrak{D}^+(g)$, the bias of $\widehat{R}_{ccUU}(g)$ can be formulated as

$$\mathbb{E}[\widehat{R}_{\mathrm{ccUU}}(g)] - R(g)$$

$$= \mathbb{E}[\widehat{R}_{\mathrm{ccUU}}(g) - \widehat{R}_{\mathrm{UU}}(g)]$$

$$= \int_{(\mathscr{X}_{\mathrm{tr}}, \mathscr{X}'_{\mathrm{tr}}) \in \mathfrak{D}^+(g)} \left(\widehat{R}_{\mathrm{ccUU}}(g) - \widehat{R}_{\mathrm{UU}}(g) \right) p_{\mathrm{tr}}(\mathscr{X}_{\mathrm{tr}}) p'_{\mathrm{tr}}(\mathscr{X}'_{\mathrm{tr}}) \mathrm{d}\mathscr{X}_{\mathrm{tr}} \mathrm{d}\mathscr{X}'_{\mathrm{tr}}$$

$$+ \int_{(\mathscr{X}_{\mathrm{tr}}, \mathscr{X}'_{\mathrm{tr}}) \in \mathfrak{D}^-(g)} \left(\widehat{R}_{\mathrm{ccUU}}(g) - \widehat{R}_{\mathrm{UU}}(g) \right) p_{\mathrm{tr}}(\mathscr{X}_{\mathrm{tr}}) p'_{\mathrm{tr}}(\mathscr{X}'_{\mathrm{tr}}) \mathrm{d}\mathscr{X}_{\mathrm{tr}} \mathrm{d}\mathscr{X}'_{\mathrm{tr}}$$

$$= \int_{(\mathscr{X}_{\mathrm{tr}}, \mathscr{X}'_{\mathrm{tr}}) \in \mathfrak{D}^-(g)} \left(\widehat{R}_{\mathrm{ccUU}}(g) - \widehat{R}_{\mathrm{UU}}(g) \right) p_{\mathrm{tr}}(\mathscr{X}_{\mathrm{tr}}) p'_{\mathrm{tr}}(\mathscr{X}'_{\mathrm{tr}}) \mathrm{d}\mathscr{X}_{\mathrm{tr}} \mathrm{d}\mathscr{X}'_{\mathrm{tr}}.$$

Thus, we have $\mathbb{E}[\widehat{R}_{\mathrm{ccUU}}(g)] - R(g) > 0$ if and only if

$$\int_{(\mathscr{X}_{\mathrm{tr}}, \mathscr{X}'_{\mathrm{tr}}) \in \mathfrak{D}^-(g)} p_{\mathrm{tr}}(\mathscr{X}_{\mathrm{tr}}) p'_{\mathrm{tr}}(\mathscr{X}'_{\mathrm{tr}}) \mathrm{d}\mathscr{X}_{\mathrm{tr}} \mathrm{d}\mathscr{X}'_{\mathrm{tr}} > 0,$$

due to the fact that $\widehat{R}_{\mathrm{ccUU}}(g) - \widehat{R}_{\mathrm{UU}}(g) > 0$ on $\mathfrak{D}^-(g)$. That is, the bias of $\widehat{R}_{\mathrm{ccUU}}(g)$ is positive if and only if the measure of $\mathfrak{D}^-(g)$ is non-zero.

Next, we study the probability measure of $\mathfrak{D}^-(g)$ by *the method of bounded differences* (McDiarmid, 1989). $R_{\mathrm{P}}^+(g) \geq \alpha_g/\pi_{\mathrm{P}}$ and $R_{\mathrm{N}}^-(g) \geq \beta_g/\pi_{\mathrm{N}}$ yield

$$\mathbb{E}[R_{\mathrm{a}} - R_{\mathrm{c}}] = \pi_{\mathrm{P}} R_{\mathrm{P}}^+(g) \geq \alpha_g,$$

$$\mathbb{E}[R_{\mathrm{d}} - R_{\mathrm{b}}] = \pi_{\mathrm{N}} R_{\mathrm{N}}^-(g) \geq \beta_g.$$

Since we have assumed that $0 \leq \ell(z) \leq C_\ell$, the change of $C_{\mathrm{a}}\widehat{R}_{\mathrm{U}}^+(g)$ and $C_{\mathrm{b}}\widehat{R}_{\mathrm{U}}^-(g)$ will be no more than $C_{\mathrm{a}}C_\ell/n$ and $C_{\mathrm{b}}C_\ell/n$ if some $x_i \in \mathscr{X}_{\mathrm{tr}}$ is replaced, or the change of $C_{\mathrm{c}}\widehat{R}_{\mathrm{U}'}^+(g)$ and $C_{\mathrm{d}}\widehat{R}_{\mathrm{U}'}^-(g)$ will be no more than $C_{\mathrm{c}}C_\ell/n'$ and $C_{\mathrm{d}}C_\ell/n'$ if some $x'_j \in \mathscr{X}'_{\mathrm{tr}}$ is replaced. Subsequently, *McDiarmid's inequality* (McDiarmid, 1989) implies

$$\Pr\{\pi_{\mathrm{P}} R_{\mathrm{P}}^+(g) - (R_{\mathrm{a}} - R_{\mathrm{c}}) \geq \alpha_g\} \leq \exp\left(-\frac{2\alpha_g^2}{n(C_{\mathrm{a}}C_\ell/n)^2 + n'(C_{\mathrm{c}}C_\ell/n')^2} \right)$$

$$= \exp\left(-\frac{2\alpha_g^2/C_\ell^2}{C_{\mathrm{a}}^2/n + C_{\mathrm{c}}^2/n'} \right),$$

and

$$\Pr\{\pi_{\mathrm{N}} R_{\mathrm{N}}^-(g) - (R_{\mathrm{d}} - R_{\mathrm{b}}) \geq \beta_g\} \leq \exp\left(-\frac{2\beta_g^2}{n'(C_{\mathrm{d}}C_\ell/n')^2 + n(C_{\mathrm{b}}C_\ell/n)^2} \right)$$

$$= \exp\left(-\frac{2\beta_g^2/C_\ell^2}{C_{\mathrm{d}}^2/n' + C_{\mathrm{b}}^2/n} \right).$$

Then the probability measure of $\mathfrak{D}^-(g)$ can be bounded as

$$\Pr(\mathfrak{D}^-(g)) \leq \Pr\{R_a - R_c \leq 0\} + \Pr\{R_d - R_b < 0\}$$

$$\leq \Pr\{R_a - R_c \leq \pi_P R_P^+(g) - \alpha_g\} + \Pr\{R_d - R_b \leq \pi_N R_N^-(g) - \beta_g\}$$

$$= \Pr\{\pi_P R_P^+(g) - (R_a - R_c) \geq \alpha_g\} + \Pr\{\pi_N R_N^-(g) - (R_d - R_b) \geq \beta_g\}$$

$$\leq \exp\left(-\frac{2\alpha_g^2/C_\ell^2}{C_a^2/n + C_c^2/n'}\right) + \exp\left(-\frac{2\beta_g^2/C_\ell^2}{C_d^2/n' + C_b^2/n}\right),$$

and thus we complete the proof. $\square$

Based on lemma 11.2, we can show the exponential decay of the bias and also the consistency.

Theorem 11.3 (Bias and consistency)   *(Lu et al., 2020) Let*

$$\Delta_g := \exp\left(-\frac{2\alpha_g^2/C_\ell^2}{C_a^2/n + C_c^2/n'}\right) + \exp\left(-\frac{2\beta_g^2/C_\ell^2}{C_d^2/n' + C_b^2/n}\right).$$

By the assumption in lemma 11.2, the bias of $\widehat{R}_{ccUU}(g)$ decays exponentially as $n, n' \to \infty$:

$$0 \leq \mathbb{E}_{\mathscr{X}_{tr}, \mathscr{X}_{tr}'}[\widehat{R}_{ccUU}(g)] - R(g) \leq (L_f + 1)(C_a + C_b + C_c + C_d)C_\ell \Delta_g. \tag{11.19}$$

Moreover, for any $\delta > 0$, let

$$C_\delta := C_\ell L_f \sqrt{\ln(2/\delta)/2},$$

$$\chi_{n,n'} := (C_a + C_b)/\sqrt{n} + (C_c + C_d)/\sqrt{n'}.$$

Then we have with probability at least $1 - \delta$,

$$|\widehat{R}_{ccUU}(g) - R(g)| \leq (L_f + 1)(C_a + C_b + C_c + C_d)C_\ell \Delta_g + C_\delta \cdot \chi_{n,n'}, \tag{11.20}$$

and with probability at least $1 - \delta - \Delta_g$,

$$|\widehat{R}_{ccUU}(g) - R(g)| \leq C_\delta \cdot \chi_{n,n'}. \tag{11.21}$$

Proof   It has been proved in lemma 11.2 that

$$\mathbb{E}[\widehat{R}_{ccUU}(g)] - R(g)$$

$$= \int_{(\mathscr{X}_{tr}, \mathscr{X}_{tr}') \in \mathfrak{D}^-(g)} \left(\widehat{R}_{ccUU}(g) - \widehat{R}_{UU}(g)\right) p_{tr}(\mathscr{X}_{tr}) p_{tr}'(\mathscr{X}_{tr}') \mathrm{d}\mathscr{X}_{tr} \mathrm{d}\mathscr{X}_{tr}'.$$

Therefore, the exponential decay of the bias can be obtained via

$$\mathbb{E}[\widehat{R}_{\text{ccUU}}(g)] - R(g) \leq \sup_{(\mathscr{X}_{\text{tr}}, \mathscr{X}'_{\text{tr}}) \in \mathfrak{D}^-(g)} \left(\widehat{R}_{\text{ccUU}}(g) - \widehat{R}_{\text{UU}}(g) \right)$$

$$\times \int_{(\mathscr{X}_{\text{tr}}, \mathscr{X}'_{\text{tr}}) \in \mathfrak{D}^-(g)} p_{\text{tr}}(\mathscr{X}_{\text{tr}}) p'_{\text{tr}}(\mathscr{X}'_{\text{tr}}) \mathrm{d}\mathscr{X}_{\text{tr}} \mathrm{d}\mathscr{X}'_{\text{tr}}$$

$$= \sup_{(\mathscr{X}_{\text{tr}}, \mathscr{X}'_{\text{tr}}) \in \mathfrak{D}^-(g)} (f_1(R_a - C_c) + f_2(R_d - R_b)$$

$$- (R_a - C_c) - (R_d - R_b)) \cdot \Pr(\mathfrak{D}^-(g))$$

$$\leq \sup_{(\mathscr{X}_{\text{tr}}, \mathscr{X}'_{\text{tr}}) \in \mathfrak{D}^-(g)} (|f_1(R_a - C_c)| + |f_2(R_d - R_b)|$$

$$+ |R_a - C_c| + |R_d - R_b|) \cdot \Pr(\mathfrak{D}^-(g))$$

$$\leq \sup_{(\mathscr{X}_{\text{tr}}, \mathscr{X}'_{\text{tr}}) \in \mathfrak{D}^-(g)} (L_f |R_a - C_c| + L_f |R_d - R_b|$$

$$+ |R_a - C_c| + |R_d - R_b|) \cdot \Pr(\mathfrak{D}^-(g))$$

$$\leq \sup_{(\mathscr{X}_{\text{tr}}, \mathscr{X}'_{\text{tr}}) \in \mathfrak{D}^-(g)} ((L_f + 1)(C_a + C_c)C_\ell$$

$$+ (L_f + 1)(C_d + C_b)C_\ell) \cdot \Pr(\mathfrak{D}^-(g))$$

$$= (L_f + 1)(C_a + C_b + C_c + C_d)C_\ell \Delta_g,$$

where we employed the Lipschitz condition, i.e., $|f_j(x) - f_j(y)| \leq L_f |x - y|$ for $j = 1, 2$, and the assumption $f_1(0) = f_2(0) = 0$ in definition 11.1. Then the deviation bound (11.20) is due to

$$|\widehat{R}_{\text{ccUU}}(g) - R(g)| \leq |\widehat{R}_{\text{ccUU}}(g) - \mathbb{E}[\widehat{R}_{\text{ccUU}}(g)]| + |\mathbb{E}[\widehat{R}_{\text{ccUU}}(g)] - R(g)|$$

$$\leq |\widehat{R}_{\text{ccUU}}(g) - \mathbb{E}[\widehat{R}_{\text{ccUU}}(g)]|$$

$$+ (L_f + 1)(C_a + C_b + C_c + C_d)C_\ell \Delta_g.$$

Denote by R'_a, R'_b, R'_c, and R'_d, which differs from R_a, R_b, R_c, and R_d on a single example. Then

$$|f_1(R_a - R_c) + f_2(R_d - R_b) - f_1(R'_a - R_c) - f_2(R_d - R'_b)|$$

$$\leq |f_1(R_a - R_c) - f_1(R'_a - R_c)| + |f_2(R_d - R_b) - f_2(R_d - R'_b)|$$

$$\leq L_f |R_a - R_c - R'_a + R_c| + L_f |R_d - R_b - R_d + R'_b|$$

$$\leq (C_a + C_b)L_f C_\ell / n. \tag{11.22}$$

Similarly, we can obtain

$$|f_1(R_a - R_c) + f_2(R_d - R_b) - f_1(R_a - R'_c) - f_2(R'_d - R_b)| \leq (C_c + C_d)L_f C_\ell / n'. \tag{11.23}$$

Therefore, the change of $\widehat{R}_{\text{ccUU}}(g)$ will be no more than $(C_a + C_b)L_f C_\ell / n$ if some $x_i \in \mathcal{X}_{\text{tr}}$ is replaced, or it will be no more than $(C_c + C_d)L_f C_\ell / n'$ if some $x'_j \in \mathcal{X}'_{\text{tr}}$ is replaced, and McDiarmid's inequality gives us

$$\Pr\{|\widehat{R}_{\text{ccUU}}(g) - \mathbb{E}[\widehat{R}_{\text{ccUU}}(g)]| \geq \epsilon\}$$

$$\leq 2\exp\left(-\frac{2\epsilon^2}{n((C_a + C_b)L_f C_\ell / n)^2 + n'((C_c + C_d)L_f C_\ell / n')^2}\right).$$

Setting the above right-hand side to be equal to δ and solving it for ϵ yields immediately the following bound: For any $\delta > 0$, with probability at least $1 - \delta$, we have

$$|\widehat{R}_{\text{ccUU}}(g) - \mathbb{E}[\widehat{R}_{\text{ccUU}}(g)]| \leq \sqrt{\frac{\ln(2/\delta)C_\ell^2 L_f^2}{2}\left(\frac{(C_a + C_b)^2}{n} + \frac{(C_c + C_d)^2}{n'}\right)}$$

$$\leq C_\delta \left(\frac{C_a + C_b}{\sqrt{n}} + \frac{C_c + C_d}{\sqrt{n'}}\right)$$

$$= C_\delta \cdot \chi_{n,n'},$$

where $C_\delta = C_\ell L_f \sqrt{\ln(2/\delta)/2}$ and $\chi_{n,n'} = (C_a + C_b)/\sqrt{n} + (C_c + C_d)/\sqrt{n'}$. Thus we obtain

$$|\widehat{R}_{\text{ccUU}}(g) - R(g)| \leq C_\delta \cdot \chi_{n,n'} + (L_f + 1)(C_a + C_b + C_c + C_d)C_\ell \Delta_g.$$

On the other hand, the deviation bound (11.21) is due to

$$|\widehat{R}_{\text{ccUU}}(g) - R(g)| \leq |\widehat{R}_{\text{ccUU}}(g) - \widehat{R}_{\text{UU}}(g)| + |\widehat{R}_{\text{UU}}(g) - R(g)|,$$

where $|\widehat{R}_{\text{ccUU}}(g) - \widehat{R}_{\text{UU}}(g)| > 0$ with probability at most Δ_g, and $|\widehat{R}_{\text{UU}}(g) - R(g)|$ shares the same concentration inequality with $|\widehat{R}_{\text{ccUU}}(g) - \mathbb{E}[\widehat{R}_{\text{ccUU}}(g)]|$. $\qquad\square$

Either (11.20) or (11.21) in theorem 11.3 indicates $\widehat{R}_{\text{ccUU}}(g) \to R(g)$ in $\mathcal{O}_p(1/\sqrt{n} + 1/\sqrt{n'})$ for fixed g. This convergence rate is optimal according to the *central limit theorem* (Chung, 1968), which means that $\widehat{R}_{\text{ccUU}}(g)$ is a biased yet optimal estimator to the risk.

11.5.2   Estimation Error

While theorem 11.3 addressed the use of (11.17) when the risk is evaluated, in what follows we study the estimation error $R(\widehat{g}_{\text{ccUU}}) - R(g^*)$ when classifiers are trained, where g^* is the true risk minimizer in the model class $\mathcal{G}$, i.e., $g^* = \text{argmin}_{g \in \mathcal{G}} R(g)$, and $\widehat{g}_{\text{ccUU}}$ is the minimizer of (11.17), i.e., $\widehat{g}_{\text{ccUU}} = \text{argmin}_{g \in \mathcal{G}} \widehat{R}_{\text{nnUU}}(g)$. Based on the analysis in section 3.1.2.5, we assume that $\|x\| \leq C_x$, and the loss function $\ell(t, y)$ is Lipschitz continuous in t for all $|t| \leq C_g$ with a Lipschitz constant L_ℓ.

Theorem 11.4 (Estimation error bound)   *(Lu et al., 2020) Assume that*

(a) $\inf_{g \in \mathscr{G}} R_{\mathrm{P}}^{+}(g) \geq \alpha/\pi_{\mathrm{P}} > 0$, $\inf_{g \in \mathscr{G}} R_{\mathrm{N}}^{-}(g) \geq \beta/\pi_{\mathrm{N}} > 0$.

(b) $\mathscr{G}$ *is closed under negation, i.e.,* $g \in \mathscr{G}$ *if and only if* $-g \in \mathscr{G}$.

Let $\Delta = \exp\left(-\dfrac{2\alpha^2/C_\ell^2}{C_{\mathrm{a}}^2/n + C_{\mathrm{c}}^2/n'}\right) + \exp\left(-\dfrac{2\beta^2/C_\ell^2}{C_{\mathrm{d}}^2/n' + C_{\mathrm{b}}^2/n}\right)$. *Then, for any* $\delta > 0$, *with probability at least* $1 - \delta$,

$$R(\widehat{g}_{\mathrm{ccUU}}) - R(g^*) \leq 8(C_{\mathrm{a}} + C_{\mathrm{b}})L_f L_\ell \mathfrak{R}_{n,p_{\mathrm{tr}}}(\mathscr{G}) + 8(C_{\mathrm{c}} + C_{\mathrm{d}})L_f L_\ell \mathfrak{R}_{n',p'_{\mathrm{tr}}}(\mathscr{G})$$

$$+ 2(L_f + 1)(C_{\mathrm{a}} + C_{\mathrm{b}} + C_{\mathrm{c}} + C_{\mathrm{d}})C_\ell \Delta + 2C'_\delta \cdot \chi_{n,n'}, \qquad (11.24)$$

where $C'_\delta = C_\ell L_f \sqrt{\ln(1/\delta)/2}$, *and* $\mathfrak{R}_{n,p_{\mathrm{tr}}}(\mathscr{G})$ *and* $\mathfrak{R}_{n',p'_{\mathrm{tr}}}(\mathscr{G})$ *are the Rademacher complexities of* $\mathscr{G}$ *for the sampling of size* n *from* $p_{\mathrm{tr}}(\boldsymbol{x})$ *and of size* n' *from* $p'_{\mathrm{tr}}(\boldsymbol{x})$, *respectively.*

Proof   The estimation error bound relies on the uniform deviation bound below.

Lemma 11.5   *Under the assumptions of theorem 11.4, for any* $\delta > 0$, *with probability at least* $1 - \delta$,

$$\sup_{g \in \mathscr{G}} |\widehat{R}_{\mathrm{ccUU}}(g) - R(g)| \leq 4(C_{\mathrm{a}} + C_{\mathrm{b}})L_f L_\ell \mathfrak{R}_{n,p_{\mathrm{tr}}}(\mathscr{G}) + 4(C_{\mathrm{c}} + C_{\mathrm{d}})L_f L_\ell \mathfrak{R}_{n',p'_{\mathrm{tr}}}(\mathscr{G})$$

$$+ (L_f + 1)(C_{\mathrm{a}} + C_{\mathrm{b}} + C_{\mathrm{c}} + C_{\mathrm{d}})C_\ell \Delta + C'_\delta \cdot \chi_{n,n'}. \qquad (11.25)$$

Proof   First, we deal with the bias of $\widehat{R}_{\mathrm{ccUU}}(g)$. Notice that the assumptions $\inf_{g \in \mathscr{G}} R_{\mathrm{P}}^{+}(g) \geq \alpha/\pi_{\mathrm{P}} > 0$ and $\inf_{g \in \mathscr{G}} R_{\mathrm{N}}^{-}(g) \geq \beta/\pi_{\mathrm{N}} > 0$ imply $\Delta = \sup_{g \in \mathscr{G}} \Delta_g$. By (11.19) we have

$$\sup_{g \in \mathscr{G}} |\widehat{R}_{\mathrm{ccUU}}(g) - R(g)|$$

$$\leq \sup_{g \in \mathscr{G}} |\widehat{R}_{\mathrm{ccUU}}(g) - \mathbb{E}[\widehat{R}_{\mathrm{ccUU}}(g)]| + \sup_{g \in \mathscr{G}} |\mathbb{E}[\widehat{R}_{\mathrm{ccUU}}(g)] - R(g)|$$

$$\leq \sup_{g \in \mathscr{G}} |\widehat{R}_{\mathrm{ccUU}}(g) - \mathbb{E}[\widehat{R}_{\mathrm{ccUU}}(g)]| + (L_f + 1)(C_{\mathrm{a}} + C_{\mathrm{b}} + C_{\mathrm{c}} + C_{\mathrm{d}})C_\ell \Delta. \qquad (11.26)$$

Second, we consider the double-sided uniform deviation $\sup_{g \in \mathscr{G}} |\widehat{R}_{\mathrm{ccUU}}(g) - \mathbb{E}[\widehat{R}_{\mathrm{ccUU}}(g)]|$. Denote by $\mathscr{X}_{\mathrm{s}} = \{(\mathscr{X}_{\mathrm{tr}}, \mathscr{X}'_{\mathrm{tr}})\}$, and $\mathscr{X}'_{\mathrm{s}}$ that differs from $\mathscr{X}_{\mathrm{s}}$ on a single example. Then we have

$$\left| \sup_{g \in \mathscr{G}} |\widehat{R}_{\mathrm{ccUU}}(g; \mathscr{X}_{\mathrm{s}}) - \mathbb{E}_{\mathscr{X}_{\mathrm{s}}}[\widehat{R}_{\mathrm{ccUU}}(g; \mathscr{X}_{\mathrm{s}})]| \right.$$

$$\left. - \sup_{g \in \mathscr{G}} |\widehat{R}_{\mathrm{ccUU}}(g; \mathscr{X}'_{\mathrm{s}}) - \mathbb{E}_{\mathscr{X}'_{\mathrm{s}}}[\widehat{R}_{\mathrm{ccUU}}(g; \mathscr{X}'_{\mathrm{s}})]| \right|$$

$$\leq \sup_{g \in \mathscr{G}} \left| |\widehat{R}_{\mathrm{ccUU}}(g; \mathscr{X}_{\mathrm{s}}) - \mathbb{E}_{\mathscr{X}_{\mathrm{s}}}[\widehat{R}_{\mathrm{ccUU}}(g; \mathscr{X}_{\mathrm{s}})]| \right.$$

$$- \left|\widehat{R}_{\mathrm{ccUU}}(g; \mathscr{X}'_{\mathrm{s}}) - \mathbb{E}_{\mathscr{X}'_{\mathrm{s}}}[\widehat{R}_{\mathrm{ccUU}}(g; \mathscr{X}'_{\mathrm{s}})]\right|\Big|$$

$$\leq \sup_{g \in \mathscr{G}} \left|\widehat{R}_{\mathrm{ccUU}}(g; \mathscr{X}_{\mathrm{s}}) - \widehat{R}_{\mathrm{ccUU}}(g; \mathscr{X}'_{\mathrm{s}})\right|,$$

where we applied the *triangle inequality*. Then, based on (11.22) and (11.23), we see that the change of $\sup_{g \in \mathscr{G}} |\widehat{R}_{\mathrm{ccUU}}(g) - \mathbb{E}[\widehat{R}_{\mathrm{ccUU}}(g)]|$ will be no more than $(C_{\mathrm{a}} + C_{\mathrm{b}})L_f C_\ell / n$ if some $x_i \in \mathscr{X}_{\mathrm{tr}}$ is replaced, or it will be no more than $(C_{\mathrm{c}} + C_{\mathrm{d}})L_f C_\ell / n'$ if some $x'_j \in \mathscr{X}'_{\mathrm{tr}}$ is replaced. Similar to the proof technique of theorem 11.3, by applying McDiarmid's inequality to the uniform deviation, we have with probability at least $1 - \delta$,

$$\sup_{g \in \mathscr{G}} |\widehat{R}_{\mathrm{ccUU}}(g) - \mathbb{E}[\widehat{R}_{\mathrm{ccUU}}(g)]| - \mathbb{E}[\sup_{g \in \mathscr{G}} |\widehat{R}_{\mathrm{ccUU}}(g) - \mathbb{E}[\widehat{R}_{\mathrm{ccUU}}(g)]|]$$

$$\leq \sqrt{\frac{\ln(1/\delta) C_\ell^2 L_f^2}{2} \left(\frac{(C_{\mathrm{a}} + C_{\mathrm{b}})^2}{n} + \frac{(C_{\mathrm{c}} + C_{\mathrm{d}})^2}{n'}\right)}$$

$$= C'_\delta \cdot \chi_{n,n'}, \tag{11.27}$$

where $C'_\delta = C_\ell L_f \sqrt{\ln(1/\delta)/2}$.

Third, we make *symmetrization* (Vapnik, 1998). Suppose that $(\mathscr{X}_{\mathrm{tr}}^{gh}, \mathscr{X}_{\mathrm{tr}}'^{gh})$ is a *ghost sample*, then

$$\mathbb{E}[\sup_{g \in \mathscr{G}} |\widehat{R}_{\mathrm{ccUU}}(g) - \mathbb{E}[\widehat{R}_{\mathrm{ccUU}}(g)]|]$$

$$= \mathbb{E}_{(\mathscr{X}_{\mathrm{tr}}, \mathscr{X}'_{\mathrm{tr}})}[\sup_{g \in \mathscr{G}} |\widehat{R}_{\mathrm{ccUU}}(g; \mathscr{X}_{\mathrm{tr}}, \mathscr{X}'_{\mathrm{tr}}) - \mathbb{E}_{(\mathscr{X}_{\mathrm{tr}}^{gh}, \mathscr{X}_{\mathrm{tr}}'^{gh})} \widehat{R}_{\mathrm{ccUU}}(g; \mathscr{X}_{\mathrm{tr}}^{gh}, \mathscr{X}_{\mathrm{tr}}'^{gh})|]$$

$$\leq \mathbb{E}_{(\mathscr{X}_{\mathrm{tr}}, \mathscr{X}'_{\mathrm{tr}})}[\sup_{g \in \mathscr{G}} \mathbb{E}_{(\mathscr{X}_{\mathrm{tr}}^{gh}, \mathscr{X}_{\mathrm{tr}}'^{gh})} |\widehat{R}_{\mathrm{ccUU}}(g; \mathscr{X}_{\mathrm{tr}}, \mathscr{X}'_{\mathrm{tr}}) - \widehat{R}_{\mathrm{ccUU}}(g; \mathscr{X}_{\mathrm{tr}}^{gh}, \mathscr{X}_{\mathrm{tr}}'^{gh})|]$$

$$\leq \mathbb{E}_{(\mathscr{X}_{\mathrm{tr}}, \mathscr{X}'_{\mathrm{tr}}),(\mathscr{X}_{\mathrm{tr}}^{gh}, \mathscr{X}_{\mathrm{tr}}'^{gh})}[\sup_{g \in \mathscr{G}} |\widehat{R}_{\mathrm{ccUU}}(g; \mathscr{X}_{\mathrm{tr}}, \mathscr{X}'_{\mathrm{tr}}) - \widehat{R}_{\mathrm{ccUU}}(g; \mathscr{X}_{\mathrm{tr}}^{gh}, \mathscr{X}_{\mathrm{tr}}'^{gh})|],$$

where we applied *Jensen's inequality* twice since the absolute value and the supremum are convex. By decomposing the difference $|\widehat{R}_{\mathrm{ccUU}}(g; \mathscr{X}_{\mathrm{tr}}, \mathscr{X}'_{\mathrm{tr}}) - \widehat{R}_{\mathrm{ccUU}}(g; \mathscr{X}_{\mathrm{tr}}^{gh}, \mathscr{X}_{\mathrm{tr}}'^{gh})|$, we can know that

$$|\widehat{R}_{\mathrm{ccUU}}(g; \mathscr{X}_{\mathrm{tr}}, \mathscr{X}'_{\mathrm{tr}}) - \widehat{R}_{\mathrm{ccUU}}(g; \mathscr{X}_{\mathrm{tr}}^{gh}, \mathscr{X}_{\mathrm{tr}}'^{gh})|$$

$$\leq \left|f_1\left(a\widehat{R}_{\mathrm{U}}^+(g; \mathscr{X}_{\mathrm{tr}}) - c\widehat{R}_{\mathrm{U}'}^+(g; \mathscr{X}'_{\mathrm{tr}})\right) - f_1\left(a\widehat{R}_{\mathrm{U}}^+(g; \mathscr{X}_{\mathrm{tr}}^{gh}) - c\widehat{R}_{\mathrm{U}'}^+(g; \mathscr{X}_{\mathrm{tr}}'^{gh})\right)\right|$$

$$+ \left|f_2\left(-b\widehat{R}_{\mathrm{U}}^-(g; \mathscr{X}_{\mathrm{tr}}) + d\widehat{R}_{\mathrm{U}'}^-(g; \mathscr{X}'_{\mathrm{tr}})\right) - f_2\left(-b\widehat{R}_{\mathrm{U}}^-(g; \mathscr{X}_{\mathrm{tr}}^{gh}) + d\widehat{R}_{\mathrm{U}'}^-(g; \mathscr{X}_{\mathrm{tr}}'^{gh})\right)\right|$$

$$\leq \left|L_f\left(a\widehat{R}_{\mathrm{U}}^+(g; \mathscr{X}_{\mathrm{tr}}) - c\widehat{R}_{\mathrm{U}'}^+(g; \mathscr{X}'_{\mathrm{tr}}) - a\widehat{R}_{\mathrm{U}}^+(g; \mathscr{X}_{\mathrm{tr}}^{gh}) + c\widehat{R}_{\mathrm{U}'}^+(g; \mathscr{X}_{\mathrm{tr}}'^{gh})\right)\right|$$

$$+ \left|L_f\left(-b\widehat{R}_{\mathrm{U}}^-(g; \mathscr{X}_{\mathrm{tr}}) + d\widehat{R}_{\mathrm{U}'}^-(g; \mathscr{X}'_{\mathrm{tr}}) + b\widehat{R}_{\mathrm{U}}^-(g; \mathscr{X}_{\mathrm{tr}}^{gh}) - d\widehat{R}_{\mathrm{U}'}^-(g; \mathscr{X}_{\mathrm{tr}}'^{gh})\right)\right|$$

$$\leq \left| aL_f \left(\widehat{R}_{\mathrm{U}}^{+}(g; \mathscr{X}_{\mathrm{tr}}) - \widehat{R}_{\mathrm{U}}^{+}(g; \mathscr{X}_{\mathrm{tr}}^{gh}) \right) \right| + \left| cL_f \left(\widehat{R}_{\mathrm{U'}}^{+}(g; \mathscr{X}_{\mathrm{tr}}') - \widehat{R}_{\mathrm{U'}}^{+}(g; \mathscr{X}_{\mathrm{tr}}'^{gh}) \right) \right|$$

$$+ \left| bL_f \left(\widehat{R}_{\mathrm{U}}^{-}(g; \mathscr{X}_{\mathrm{tr}}) - \widehat{R}_{\mathrm{U}}^{-}(g; \mathscr{X}_{\mathrm{tr}}^{gh}) \right) \right| + \left| dL_f \left(\widehat{R}_{\mathrm{U'}}^{-}(g; \mathscr{X}_{\mathrm{tr}}') - \widehat{R}_{\mathrm{U'}}^{-}(g; \mathscr{X}_{\mathrm{tr}}'^{gh}) \right) \right|,$$

where we employed the Lipschitz condition. This decomposition results in

$$\mathbb{E}[\sup_{g \in \mathscr{G}} |\widehat{R}_{\mathrm{ccUU}}(g) - \mathbb{E}[\widehat{R}_{\mathrm{ccUU}}(g)]|]$$

$$\leq aL_f \mathbb{E}_{\mathscr{X}_{\mathrm{tr}}, \mathscr{X}_{\mathrm{tr}}^{gh}} \left[\sup_{g \in \mathscr{G}} \left| \left(\widehat{R}_{\mathrm{U}}^{+}(g; \mathscr{X}_{\mathrm{tr}}) - \widehat{R}_{\mathrm{U}}^{+}(g; \mathscr{X}_{\mathrm{tr}}^{gh}) \right) \right| \right]$$

$$+ cL_f \mathbb{E}_{\mathscr{X}_{\mathrm{tr}}', \mathscr{X}_{\mathrm{tr}}'^{gh}} \left[\sup_{g \in \mathscr{G}} \left| \left(\widehat{R}_{\mathrm{U'}}^{+}(g; \mathscr{X}_{\mathrm{tr}}') - \widehat{R}_{\mathrm{U'}}^{+}(g; \mathscr{X}_{\mathrm{tr}}'^{gh}) \right) \right| \right]$$

$$+ bL_f \mathbb{E}_{\mathscr{X}_{\mathrm{tr}}, \mathscr{X}_{\mathrm{tr}}^{gh}} \left[\sup_{g \in \mathscr{G}} \left| \left(\widehat{R}_{\mathrm{U}}^{-}(g; \mathscr{X}_{\mathrm{tr}}) - \widehat{R}_{\mathrm{U}}^{-}(g; \mathscr{X}_{\mathrm{tr}}^{gh}) \right) \right| \right]$$

$$+ dL_f \mathbb{E}_{\mathscr{X}_{\mathrm{tr}}', \mathscr{X}_{\mathrm{tr}}'^{gh}} \left[\sup_{g \in \mathscr{G}} \left| \left(\widehat{R}_{\mathrm{U'}}^{-}(g; \mathscr{X}_{\mathrm{tr}}') - \widehat{R}_{\mathrm{U'}}^{-}(g; \mathscr{X}_{\mathrm{tr}}'^{gh}) \right) \right| \right].$$

Fourth, we relax those expectations to Rademacher complexities (see section 3.1.2.3 for the definition of Rademacher complexity with an additional absolute value of the average before taking the supremum). The original ℓ may miss the origin, i.e., $\ell(0, y) \neq 0$, with which we need to cope. Let

$$\bar{\ell}(t, y) = \ell(t, y) - \ell(0, y)$$

be a *shifted loss* so that $\bar{\ell}(0, y) = 0$. Hence,

$$\widehat{R}_{\mathrm{U}}^{+}(g; \mathscr{X}_{\mathrm{tr}}) - \widehat{R}_{\mathrm{U}}^{+}(g; \mathscr{X}_{\mathrm{tr}}^{gh})$$

$$= \frac{1}{n} \sum_{x_i \in \mathscr{X}_{\mathrm{tr}}} \ell(g(x_i), +1) - \frac{1}{n} \sum_{x_i^{gh} \in \mathscr{X}_{\mathrm{tr}}^{gh}} \ell(g(x_i^{gh}), +1)$$

$$= \frac{1}{n} \sum_{i=1}^{n} (\ell(g(x_i), +1) - \ell(g(x_i^{gh}), +1))$$

$$= \frac{1}{n} \sum_{i=1}^{n} (\bar{\ell}(g(x_i), +1) - \bar{\ell}(g(x_i^{gh}), +1)).$$

This is already a standard form where we can attach Rademacher variables to every $\bar{\ell}(g(x_i), +1) - \bar{\ell}(g(x_i^{gh}), +1)$, so we have

$$\mathbb{E}_{\mathscr{X}_{\mathrm{tr}}, \mathscr{X}_{\mathrm{tr}}^{gh}}[\sup_{g \in \mathscr{G}} |\widehat{R}_{\mathrm{U}}^{+}(g; \mathscr{X}_{\mathrm{tr}}) - \widehat{R}_{\mathrm{U}}^{+}(g; \mathscr{X}_{\mathrm{tr}}^{gh})|]$$

$$= \mathbb{E}_{\mathscr{X}_{\mathrm{tr}}, \mathscr{X}_{\mathrm{tr}}^{gh}} \left[\sup_{g \in \mathscr{G}} \left| (1/n) \sum_{i=1}^{n} (\bar{\ell}(g(x_i), +1) - \bar{\ell}(g(x_i^{gh}), +1)) \right| \right]$$

$$= \mathbb{E}_{\boldsymbol{\varepsilon},\mathscr{X}_{\mathrm{tr}},\mathscr{X}^{gh}_{\mathrm{tr}}} \left[\sup_{g\in\mathscr{G}} \left| (1/n) \sum_{i=1}^{n} \varepsilon_i (\bar{\ell}(g(\boldsymbol{x}_i),+1) - \bar{\ell}(g(\boldsymbol{x}^{gh}_i),+1)) \right| \right]$$

$$\leq \mathbb{E}_{\boldsymbol{\varepsilon},\mathscr{X}_{\mathrm{tr}}} \left[\sup_{g\in\mathscr{G}} \left| (1/n) \sum_{i=1}^{n} \varepsilon_i (\bar{\ell}(g(\boldsymbol{x}_i),+1)) \right| \right]$$

$$+ \mathbb{E}_{\boldsymbol{\varepsilon},\mathscr{X}^{gh}_{\mathrm{tr}}} \left[\sup_{g\in\mathscr{G}} \left| (1/n) \sum_{i=1}^{n} \varepsilon_i (\bar{\ell}(g(\boldsymbol{x}^{gh}_i),+1)) \right| \right]$$

$$= 2\mathbb{E}_{\boldsymbol{\varepsilon},\mathscr{X}_{\mathrm{tr}}} \left[\sup_{g\in\mathscr{G}} \left| (1/n) \sum_{i=1}^{n} \varepsilon_i (\bar{\ell}(g(\boldsymbol{x}_i),+1)) \right| \right]$$

$$= 2\mathfrak{R}'_{n,p_{\mathrm{tr}}}(\bar{\ell}(\cdot,+1)\circ\mathscr{G}).$$

The other three expectations can be handled analogously. As a result,

$$\mathbb{E}[\sup_{g\in\mathscr{G}} |\widehat{R}_{\mathrm{ccUU}}(g) - \mathbb{E}[\widehat{R}_{\mathrm{ccUU}}(g)]|]$$

$$\leq 2C_{\mathrm{a}}L_f \mathfrak{R}'_{n,p_{\mathrm{tr}}}(\bar{\ell}(\cdot,+1)\circ\mathscr{G}) + 2C_{\mathrm{c}}L_f \mathfrak{R}'_{n',p'_{\mathrm{tr}}}(\bar{\ell}(\cdot,+1)\circ\mathscr{G})$$

$$+ 2C_{\mathrm{b}}L_f \mathfrak{R}'_{n,p_{\mathrm{tr}}}(\bar{\ell}(\cdot,-1)\circ\mathscr{G}) + 2C_{\mathrm{d}}L_f \mathfrak{R}'_{n',p'_{\mathrm{tr}}}(\bar{\ell}(\cdot,-1)\circ\mathscr{G}).$$

Finally, we transform the Rademacher complexities of composite function classes to the original function class. It is obvious that $\bar{\ell}$ shares the same Lipschitz constant L_ℓ with ℓ, and consequently

$$\mathfrak{R}'_{n,p_{\mathrm{tr}}}(\bar{\ell}(\cdot,+1)\circ\mathscr{G}) \leq 2L_\ell \mathfrak{R}'_{n,p_{\mathrm{tr}}}(\mathscr{G}) = 2L_\ell \mathfrak{R}_{n,p_{\mathrm{tr}}}(\mathscr{G})$$

$$\mathfrak{R}'_{n',p'_{\mathrm{tr}}}(\bar{\ell}(\cdot,+1)\circ\mathscr{G}) \leq 2L_\ell \mathfrak{R}'_{n',p'_{\mathrm{tr}}}(\mathscr{G}) = 2L_\ell \mathfrak{R}_{n',p'_{\mathrm{tr}}}(\mathscr{G})$$

$$\mathfrak{R}'_{n,p_{\mathrm{tr}}}(\bar{\ell}(\cdot,-1)\circ\mathscr{G}) \leq 2L_\ell \mathfrak{R}'_{n,p_{\mathrm{tr}}}(\mathscr{G}) = 2L_\ell \mathfrak{R}_{n,p_{\mathrm{tr}}}(\mathscr{G})$$

$$\mathfrak{R}'_{n',p'_{\mathrm{tr}}}(\bar{\ell}(\cdot,-1)\circ\mathscr{G}) \leq 2L_\ell \mathfrak{R}'_{n',p'_{\mathrm{tr}}}(\mathscr{G}) = 2L_\ell \mathfrak{R}_{n',p'_{\mathrm{tr}}}(\mathscr{G}),$$

where we used the results in section 3.1.2.3 that $\mathfrak{R}'_{n,q}(f\circ\mathscr{G}) \leq 2L_f \mathfrak{R}'_{n,q}(\mathscr{G})$, where $f: \mathbb{R} \to \mathbb{R}$ is a Lipschitz continuous function with a Lipschitz constant L_f and $f(0)=0$, and $\mathfrak{R}'_{n,p}(\mathscr{G}) = \mathfrak{R}_{n,p}(\mathscr{G})$ if $\mathscr{G}$ is closed under negation. So we have

$$\mathbb{E}[\sup_{g\in\mathscr{G}} |\widehat{R}_{\mathrm{ccUU}}(g) - \mathbb{E}[\widehat{R}_{\mathrm{ccUU}}(g)]|]$$

$$\leq 4(C_{\mathrm{a}} + C_{\mathrm{b}})L_f L_\ell \mathfrak{R}_{n,p_{\mathrm{tr}}}(\mathscr{G}) + 4(C_{\mathrm{c}} + C_{\mathrm{d}})L_f L_\ell \mathfrak{R}_{n',p'_{\mathrm{tr}}}(\mathscr{G}). \tag{11.28}$$

Combining equations (11.26), (11.27), and (11.28) finishes the proof of the uniform deviation bound (11.25). $\qquad\square$

We are now ready to prove our estimation error bound based on the uniform deviation bound in lemma 11.5 as follows:

$$R(\widehat{g}_{\text{ccUU}}) - R(g^*) = \left(\widehat{R}_{\text{ccUU}}(\widehat{g}_{\text{ccUU}}) - \widehat{R}_{\text{ccUU}}(g^*)\right)$$

$$+ \left(R(\widehat{g}_{\text{ccUU}}) - \widehat{R}_{\text{ccUU}}(\widehat{g}_{\text{ccUU}})\right) + \left(\widehat{R}_{\text{ccUU}}(g^*) - R(g^*)\right)$$

$$\leq 0 + 2 \sup_{g \in \mathcal{G}} |\widehat{R}_{\text{ccUU}}(g) - R(g)|$$

$$\leq 8(C_a + C_b)L_f L_\ell \mathfrak{R}_{n,p_{\text{tr}}}(\mathcal{G}) + 8(C_c + C_d)L_f L_\ell \mathfrak{R}_{n',p'_{\text{tr}}}(\mathcal{G})$$

$$+ 2(L_f + 1)(C_a + C_b + C_c + C_d)C_\ell \Delta + 2C'_\delta \cdot \chi_{n,n'},$$

where $\widehat{R}_{\text{ccUU}}(\widehat{g}_{\text{ccUU}}) \leq \widehat{R}_{\text{ccUU}}(g^*)$ by the definition of g^* and $\widehat{g}_{\text{ccUU}}$. $\qquad\square$

Theorem 11.4 ensures that classification with (11.17) is also consistent: as $n, n' \to \infty$, $R(\widehat{g}_{\text{ccUU}}) \to R(g^*)$, since $\mathfrak{R}_{n,p_{\text{tr}}}(\mathcal{G})$, $\mathfrak{R}_{n',p'_{\text{tr}}}(\mathcal{G}) \to 0$ for all parametric models with a bounded norm and $\Delta \to 0$. For linear-in-parameter models with a bounded norm, $\mathfrak{R}_{n,p_{\text{tr}}}(\mathcal{G}) = \mathcal{O}(1/\sqrt{n})$ and $\mathfrak{R}_{n',p'_{\text{tr}}}(\mathcal{G}) = \mathcal{O}(1/\sqrt{n'})$, and thus $R(\widehat{g}_{\text{ccUU}}) \to R(g^*)$ in $\mathcal{O}_p(1/\sqrt{n} + 1/\sqrt{n'})$.

11.6 Experiments

Now, we compare the corrected non-negative classification methods with their unbiased counterparts.

11.6.1 Comparison of PN, uPU, and nnPU Classification

In this section, we compare PN, uPU, and nnPU classification experimentally. We focus on training deep neural networks, as uPU classification usually does not overfit if a linear-in-parameter model is used (Niu et al., 2016).

Table 11.1 describes the specification of benchmark datasets. *MNIST*[5] (LeCun et al., 1998), *20News*[6] (Lang, 1995), and *CIFAR-10*[7] (Krizhevsky, 2009) have 10, 7, and 10 classes originally, and we constructed the P and N classes from them as follows: MNIST was preprocessed in such a way that "0," "2," "4," "6," and "8" constitute the P class, while "1," "3," "5," "7," and "9" constitute the N class. For 20news, "alt," "comp," "misc," and "rec" make up the P class, and "sci," "soc," and "talk" make up the N class. For CIFAR-10, the P class is formed by "airplane," "automobile," "ship," and "truck," and the N class is formed by "bird," "cat," "deer," "dog," "frog," and "horse." The dataset *epsilon*[8] (Yuan et al., 2012) has only 2 classes and we just used them directly.

Table 11.1
Specification of benchmark datasets.

Name	# Train	# Test	# Feature	π_P
MNIST	60,000	10,000	784	0.49
epsilon	400,000	100,000	2,000	0.50
20News	11,314	7,532	61,188	0.44
CIFAR-10	50,000	10,000	3,072	0.40

The model $g(x)$ and optimization algorithm $\mathscr{A}$ are chosen as follows:

MNIST: We used a six-layer MLP d-300-300-300-300-1 with ReLUs (Nair and Hinton, 2010) and the Adam algorithm (Kingma and Ba, 2015).

epsilon: The model and optimization algorithm were similar to MNIST, but the activation was replaced with Softsign (Glorot and Bengio, 2010) for better performance.

20News: We borrowed the pretrained word embeddings from GloVe (Pennington et al., 2014), and the model can be written as d-avg_pool(word_emb(d,300))-300-300-1, where word_emb(d,300) retrieves 300-dimensional word embeddings for all words in a document, avg_pool executes average pooling, and the resulting vector is fed to a four-layer MLP with Softsign. AdaGrad (Duchi et al., 2011) was used as an optimization algorithm.

CIFAR-10: The model was an *all convolutional net* (Springenberg et al., 2015): (32*32*3)-[C(3*3,96)]*2-C(3*3,96,2)-[C(3*3,192)]*2-C(3*3,192,2)-C(3*3,192)-C(1*1,192)-C(1*1,10)-1000-1000-1, where the input is a 32*32 RGB image, C(3*3,96) means 96 channels of 3*3 convolutions followed by ReLU, [·]*2 means there are two such layers, C(3*3,96,2) means a similar layer but with stride 2; it is one of the best architectures for CIFAR-10. Batch normalization (Ioffe and Szegedy, 2015) was applied before hidden layers. Furthermore, the sigmoid loss was used as the surrogate loss and an ℓ_2-regularization was also added.

We fixed $\beta = 0$ and $\gamma = 1$ for simplicity (see algorithm 1). Three classification methods were set up as follows:

PN classification: $n_P = 1,000$ and $n_N = (\pi_N/2\pi_P)^2 n_P$.

uPU classification: $n_P = 1,000$ and n_U is the total number of training data.

nnPU classification: n_P and n_U are exactly the same as uPU classification.

For uPU and nnPU classification, P and U data were dependent, because neither $\widehat{R}_{PU}(g)$ in (4.3) nor $\widehat{R}_{nnPU}(g)$ in (11.14) requires them to be independent. The choice of n_N was motivated by Niu et al. (2016) and may make nnPU classification potentially better than PN classification as $n_U \rightarrow \infty$ (whether $n_P < \infty$ or $n_P \leq n_U$).

The experimental results are reported in figure 11.6, where means and standard deviations of training and test risks based on the same 10 random samplings are shown. We can see

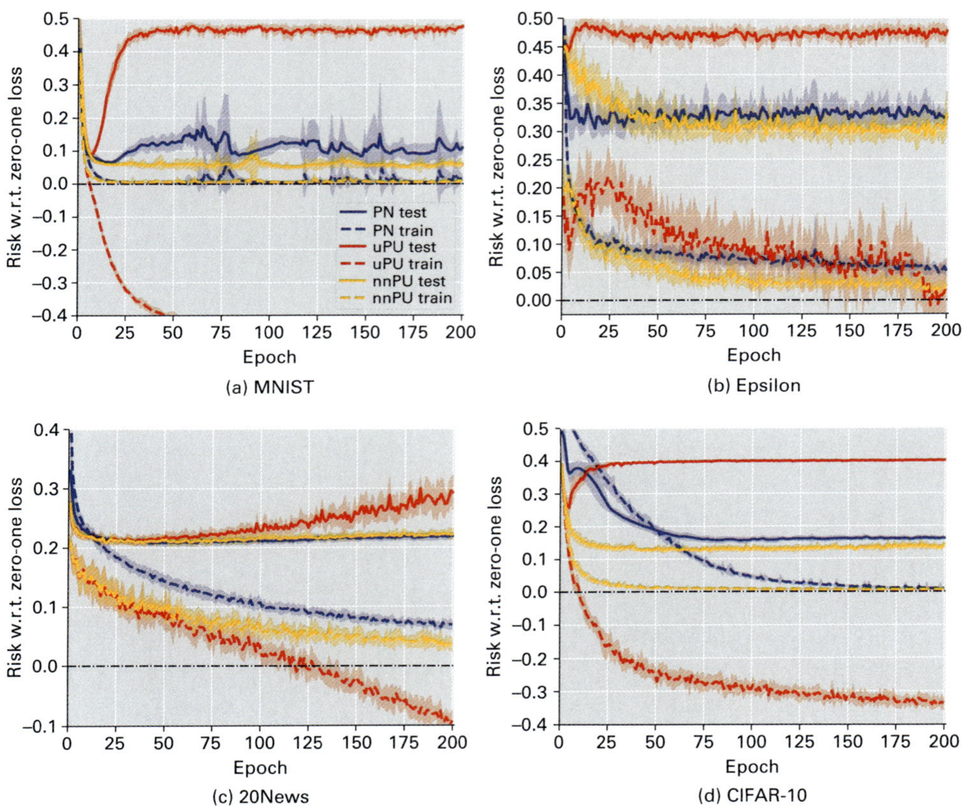

Figure 11.6
Comparison of PN, uPU, and nnPU classification.

that uPU classification overfitted training data and nnPU classification fixed this problem. Additionally, given limited N data, nnPU classification outperformed PN classification on MNIST, epsilon, and CIFAR-10 and was comparable to it on 20News. In summary, with the non-negative risk estimator, we are able to use very flexible models given limited P data.

We further tried some cases where π_P is misspecified, in order to simulate PU classification in the wild, where we must suffer from errors in estimating π_P. More specifically, we tested nnPU classification by replacing π_P with $\pi_P' \in \{0.8\pi_P, 0.9\pi_P, \ldots, 1.2\pi_P\}$ and giving π_P' to the classification method so that it would regard π_P' as π_P during the entire training process. The experimental setup was exactly the same as before except for the replacement of π_P.

The experimental results are reported in figure 11.7, where means of test risks of nnPU classification based on the same 10 random samplings are shown, and the best test risks

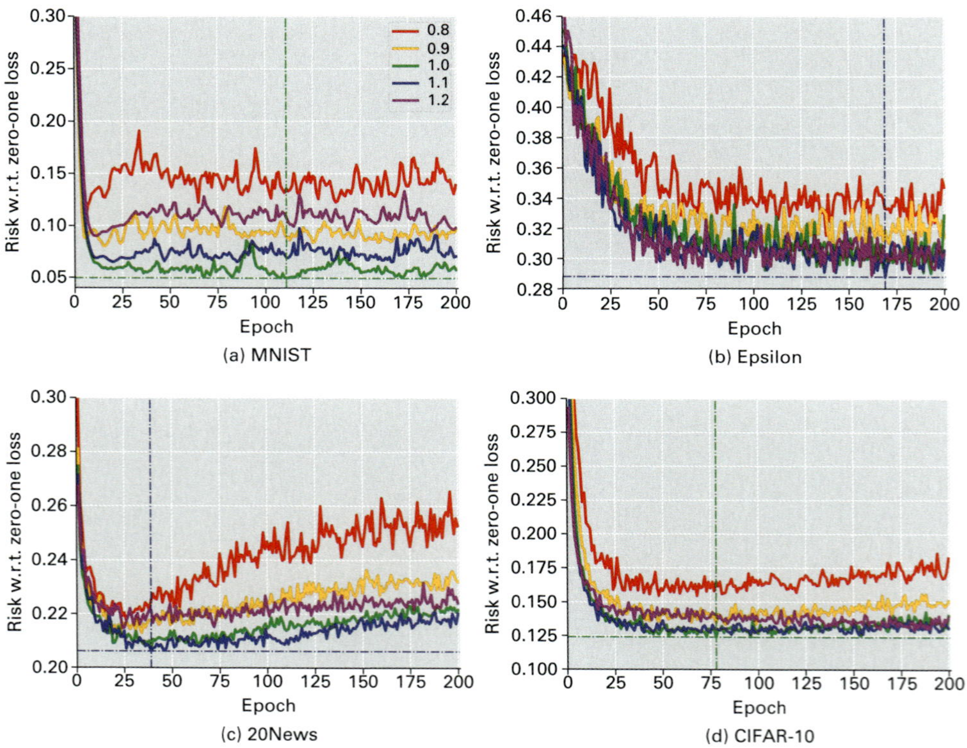

Figure 11.7
Experimental results given $\pi_{\mathrm{P}}' \in \{0.8\pi_{\mathrm{P}}, 0.9\pi_{\mathrm{P}}, \ldots, 1.2\pi_{\mathrm{P}}\}$.

are identified (horizontal lines are the best mean test risks and vertical lines are the epochs when they were achieved). We can see that on MNIST, the more the misspecification was, the worse nnPU classification performed, while under-misspecification hurt more than over-misspecification. On epsilon, when π_{P}' equals π_{P}, $1.1\pi_{\mathrm{P}}$ and $1.2\pi_{\mathrm{P}}$ were comparable, but the best was $\pi_{\mathrm{P}}' = 1.1\pi_{\mathrm{P}}$ rather than $\pi_{\mathrm{P}}' = \pi_{\mathrm{P}}$. On 20News, these three cases became different, such that $\pi_{\mathrm{P}}' = \pi_{\mathrm{P}}$ was superior to $\pi_{\mathrm{P}}' = 1.2\pi_{\mathrm{P}}$ but inferior to $\pi_{\mathrm{P}}' = 1.1\pi_{\mathrm{P}}$. At last, on CIFAR-10, $\pi_{\mathrm{P}}' = \pi_{\mathrm{P}}$ was the winner and $\pi_{\mathrm{P}}' = 1.1\pi_{\mathrm{P}}$ was comparable again.

In all the experiments, we have fixed $\beta = 0$, which may explain this phenomenon. Recall that uPU classification overfitted seriously on all the benchmark datasets, and note that the larger π_{P}' is, the more different nnPU classification is from uPU classification. Therefore, the replacement of π_{P} with some $\pi_{\mathrm{P}}' > \pi_{\mathrm{P}}$ introduces additional bias of $\widehat{R}_{\mathrm{nnPU}}(g)$ in estimating $R(g)$, but it also pushes $\widehat{R}_{\mathrm{nnPU}}(g)$ away from $\widehat{R}_{\mathrm{PU}}(g)$ and then pushes nnPU classification away from uPU classification. This may result in lower test risks given some π_{P}' slightly larger than π_{P} as shown in figure 11.7. This is also why under-misspecified π_{P}' hurt more than over-misspecified π_{P}'.

Table 11.2
Specification of benchmark datasets and models.

Dataset	# Train	# Test	# Feature	# Class	Models
MNIST	60,000	10,000	784	10	Linear, MLP
Fashion-MNIST	60,000	10,000	784	10	Linear, MLP
Kuzushiji-MNIST	60,000	10,000	784	10	Linear, MLP
CIFAR-10	50,000	10,000	3,072	10	DenseNet, ResNet

An implementation of the nnPU classification algorithm using Chainer (Tokui et al., 2015) is available from https://github.com/kiryor/nnPUlearning.

11.6.2   Comparison of uCL and nnCL Classification

In this section, we compare uCL classification (9.8) (denoted by Free), nnCL classification (11.16) (denoted by Max Operator), and nnCL classification using the gradient ascent trick (denoted by Gradient Ascent) where $\gamma = 1$. We also add two more baseline methods: Pairwise comparison (denoted by PC) with ramp loss (see section 9.2) and Forward correction from Yu et al. (2018).

Table 11.2 summarizes the benchmark datasets. MNIST[9] (LeCun et al., 1998) is a 10-class dataset of handwritten digits: $0, 1, 2, \ldots, 9$. Each sample is a 28×28 gray-scale image. Fashion-MNIST[10] (Xiao et al., 2017) is a 10-class dataset of fashion items: T-shirt/top, Trouser, Pullover, Dress, Coat, Sandal, Shirt, Sneaker, Bag, and Ankle boot. Each sample is a 28×28 gray-scale image. Kuzushiji-MNIST[11] (Clanuwat et al., 2018) is a 10-class dataset of cursive Japanese ("Kuzushiji") characters. Each sample is a 28×28 gray-scale image. CIFAR-10[12] is a 10-class dataset of various objects: airplane, automobile, bird, cat, deer, dog, frog, horse, ship, and truck. Each sample is a color image in $32 \times 32 \times 3$ RGB format. It is a subset of the 80 million tiny images dataset (Torralba et al., 2008).

The model $\boldsymbol{g}(\boldsymbol{x})$ and optimization algorithm $\mathscr{A}$ are chosen as follows:

MNIST, Fashion-MNIST, and Kuzushiji-MNIST: For models, a linear-in-input model with a bias term and a MLP model (d-500-1) with softmax cross-entropy loss function (except PC) were used. Weight decay was also applied. For optimization, Adam (Kingma and Ba, 2015) was used.

CIFAR-10: For models, DenseNet (Huang et al., 2017) and ResNet-34 (He et al., 2016) were used with weight decay. For optimization, stochastic gradient descent with momentum was used.

We show the accuracy for all 300 epochs on test data to demonstrate how the overfitting issues appear and how different implementations of section 11.4.4 are effective. The experimental results are reported in figure 11.8, where means and standard deviations of test accuracy for five trials on test data evaluated with ordinary labels are shown.

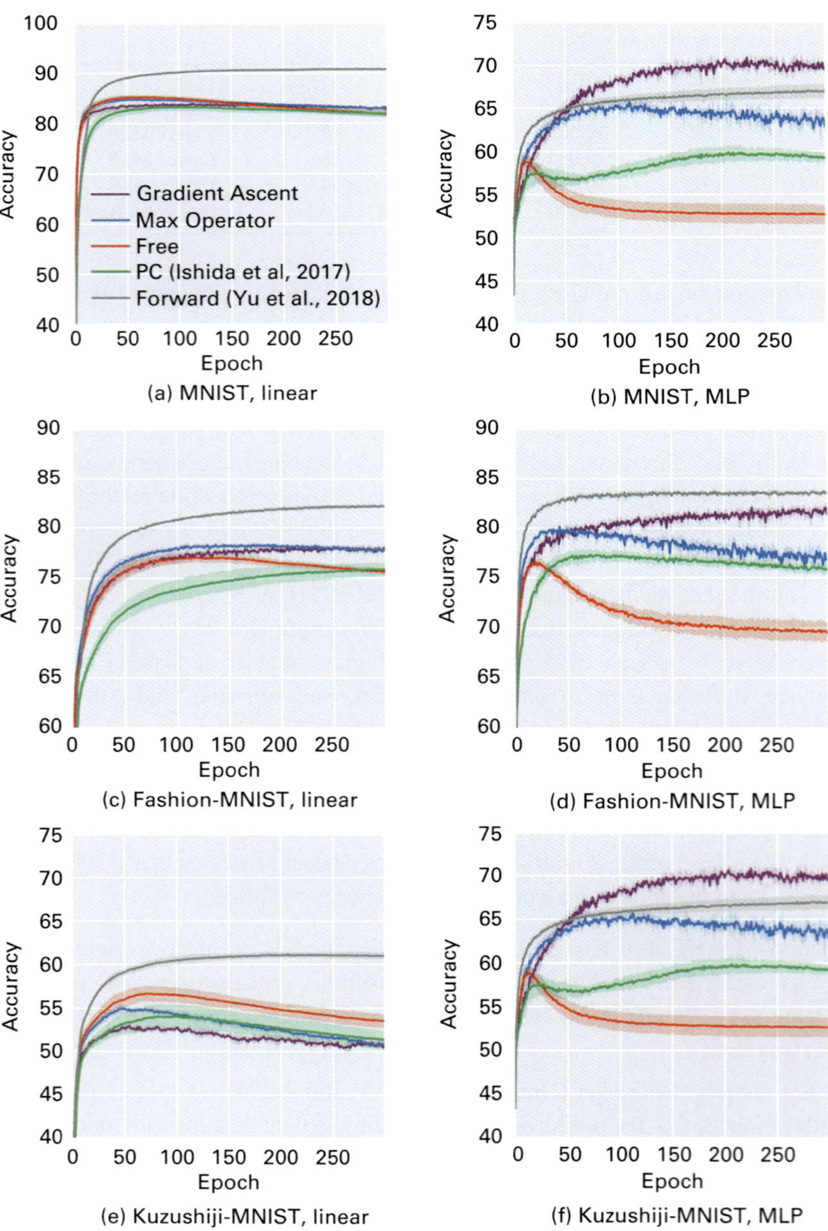

Figure 11.8
Comparison of uCL and nnCL classification. Dark colors show the mean accuracy of five trials and light colors show standard deviation.

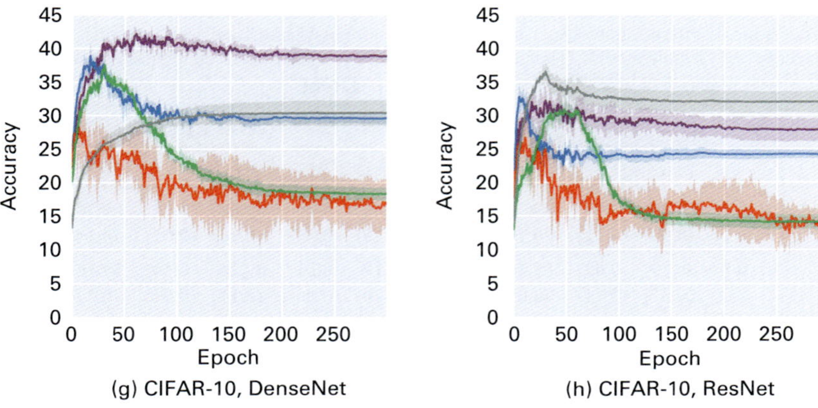

(g) CIFAR-10, DenseNet   (h) CIFAR-10, ResNet

Figure 11.8
(continued)

First, we compare uCL classification (Free) and nnCL classification (Max Operator and Gradient Ascent). For linear models in MNIST, Fashion-MNIST, and Kuzushiji-MNIST, the three methods work similarly. However, in the case of using a more flexible MLP model or using DenseNet/ResNet in CIFAR-10, we can see that Free is the worst, Max Operator is better, and Gradient Ascent is the best out of the proposed three methods for most of the epochs (Free < Max Operator < Gradient Ascent).

Next, we compare with the baseline methods. For linear models, the Forward method seems to work well. However, for deep models (MLP, DenseNet, and ResNet), the superiority stands out for Gradient Ascent for many datasets, but in some cases, the Forward method seems to be a good choice.

An implementation of the nnCL classification algorithm using PyTorch (Paszke et al., 2019) is available from https://github.com/takashiishida/comp.

11.6.3   Comparison of uUU and ccUU Classification

In this section, we experimentally compare uUU classification with three implementations of ccUU classification, i.e., using ReLU function (ccUU-ReLU), this corresponds to nnUU classification (11.15); using absolute value function (ccUU-ABS); and using generalized leaky ReLU function (ccUU-LReLU). We add a biased UU (bUU) classification baseline (i.e., supervised classification taking the U set with larger class-prior as P data and the other U set with smaller class-prior as N data), which is a straightforward method to handle the UU classification problem. We test them on simple and deep models, using different datasets and class-prior setups.

Again, MNIST, Fashion-MNIST, Kuzushiji-MNIST, and CIFAR-10 benchmark datasets were used. As shown in table 11.2, they have 10 classes originally, and we constructed

Table 11.3
Means (standard deviations) of the accuracy (Acc) and the drop (Δ_A) for bUU, uUU, and ccUU classification over five trials in percentage with simple models. The best and comparable methods based on the paired t-test at the significance level 5% are highlighted in boldface.

Dataset	θ, θ'	bUU Acc	Δ_A	uUU Acc	Δ_A	ccUU-ABS Acc	Δ_A	ccUU-ReLU Acc	Δ_A	ccUU-LReLU Acc	Δ_A
MNIST	0.8, 0.2	89.30 (0.09)	0.28 (0.12)	**89.76** **(0.13)**	0.10 (0.04)	**89.80** **(0.20)**	0.10 (0.03)	**89.68** **(0.14)**	0.14 (0.07)	**89.70** **(0.14)**	0.12 (0.07)
	0.7, 0.3	88.59 (0.22)	0.42 (0.11)	**89.26** **(0.09)**	0.11 (0.07)	**89.19** **(0.20)**	0.21 (0.06)	**89.24** **(0.11)**	0.14 (0.07)	**89.15** **(0.27)**	0.20 (0.06)
	0.6, 0.4	84.64 (0.42)	2.03 (0.15)	**87.15** **(0.34)**	0.54 (0.22)	**87.28** **(0.38)**	0.49 (0.23)	**87.13** **(0.33)**	0.60 (0.25)	**87.26** **(0.37)**	0.40 (0.08)
Fashion-MNIST	0.8, 0.2	**87.27** **(0.83)**	0.82 (0.73)	**87.73** **(0.11)**	0.43 (0.06)	**87.72** **(0.11)**	0.44 (0.06)	**87.78** **(0.20)**	0.35 (0.14)	**87.78** **(0.20)**	0.39 (0.13)
	0.7, 0.3	85.53 (0.93)	2.02 (0.77)	**86.99** **(0.17)**	0.73 (0.14)	**87.02** **(0.35)**	0.71 (0.26)	**87.07** **(0.28)**	0.72 (0.20)	**86.84** **(0.56)**	0.97 (0.52)
	0.6, 0.4	80.66 (2.22)	4.59 (1.91)	**83.69** **(0.53)**	2.70 (0.46)	**84.18** **(0.57)**	2.41 (0.57)	**84.20** **(0.44)**	2.08 (0.63)	**83.92** **(1.07)**	2.59 (1.09)
Kuzushiji-MNIST	0.8, 0.2	72.73 (0.39)	1.59 (0.45)	**79.19** **(0.29)**	0.39 (0.18)	**79.28** **(0.29)**	0.39 (0.18)	**79.28** **(0.29)**	0.39 (0.18)	**79.32** **(0.19)**	0.35 (0.20)
	0.7, 0.3	72.21 (0.74)	1.91 (0.52)	**78.67** **(0.34)**	0.75 (0.25)	**78.89** **(0.40)**	0.75 (0.23)	**78.79** **(0.21)**	0.63 (0.25)	**78.90** **(0.40)**	0.63 (0.28)
	0.6, 0.4	69.76 (0.46)	3.14 (0.70)	**77.73** **(0.37)**	1.24 (0.24)	**77.95** **(0.71)**	1.20 (0.43)	**77.84** **(0.65)**	1.26 (0.36)	**77.86** **(0.72)**	1.19 (0.29)
CIFAR-10	0.8, 0.2	76.94 (5.49)	4.62 (5.35)	**80.50** **(1.20)**	1.49 (1.22)	**80.48** **(1.19)**	1.50 (1.21)	**80.82** **(0.69)**	1.07 (0.58)	**81.13** **(0.51)**	0.76 (0.41)
	0.7, 0.3	**78.04** **(2.02)**	2.22 (2.18)	**79.68** **(0.66)**	1.54 (0.56)	**80.12** **(0.42)**	1.20 (0.35)	**80.28** **(0.14)**	1.03 (0.21)	**79.95** **(0.67)**	1.32 (0.59)
	0.6, 0.4	67.23 (6.68)	9.05 (6.77)	**76.34** **(1.41)**	3.74 (1.51)	**75.21** **(1.95)**	4.81 (1.93)	**76.24** **(0.96)**	3.85 (0.99)	**76.28** **(0.92)**	3.72 (1.06)

the P and N classes from them as follows: MNIST was preprocessed in such a way that even digits constitute the P class and the odd digits constitute the N class. For Fashion-MNIST, the P class was formed by "T-shirt," "Trouser," "Shirt," and "Sneaker," and the N class was formed by "Pullover," "Dress," "Coat," "Sandal," "Bag," and "Ankle boot." For Kuzushiji-MNIST, "ki," "re," and "wo" made up the P class, and "o," "su," "tsu," "na," "ha," "ma," and "ya" made up the N class. For CIFAR-10, the P class was composed of "bird," "deer," "dog," "frog," "ship," and "truck," and the N class was composed of "airplane," "automobile," "cat," and "horse." $\mathcal{X}_{\text{tr}}$ and $\mathcal{X}'_{\text{tr}}$ of the same sample size were drawn,[13] where the class-priors (θ, θ') were chosen as $(0.9, 0.1)$, $(0.8, 0.2)$, or $(0.7, 0.3)$.

The following models $g(\boldsymbol{x})$ are chosen:

Table 11.4

Means (standard deviations) of the accuracy (Acc) and the drop (Δ_A) for bUU, uUU, and ccUU classification over five trials in percentage with deep models. The best and comparable methods based on the paired t-test at the significance level 5% are highlighted in boldface.

Dataset	θ, θ'	bUU Acc	Δ_A	uUU Acc	Δ_A	ccUU-ABS Acc	Δ_A	ccUU-ReLU Acc	Δ_A	ccUU-LReLU Acc	Δ_A
MNIST	0.8, 0.2	80.56 (0.62)	14.99 (0.74)	78.01 (0.45)	18.07 (0.55)	**95.19** **(0.12)**	0.86 (0.13)	**95.15** **(0.43)**	1.11 (0.36)	**95.21** **(0.42)**	1.06 (0.42)
	0.7, 0.3	70.55 (0.66)	20.80 (0.75)	64.74 (0.78)	29.78 (0.84)	91.69 (1.13)	2.77 (1.08)	**93.01** **(0.39)**	1.88 (0.53)	**93.29** **(0.81)**	1.60 (0.76)
	0.6, 0.4	59.85 (0.59)	19.37 (1.11)	53.34 (0.88)	37.74 (1.31)	78.54 (1.19)	11.73 (1.27)	**88.11** **(1.48)**	3.37 (1.38)	**90.34** **(0.84)**	1.13 (0.66)
Fashion-MNIST	0.8, 0.2	81.51 (0.77)	9.56 (0.65)	80.19 (0.81)	11.40 (0.81)	**90.41** **(0.56)**	1.13 (0.43)	90.20 (0.53)	1.35 (0.37)	**90.90** **(0.26)**	0.64 (0.25)
	0.7, 0.3	72.07 (0.94)	16.23 (0.63)	71.93 (1.42)	18.48 (1.29)	87.84 (0.80)	2.43 (0.70)	**88.14** **(0.90)**	2.22 (1.00)	**89.39** **(0.18)**	0.97 (0.15)
	0.6, 0.4	61.58 (1.30)	17.89 (0.75)	63.01 (1.07)	25.10 (1.17)	80.86 (1.38)	6.83 (1.59)	83.78 (1.00)	4.11 (0.84)	**86.25** **(0.32)**	1.63 (0.24)
Kuzushiji-MNIST	0.8, 0.2	78.10 (0.69)	8.22 (0.83)	74.60 (0.71)	14.76 (0.77)	**86.62** **(1.11)**	2.85 (1.19)	**87.13** **(0.99)**	2.28 (0.87)	**87.56** **(0.62)**	1.85 (0.45)
	0.7, 0.3	70.77 (0.58)	10.95 (0.63)	66.40 (0.49)	21.12 (0.48)	83.79 (0.66)	3.81 (0.55)	**85.35** **(0.60)**	2.20 (0.41)	**85.65** **(0.29)**	1.60 (0.23)
	0.6, 0.4	61.70 (0.76)	11.44 (1.41)	60.12 (0.90)	23.59 (1.12)	77.82 (1.12)	5.79 (1.19)	80.52 (1.35)	3.32 (1.10)	**82.22** **(0.52)**	1.61 (0.27)
CIFAR-10	0.8, 0.2	74.28 (0.94)	10.76 (1.37)	76.12 (3.51)	11.48 (3.21)	**84.39** **(1.34)**	3.22 (1.04)	**84.47** **(1.68)**	3.18 (1.32)	**84.51** **(1.33)**	3.11 (0.95)
	0.7, 0.3	65.06 (0.46)	14.09 (1.59)	67.52 (3.07)	17.84 (2.85)	**80.53** **(1.52)**	4.84 (0.72)	**81.64** **(1.46)**	3.73 (1.19)	**81.26** **(2.51)**	4.08 (2.44)
	0.6, 0.4	57.12 (0.46)	12.52 (2.25)	57.26 (1.33)	23.95 (1.35)	71.53 (1.40)	9.30 (0.66)	76.83 (1.26)	4.10 (0.91)	**78.34** **(1.00)**	2.62 (0.69)

MNIST, Fashion-MNIST, and Kuzushiji-MNIST: The simple model used was a linear-in-input model $g(x) = \omega^T x + b$ where $\omega \in R^{784}$ and $b \in R$ with ℓ_2-regularization. The deep model used was a six-layer MLP with ReLU as the activation function: d-300-300-300-300-1 with ℓ_2-regularization.

CIFAR-10: The simple model used for training CIFAR-10 was also a linear-in-input model $g(x) = \omega^T x + b$ where $\omega \in R^{3072}$ and $b \in R$ with ℓ_2-regularization. The deep model was a 32-layer *residual net* (ResNet) (He et al., 2016): (32*32*3)-C(3*3, 16)-[C(3*3, 16), C(3*3, 16)]*5-[C(3*3, 32), C(3*3, 32)]*5-[C(3*3, 64), C(3*3, 64)]*5-Global Average Pooling-1, where [·, ·] means a building block (He et al., 2016). Furthermore, batch normalization and ℓ_2-regularization were also applied.

As a common practice, Adam (Kingma and Ba, 2015) with logistic loss $\ell_{\log}(z) = \ln(1 + \exp(-z))$ was used for optimization. For all the experiments, the generalized leaky ReLU hyper-parameter γ was selected from -0.01 to 1. We trained 200 epochs, and besides the final classification accuracy (Acc), we also report the classification accuracy drop (Δ_A), which demonstrates the overfitting by quantifying the fall of performance during training.

We first test on simple models and report our results in table 11.3. We can see that the uUU method and three ccUU methods outperform the bUU method. The advantage increases as the classification task becomes harder, that is, the class-priors move closer.[14] Moreover, the overfitting issue in uUU method is not severe for linear models, but we can see the tendency that overfitting gets slightly worse when class-priors are closer.

Then we test on deep neural networks that are more flexible than the aforementioned simple models. Observations of the experimental results in table 11.4 are as follows. First, compared to simple model experiments, overfitting of the bUU and uUU methods become catastrophic: The performance drops behind their linear counterparts. An explanation may be that flexible models have larger capacity to fit patterns, and thus the empirical risk tends to be negative. And we observe that the closer the class-priors are, the more severe the overfitting is. Second, the non-negative correction methods significantly alleviate the overfitting even in the hardest classification scenario, and their classification accuracy improves compared to the simple model experiments. Among all methods, the ccUU-LReLU method achieves the best performance for all the datasets and class-prior settings and has the smallest performance drop when the class-priors get closer, which implies that it is relatively robust against the closeness of class-priors.

12   Class-Prior Estimation

In part II, various weakly supervised classification methods for *binary* classification have been introduced. However, there is still a missing component: When the risk estimators were designed, we assumed that the class-prior probabilities $\pi_\mathrm{P} := p(y = +1)$ and $\pi_\mathrm{N} := p(y = -1)$ were known; to implement some of those methods, estimates of the class-prior probabilities are needed. In this chapter, we show how to estimate them for those weakly supervised classification methods. Note that weakly supervised classification methods for *multi-class* classification explored in part III do not require class-prior estimation.

12.1   Introduction

The purpose of this chapter is to introduce class-prior estimation methods for weakly supervised classification. Nevertheless, before addressing this topic, we would like to emphasize that the supervised classification methods relying on multi-sample empirical risk estimators suffer from the same issue, and it may be even worse. For example, our starting point, PN classification discussed in chapter 3, is based on $\mathcal{X}_\mathrm{P}$ of size n_P from $p_\mathrm{P}(\boldsymbol{x}) := p(\boldsymbol{x} \mid y = +1)$ and $\mathcal{X}_\mathrm{N}$ of size n_N from $p_\mathrm{N}(\boldsymbol{x}) := p(\boldsymbol{x} \mid y = -1)$. In this setting, the class-conditional densities $p_\mathrm{P}(\boldsymbol{x})$ and $p_\mathrm{N}(\boldsymbol{x})$ are completely independent of the class-priors π_P and π_N. Therefore, from $\mathcal{X}_\mathrm{P}$ and $\mathcal{X}_\mathrm{N}$, we cannot extract any information on the class-prior probabilities. As a result, in the PN setting, class-prior estimation is theoretically impossible, which implies that classifier training is actually an easier component than class-prior estimation.

Class-prior estimation for PN classification is usually carried out with the help of unlabeled data $\mathcal{X}_\mathrm{U}$ of size n_U from $p_\mathrm{U}(\boldsymbol{x}) := \pi_\mathrm{P} p_\mathrm{P}(\boldsymbol{x}) + \pi_\mathrm{N} p_\mathrm{N}(\boldsymbol{x})$, in the same way as that for PNU classification discussed in chapter 5. Although it seems that the marginal density $p_\mathrm{U}(\boldsymbol{x})$ is independent of the class-prior probability $p(y)$ and thus does not include any information on the class-prior probabilities, $p_\mathrm{U}(\boldsymbol{x})$ in fact does contain sufficient information for class-prior estimation. In the PNU setting, class-prior estimation is a mature research topic and there are many *distribution matching* methods for the purpose (e.g., Saerens et al., 2002; du Plessis and Sugiyama, 2014b; Sugiyama et al., 2013; Iyer et al., 2014). The advantage of distribution matching is that no distributional assumption is required—for example, we do not need to assume that $p_{\mathrm{P}}(\boldsymbol{x})$ and $p_{\mathrm{N}}(\boldsymbol{x})$ are Gaussian and $p_{\mathrm{U}}(\boldsymbol{x})$ is a Gaussian mixture. We will discuss distribution matching in section 12.2.

On the other hand, in the PU and NU settings discussed in chapter 4, we cannot perform distribution matching naively, since either $\mathcal{X}_{\mathrm{N}}$ or $\mathcal{X}_{\mathrm{P}}$ is unavailable. Fortunately, class-prior estimation by *mixture proportion estimation* (Scott, 2015; Blanchard et al., 2010) and *partial distribution matching* (du Plessis et al., 2017) is possible, which will be discussed in section 12.3 and section 12.4, respectively. We also introduce a specific implementation of partial distribution matching based on a penalized divergence in section 12.5. Even though mixture proportion estimation has no distributional assumption, it instead requires strong assumptions that can never be hypothesis-tested from data; if these assumptions are violated, the class-prior probabilities are systematically overestimated. To systematically reduce the positive bias of any mixture proportion estimation method, there is a general technique called *regrouping* (Yao et al., 2022), which will be introduced in section 12.6. Employing this regrouping technique before mixture proportion estimation is promising, as we cannot know whether the assumptions hold or not in practice.

The cases of SD, SU, DU, and SDU settings explored in chapter 7 are analogous to PN, PU, NU, and PNU settings, and hence their class-prior estimation can also be performed by full/partial distribution matching and mixture proportion estimation. Consequently, we can systematically estimate the class-prior probabilities for PN classification, PU classification, NU classification, PNU classification, and pairwise-constraint (SD, SU, DU, and SDU) classification. Note that we do not need to estimate the class-prior probabilities for Pconf classification in chapter 6, complementary-label classification in chapter 9, or partial-label classification in chapter 10. We should also mention that class-prior estimation for UU classification in chapter 8 is only possible with stronger assumptions than those for PU class-prior estimation (Scott, 2015; Menon et al., 2015). The assumptions cannot be hypothesis-tested from data, either; when the assumptions are violated, the direction of the bias is unclear, and we do not have practically reliable estimation methods (such as regrouping) for it up to now. Thus, for UU classification, we have to obtain the class-prior probabilities from other sources such as domain knowledge.

In fact, it should be a good idea to obtain the class-prior probabilities from other sources even for the settings where we can estimate them from given data. This allows us to obtain (estimates of) the class-prior probabilities without additional mathematical assumptions; moreover, we do not need to incur the computational cost for class-prior estimation. If we are lucky, we may find the necessary information from public sources for free—for instance, governments frequently release statistics about economics and social science. Otherwise, we should prepare a tiny subset of $\mathcal{X}_{\mathrm{U}}$, denoted by $\mathcal{X}_{\mathrm{U}}'$, and ask domain experts to label $\mathcal{X}_{\mathrm{U}}'$ or give us the class-prior probabilities of $\mathcal{X}_{\mathrm{U}}'$ directly. The tiny subset $\mathcal{X}_{\mathrm{U}}'$ with labels should be enough for estimating the class-prior probabilities that are just a few scalars, and labeling such a tiny subset would not cost much.

12.2 Full Distribution Matching

Distribution matching is the foundation of class-prior estimation for PN classification (discussed in chapter 3) and PNU classification (discussed in Chapter 5). The intuition behind it can easily be explained as follows.

Suppose that we can directly access $p_P(x)$, $p_N(x)$, and $p_U(x)$ rather than extracting the information about them from $\mathcal{X}_P$, $\mathcal{X}_N$, and $\mathcal{X}_U$. Given a variable θ such that $0 \leq \theta \leq 1$, let us define a *mixture model*, i.e., a mixture distribution with θ as the mixture proportion on the sample space $\mathcal{X}$, as

$$q_\theta(x) := \theta p_P(x) + (1 - \theta)p_N(x).$$

We would like $q_\theta(x)$ to be a model of $p_U(x)$ and θ to be a model of π_P. Let $\mathrm{Div}(p\|q)$ be a proper divergence measure such that $\mathrm{Div}(p\|q) = 0$ if and only if $p = q$, where $p = q$ means that $p(x) = q(x)$ for every $x \in \mathcal{X}$ except a set of zero probability measure according to $p(x)$:[1]

$$\int (p(x) - q(x))^2 p(x)\mathrm{d}x = 0.$$

Then, we can fit the model $q_\theta(x)$ to the ground-truth $p_U(x)$ by either finding the root of $\mathrm{Div}(p_U\|q) = 0$ or minimizing the divergence as

$$\theta^* := \underset{0 \leq \theta \leq 1}{\mathrm{argmin}}\, \mathrm{Div}(p_U\|q_\theta).$$

Thus, the true class-prior probability π_P is just the divergence minimizer θ^*.

In practice, we have to perform distribution matching on the data instead of the distributions. A naive idea is to employ some *density estimation* methods, such as kernel density estimation (Rosenblatt, 1956; Parzen, 1962), for $p_P(x)$, $p_N(x)$, and $p_U(x)$, and try

$$\underset{0 \leq \theta \leq 1}{\mathrm{argmin}}\, \mathrm{Div}(\widehat{p}_U\|\widehat{q}_\theta), \quad \text{where} \quad \widehat{q}_\theta(x) := \theta\widehat{p}_P(x) + (1 - \theta)\widehat{p}_N(x).$$

Here, $\widehat{p}_P(x)$, $\widehat{p}_N(x)$, and $\widehat{p}_U(x)$ are corresponding density estimators, respectively. However, this density estimation approach is actually not a good idea: We suffer from the estimation errors in density estimation, and they may be enlarged in the computation of $\mathrm{Div}(\widehat{p}_U\|\widehat{q}_\theta)$ and its minimization with respect to θ, especially when $\widehat{p}_U(x)$ or $\widehat{q}_\theta(x)$ appears in the denominator. Hence, we strongly recommend employing some *divergence estimation* methods[2] to directly obtain $\widehat{\mathrm{Div}(p_U\|q_\theta)}$ without explicitly estimating $p_P(x)$, $p_N(x)$, and $p_U(x)$, so that the minimization of $\widehat{\mathrm{Div}(p_U\|q_\theta)}$ with respect to θ is more reliable. After we employ divergence estimation, we may encounter that $\widehat{\mathrm{Div}(p_U\|q_\theta)} < 0$ or $\min_q \widehat{\mathrm{Div}(p_U\|q_\theta)} > 0$, and thus we should minimize $\widehat{\mathrm{Div}(p_U\|q_\theta)}$ but not find the root of $\widehat{\mathrm{Div}(p_U\|q_\theta)} = 0$.

For the purpose of class-prior estimation equipped with divergence estimation, various divergence measures can be adopted, such as the Kullback-Leibler divergence

(Saerens et al., 2002), the Pearson divergence (du Plessis and Sugiyama, 2014b), the L_2-distance (Sugiyama et al., 2013), and the maximum mean discrepancy (Iyer et al., 2014).

Specifically, a standard choice of the divergence measure here would be the f-*divergence* (Ali and Silvey, 1966; Csiszár, 1967):

$$\mathrm{Div}(p\|q) := \int f\left(\frac{q(\boldsymbol{x})}{p(\boldsymbol{x})}\right) p(\boldsymbol{x})\mathrm{d}\boldsymbol{x},$$

where $f\colon [0, +\infty) \to \mathbb{R}$ is a convex function such that $f(1) = 0$. The equation requires that $q(\boldsymbol{x})$ is *absolutely continuous* with respect to $p(\boldsymbol{x})$, i.e., $p(\boldsymbol{x}) = 0$ implies $q(\boldsymbol{x}) = 0$. The requirement must be automatically satisfied in the context of class-prior estimation: We can naturally assume $0 < \pi_{\mathrm{P}} < 1$, so that $p_{\mathrm{U}}(\boldsymbol{x}) = 0$ means both $p_{\mathrm{P}}(\boldsymbol{x}) = 0$ and $p_{\mathrm{N}}(\boldsymbol{x}) = 0$ and thus $q_\theta(\boldsymbol{x}) = 0$. When $f(t) = -\ln(t)$, the f-divergence is reduced to the *Kullback-Leibler divergence* (Kullback and Leibler, 1951):

$$\int \ln\left(\frac{p(\boldsymbol{x})}{q(\boldsymbol{x})}\right) p(\boldsymbol{x})\mathrm{d}\boldsymbol{x}.$$

When $f(t) = \frac{1}{2}(t-1)^2$, the f-divergence corresponds to the *Pearson divergence* (Pearson, 1900):

$$\frac{1}{2}\int \left(\frac{q(\boldsymbol{x})}{p(\boldsymbol{x})} - 1\right)^2 p(\boldsymbol{x})\mathrm{d}\boldsymbol{x} = \frac{1}{2}\int \frac{q(\boldsymbol{x})}{p(\boldsymbol{x})} q(\boldsymbol{x})\mathrm{d}\boldsymbol{x} - \frac{1}{2}.$$

For $f(t) = |t-1|$, the f-divergence agrees with the L_1-*distance*:

$$\int |p(\boldsymbol{x}) - q(\boldsymbol{x})|\mathrm{d}\boldsymbol{x}.$$

Another popular choice is the L_2-*distance* (Sugiyama et al., 2013):

$$\int (p(\boldsymbol{x}) - q(\boldsymbol{x}))^2\mathrm{d}\boldsymbol{x},$$

which looks similar to the Pearson divergence and the L_1-distance, but it is not an f-divergence.

To be clear, a divergence measure and a distance metric are mathematically different. A divergence measure may only be a positive-definite function, i.e., $\mathrm{Div}(p\|q) \geq 0$, and $\mathrm{Div}(p\|q) = 0$ if and only if $p = q$. A distance metric must also satisfy the symmetry and the triangle inequality. The L_1-distance and L_2-distance are indeed distance metrics, and thus we have $\mathrm{Div}(p\|q) = \mathrm{Div}(q\|p)$ and $\mathrm{Div}(p\|q) \leq \mathrm{Div}(p\|r) + \mathrm{Div}(r\|q)$ for any $p(\boldsymbol{x})$, $q(\boldsymbol{x})$, and $r(\boldsymbol{x})$ on $\mathscr{X}$.

12.3   Mixture Proportion Estimation

Now, let us focus on class-prior estimation for PU classification. Since no negative data is available in the PU setting, we cannot perform full distribution matching anymore.

Fortunately, we have two mathematical tools for this task: *mixture proportion estimation* and *partial distribution matching*. The former can solve the problem under strong assumptions without any bias, while the latter is assumption-free but has a positive bias. In this section, we explain the mixture proportion estimation approach; partial distribution matching will be explored in the next section. The *mixture proportion estimation* problem (Scott, 2015) looks similar to class-prior estimation for PU classification, but it actually has a fairly different goal.

12.3.1   Estimation Goal and Optimization Goal

In mixture proportion estimation, there are a mixture density $\mu_F(x)$ and two component densities $\mu_G(x)$ and $\mu_H(x)$ on the sample space $\mathcal{X}$, such that

$$\mu_F = (1 - \kappa)\mu_G + \kappa\mu_H, \tag{12.1}$$

where the mixture proportion κ satisfies $0 \leq \kappa \leq 1$. In (12.1), μ_F and μ_H are observable, while μ_G is unobservable. Note that μ_F, μ_G, and μ_H are not related by any joint distribution of x and y in the original context of mixture proportion estimation. Under this setting, we hope to estimate κ given the samples from μ_F and μ_H. However, this is unfortunately not possible: κ is *unidentifiable* even if we can access μ_F and μ_H, since μ_F, μ_G, and μ_H are not related and μ_G is free. Specifically, as we will see in section 12.3.3, any κ' such that $0 < \kappa' < \kappa$ is also a valid mixture proportion for (12.1).

Therefore, an *achievable* goal in mixture proportion estimation is to find the maximum proportion of μ_H in μ_F (cf. Blanchard et al., 2010):

$$\kappa^* := \sup\{\kappa \mid \exists\mu_G : \mu_F = (1 - \kappa)\mu_G + \kappa\mu_H\}. \tag{12.2}$$

Under certain conditions, κ may become identifiable (see section 12.3.3 for details); in such cases, we ensure $\kappa = \kappa^*$. Thus, under those conditions, the estimation goal (i.e., trying to estimate the true κ as appeared in (12.1)) and the optimization goal (i.e., trying to find κ^* as defined in (12.2)) are compatible with each other.

It is clear that (12.1) coincides with the equation

$$p_U(x) = \pi_P p_P(x) + \pi_N p_N(x), \tag{12.3}$$

if we let $\mu_F = p_U$, $\mu_G = p_N$, $\mu_H = p_P$, and $\kappa = \pi_P$. This would assign specific meanings to the variables in (12.1) in the context of PU classification, and therefore (12.1) can serve as a model of PU class-prior estimation. However, (12.1) can also serve as a model for other problems, while PU class-prior estimation may also have other models than mixture proportion estimation. Under the same conditions of the identifiability of κ (see section 12.3.3), π_P is equal to κ^* in (12.2); otherwise, π_P is strictly smaller than κ^* due to unidentifiability.

Here, we adopt the optimization goal (12.2) because it is achievable and guarantees that mixture proportion estimation always possesses outputs. If we adopt the estimation goal and try to obtain the true κ in (12.1), it may be unidentifiable. In this case, mixture proportion estimation has no output; it is even worse that we can never know whether the identifiability conditions are valid or not. As a consequence, it is more sensible to adopt the optimization goal, where μ_G and κ in (12.1) are regarded as two optimization variables as shown in (12.2).

In contrast to mixture proportion estimation (12.1), p_{N} and π_{P} are objective existence in PU class-prior estimation: They are fixed but unknown, and they cannot be regarded as optimization variables since all of p_{P}, p_{N}, p_{U}, and π_{P} are tightly related by the joint probability density $p(\boldsymbol{x}, y)$. Even when κ in (12.1) is unidentifiable, π_{P} is still estimable by, for example, *partial distribution matching* (shown in section 12.4). However, the estimate can be inaccurate and we will suffer a considerable positive bias, in the same way as κ^* in (12.2).

12.3.2　Redefinition of Optimization Goal

Equation (12.2) is not computationally friendly, as taking the supremum of an infinite set is not really a solvable optimization problem. A naive idea to overcome this issue is to express

$$\kappa = \frac{\mu_F - \mu_G}{\mu_H - \mu_G},$$

which is obtained by solving (12.1) for κ. Then, we can easily confirm that the optimal solution κ^* is given by

$$\kappa^* = \sup_{\mu_G} \frac{\mu_F - \mu_G}{\mu_H - \mu_G}.$$

Unfortunately, this is as difficult to solve as (12.2).

Recall that in (12.2), we are finding the maximum proportion of μ_H in μ_F subject to $\mu_G(\boldsymbol{x}) \geq 0$ for all $\boldsymbol{x} \in \mathcal{X}$ to ensure that μ_G is a valid density. As a result, κ^* must satisfy the following two conditions:

- $\kappa^* \mu_H(\boldsymbol{x}) \leq \mu_F(\boldsymbol{x})$ for all $\boldsymbol{x} \in \mathcal{X}$.

- $\kappa^* \mu_H(\boldsymbol{x}_0) = \mu_F(\boldsymbol{x}_0)$ for some special $\boldsymbol{x}_0$.

Thus, if both of μ_F and μ_H are directly accessible, κ^* can be obtained as

$$\kappa^* = \inf_{\boldsymbol{x} \in \mathcal{X}} \frac{\mu_F(\boldsymbol{x})}{\mu_H(\boldsymbol{x})}.$$

That being said, the above optimization is still computationally difficult since it is an optimization of $\boldsymbol{x}$ over $\mathcal{X}$ and $\mu_F(\boldsymbol{x})/\mu_H(\boldsymbol{x})$ can be discontinuous. To overcome this issue, we need formal probability theory. Let $\mathfrak{S}$ be a σ-algebra on $\mathcal{X}$ so that $(\mathcal{X}, \mathfrak{S})$ defines a measurable space (e.g., $\mathfrak{S}$ can be the power set $2^{\mathcal{X}}$), and let μ_F, μ_G, and μ_H be probability measures on $(\mathcal{X}, \mathfrak{S})$. Then,

$$\kappa^* = \inf_{S \in \mathfrak{S}, \mu_H(S) > 0} \frac{\mu_F(S)}{\mu_H(S)}, \tag{12.4}$$

which is an optimization of a set S over $\mathfrak{S}$. For (12.4), we can set $\mu_H(S) = \epsilon$ for $\epsilon > 0$, compute the infimum given $\mu_H(S) = \epsilon$, and let $\epsilon \to 0$ to obtain κ^*, where $\mu_F(S)/\mu_H(S)$ must be continuous as long as S has a non-zero measure. Hence, (12.4) is computationally friendly and can be solved stably, if μ_F and μ_H are directly accessible.

12.3.3   Irreducibility Assumption

We have claimed in section 12.3.1 that, under certain conditions, κ in (12.1) is identifiable and κ^* in (12.2) is equal to π_P. Here, we explain those conditions. Without loss of generality, we focus on PU class-prior estimation with $\mu_F = p_U$, $\mu_G = p_N$, and $\mu_H = p_P$ and study when $\kappa^* = \pi_P$ holds.

To see why $\kappa^* = \pi_P$ may not hold, suppose that there exists δ such that $0 < \delta < \pi_N$ and $\pi_N p_N(x) \geq \delta p_P(x)$ over $\mathcal{X}$. Let

$$\mu_G' = \frac{\pi_N p_N - \delta p_P}{\pi_N - \delta}.$$

Then, p_U can also be expressed as

$$p_U = (\pi_N - \delta)\mu_G' + (\pi_P + \delta)p_P,$$

and if we find the maximum proportion $\kappa^* = \inf_{x \in \mathcal{X}} p_U(x)/p_P(x)$, it would be at least $\pi_P + \delta$. As

$$\pi_P p_P \leq p_U \quad \text{with} \quad 0 < \pi_P < 1$$

means that p_P is a component of p_U,

$$\frac{\delta}{\pi_N} p_P \leq p_N \quad \text{with} \quad 0 < \frac{\delta}{\pi_N} < 1$$

means that p_P is a component of p_N. This is why $\kappa^* > \pi_P$ can hold, and this is what we want to avoid.

The *irreducibility assumption* (Blanchard et al., 2010) guarantees that p_P is not a component of p_N. More specifically, if p_N is irreducible with respect to p_P, p_N is not a mixture distribution containing p_P. Thus, it is impossible to write $p_N = \delta p_P + (1 - \delta)\mu_G'$, where μ_G' is a density on $\mathcal{X}$ and $0 < \delta \leq 1$. For p_P and p_N satisfying the irreducibility assumption,

$$\inf_{x \in \mathcal{X}} \frac{p_N(x)}{p_P(x)} = 0,$$

and with (12.3), we have

$$\inf_{x \in \mathcal{X}} \frac{p_U(x)}{p_P(x)} = \pi_P + \inf_{x \in \mathcal{X}} \frac{\pi_N p_N(x)}{p_P(x)} = \pi_P.$$

The irreducibility assumption is equivalent to that there are anchor points in the sample space $\mathcal{X}$. A data point x_0 is said to be an *anchor point* (Liu and Tao, 2015) for the positive class if $p(y=+1\,|\,x=x_0)=1$. Namely, x_0 belongs to the positive class in a deterministic manner and it serves as an anchor to "moor" the positive class to its neighborhood. For an anchor point x_0,

$$p_\mathrm{P}(x_0) = \frac{p(x=x_0, y=+1)}{p(y=+1)} = \frac{p(x_0)p(y=+1\,|\,x=x_0)}{\pi_\mathrm{P}} = \frac{p(x_0)}{\pi_\mathrm{P}} > 0,$$

$$p_\mathrm{N}(x_0) = \frac{p(x=x_0, y=-1)}{p(y=-1)} = \frac{p(x_0)p(y=-1\,|\,x=x_0)}{\pi_\mathrm{N}} = 0.$$

Thus, we can see that x_0 is the optimal solution of $\inf_{x\in\mathcal{X}} p_\mathrm{N}(x)/p_\mathrm{P}(x)$ and is also where $p_\mathrm{N}(x)/p_\mathrm{P}(x)=0$ is attained. Therefore, p_N is irreducible with respect to p_P if and only if there is at least one anchor point for the positive class.

12.3.4   Anchor Set/Point Assumption

Nevertheless, the irreducibility assumption is not enough to design computationally efficient algorithms for class-prior estimation. Let us take a closer look at the nuance of the infima in two equivalent expressions:

$$\inf_{x\in\mathcal{X}} \frac{p_\mathrm{N}(x)}{p_\mathrm{P}(x)} = 0 \quad \text{and} \quad \inf_{S\in\mathfrak{S}, p_\mathrm{P}(S)>0} \frac{p_\mathrm{N}(S)}{p_\mathrm{P}(S)} = 0.$$

The former infimum can be replaced with $\min_{x\in\mathcal{X}}$, whereas the latter cannot be replaced with $\min_{S\in\mathfrak{S}, p_\mathrm{P}(S)>0}$ since the feasible region is an open set and the minimizer may be outside the feasible region. Denote by

$$S_0 := \{x_0\,|\,p(y=+1\,|\,x=x_0)=1\}$$

the set of all anchor points for the positive class. If $p_\mathrm{P}(S_0)=0$, meaning that there are finite or countably infinite anchor points, we will have no chance to find a set S such that $p_\mathrm{P}(S)>0$ and $p_\mathrm{N}(S)=0$. In other words, if $p_\mathrm{P}(S_0)=0$, the existing witnesses to $p_\mathrm{N}(x)/p_\mathrm{P}(x)=0$ are too few. Thus, it is extremely hard to find a witness, and the convergence rate of

$$\lim_{\epsilon\to 0} \inf_{S\in\mathfrak{S}, p_\mathrm{P}(S)=\epsilon} \frac{p_\mathrm{N}(S)}{p_\mathrm{P}(S)}$$

can be arbitrarily slow with respect to ϵ, since the infimum can be continuous but not Lipschitz continuous in ϵ.

To overcome this issue, stronger assumptions are needed to increase the chance of finding a witness. The *anchor set assumption* (Liu and Tao, 2015) suffices: There is an anchor set for the positive class of a non-zero measure in the sample space $\mathcal{X}$, i.e., there is $S_0 \subset \mathcal{X}$ such that $p(y=+1\,|\,x=x_0)=1$ for all $x_0 \in S_0$ and $p_\mathrm{P}(S_0)>0$. Thanks to the existence of the anchor set, we can enjoy

$$\min_{S\in\mathfrak{S}, p_\mathrm{P}(S)>0} \frac{p_\mathrm{N}(S)}{p_\mathrm{P}(S)} = 0,$$

where the anchor set is a minimizer inside the feasible region. The anchor set assumption guarantees that we can easily sample an anchor point from $\pi_P(x)$. Suppose that $p_P(S_0) = \epsilon$, where S_0 is the largest possible anchor set. Then, there is at least one anchor point in $\mathscr{X}_P$ with probability at least $1 - (1 - \epsilon)^{n_P}$.

The *anchor point assumption* (Liu and Tao, 2015) is stronger and guarantees that we have already sampled an anchor point and it is now in $\mathscr{X}_P$. Under this assumption, as $\mathscr{X}_P$ possesses at least one anchor point, we have

$$\min_{x_i^P \in \mathscr{X}_P} \frac{p_U(x_i^P)}{p_P(x_i^P)} = \pi_P,$$

where the feasible region is only a finite set of size n_P, and this optimization must be computationally extremely efficient.

In practice, we cannot access $p_P(x)$ and $p_U(x)$ directly, and we have to estimate the density ratio function,

$$\frac{p_P(x)}{p_U(x)},$$

based on $\mathscr{X}_P$ and $\mathscr{X}_U$.[3] This is a fundamental problem in machine learning and artificial intelligence known as *density ratio estimation* (Sugiyama et al., 2012). Following the representative *least-squares importance fitting* method (Kanamori et al., 2009), it holds that

$$\frac{p_P(x)}{p_U(x)} = \underset{r}{\mathrm{argmin}} \; \frac{1}{2} \int \left(\frac{p_P(x)}{p_U(x)} - r(x) \right)^2 p_U(x) \mathrm{d}x$$

$$= \underset{r}{\mathrm{argmin}} \; \frac{1}{2} \int r(x)^2 p_U(x) \mathrm{d}x - \int r(x) p_P(x) \mathrm{d}x.$$

Thus, in order to estimate the density ratio $p_P(x)/p_U(x)$, we can practically obtain a solution through

$$\widehat{r} := \underset{r \in \mathscr{R}}{\mathrm{argmin}} \; \frac{1}{n_U} \sum_{i=1}^{n_U} r(x_i^U)^2 - \frac{2\pi_P}{n_P} \sum_{i=1}^{n_P} r(x_i^P) + \lambda \Omega(r), \tag{12.5}$$

where $\mathscr{R}$ is some function class, $\Omega(\cdot)$ is a regularizer, and $\lambda \geq 0$ is a regularization parameter. With a density ratio estimator $\widehat{r}(x)$, we can simply estimate π_P by

$$\widehat{\pi}_P := \min_{x_i^P \in \mathscr{X}_P} \frac{1}{\widehat{r}(x_i^P)}. \tag{12.6}$$

Without loss of generality, assume that $n_U > n_P$.[4] Since the estimated density ratio function $\widehat{r}(x)$ in (12.5) converges to the true density ratio function $p_P(x)/p_U(x)$ in $\mathcal{O}_p(1/\sqrt{n_P})$, the estimated class-prior probability $\widehat{\pi}_P$ in (12.6) also converges to the true class-prior probability π_P in $\mathcal{O}_p(1/\sqrt{n_P})$, under the anchor point assumption. Given that class-prior estimation is essentially an estimation problem but not an optimization problem, this order, $\mathcal{O}_p(1/\sqrt{n_{\mathrm{P}}})$, is already the optimal parametric rate. Therefore, the density-ratio-estimation-based estimator (12.6) is already one of the best possible PU class-prior estimators we can have, as long as the anchor point assumption holds.

12.3.5   Remarks

Before ending this section, we would like to give a few remarks on this mixture proportion estimation approach to PU class-prior estimation.

The irreducibility and anchor set/point assumptions cannot be verified in practice: Since p_{N} is unobservable, we can never know whether they are satisfied or not given p_{P} and p_{U}, let alone hypothesis-test them given $\mathcal{X}_{\mathrm{P}}$ and $\mathcal{X}_{\mathrm{U}}$. Those assumptions are also indispensable to the one-sample case of PU classification for estimating the probability of converting unlabeled positive data into labeled positive data (see section 4.2). Let $s \in \{0, 1\}$ be the positive-vs.-unlabeled label, while $y \in \{-1, 1\}$ is the positive-vs.-negative label. Then, if we can find an anchor point x_0 for the positive class in $\mathcal{X}$ or the set of labeled positive data, we can enjoy

$$p(s = 1 \mid y = 1) = p(s = 1 \mid x = x_0),$$

where $p(s = 1 \mid x)$ can be estimated from labeled positive and unlabeled data; if we cannot find any anchor point, $p(s = 1 \mid y = 1)$ is likewise unidentifiable.

Last but not least, recall that $p_{\mathrm{U}}(S)/p_{\mathrm{P}}(S)$ may be more stable with respect to $S \in \mathfrak{S}$ than $p_{\mathrm{U}}(x)/p_{\mathrm{P}}(x)$ with respect to $x \in \mathcal{X}$, since $p_{\mathrm{P}}(x)$ and $p_{\mathrm{U}}(x)$ may only be Lebesgue-integrable. However, when $p_{\mathrm{U}}(x)/p_{\mathrm{P}}(x)$ is to be estimated, $p_{\mathrm{P}}(x)$ and $p_{\mathrm{U}}(x)$ must be continuous, and then there is no difference in terms of stability between $\inf_{x \in \mathcal{X}} p_{\mathrm{U}}(x)/p_{\mathrm{P}}(x)$ and $\inf_{S \in \mathfrak{S}, p_{\mathrm{P}}(S) > 0} p_{\mathrm{U}}(S)/p_{\mathrm{P}}(S)$.

12.4   Partial Distribution Matching

In the previous section, we detailed the mixture proportion estimation approach to PU class-prior estimation. In this section, we introduce another approach called *partial distribution matching*. While mixture proportion estimation can solve the PU class-prior estimation problem under strong assumptions without any bias, partial distribution matching is assumption-free but has a positive bias.

12.4.1   Formulation

In the PNU setting discussed in chapter 5, we have $\mathcal{X}_{\mathrm{P}}$, $\mathcal{X}_{\mathrm{N}}$, and $\mathcal{X}_{\mathrm{U}}$ to perform *full* distribution matching:

$$\widehat{\pi}_{\mathrm{P}} := \underset{0 \leq \theta \leq 1}{\operatorname{argmin}} \ \widehat{\operatorname{Div}}(p_{\mathrm{U}} \| q_\theta),$$

where $\widehat{\mathrm{Div}}(p\|q)$ is an empirical divergence from q to p and

$$q_\theta(\boldsymbol{x}) := \theta p_{\mathrm{P}}(\boldsymbol{x}) + (1-\theta)p_{\mathrm{N}}(\boldsymbol{x})$$

is a model of $p_{\mathrm{U}}(\boldsymbol{x})$, and θ is a model of π_{P}. Since we do not have $\mathscr{X}_{\mathrm{N}}$ in the PU setting, we cannot perform full distribution matching directly. Here, we employ a *partial model* instead (du Plessis and Sugiyama, 2014a; du Plessis et al., 2017):

$$q_\theta(\boldsymbol{x}) := \theta p_{\mathrm{P}}(\boldsymbol{x}),$$

where $(1-\theta)p_{\mathrm{N}}(\boldsymbol{x})$ has been dropped from the mixture model. Suppose that we can access $p_{\mathrm{P}}(\boldsymbol{x})$ and $p_{\mathrm{U}}(\boldsymbol{x})$. Then, we can perform distribution matching in a partial manner as

$$\widehat{\pi}_{\mathrm{P}} := \underset{0 \le \theta \le 1}{\operatorname{argmin}}\, \mathrm{Div}(p_{\mathrm{U}}\|q_\theta),$$

which is called *partial distribution matching*. In particular, if we adopt the f-divergences mentioned in section 12.2, the objective function $\mathrm{Div}(\theta)$ is given by

$$\mathrm{Div}(\theta) := \mathrm{Div}(p_{\mathrm{U}}\|q_\theta) = \int f\left(\frac{\theta p_{\mathrm{P}}(\boldsymbol{x})}{p_{\mathrm{U}}(\boldsymbol{x})}\right) p_{\mathrm{U}}(\boldsymbol{x})\mathrm{d}\boldsymbol{x}, \tag{12.7}$$

where $f : [0, +\infty) \to \mathbb{R}$ is a convex function such that $f(1) = 0$.

Although partial matching looks very similar to full matching, their meanings are different due to the change of the model $q_\theta(\boldsymbol{x})$. In full distribution matching, the objective function $\mathrm{Div}(p_{\mathrm{U}}\|\theta p_{\mathrm{P}} + (1-\theta)p_{\mathrm{N}})$ attains zero if $\theta = \pi_{\mathrm{P}}$. On the other hand, in partial distribution matching, $\mathrm{Div}(p_{\mathrm{U}}\|\theta p_{\mathrm{P}})$ is still non-zero even if $\theta = \pi_{\mathrm{P}}$. Actually, achieving $\mathrm{Div}(p_{\mathrm{U}}\|\theta p_{\mathrm{P}}) = 0$ is generally impossible since it implies $p_{\mathrm{U}}(\boldsymbol{x}) = \theta p_{\mathrm{P}}(\boldsymbol{x})$; this could only be satisfied by $\theta = 1$, $p_{\mathrm{U}}(\boldsymbol{x}) = p_{\mathrm{P}}(\boldsymbol{x})$, and $p_{\mathrm{N}}(\boldsymbol{x}) = 0$, but this is not a valid binary classification problem. Therefore, $\theta = \pi_{\mathrm{P}}$ is not a solution of $\mathrm{Div}(\theta) = 0$ as this equation has no solution.

12.4.2 Differentiable Divergences

The next natural question is whether $\theta = \pi_{\mathrm{P}}$ is a solution of $\min_\theta \mathrm{Div}(\theta)$. For f-divergences, $f(t)$ is convex and $t = \theta p_{\mathrm{P}}(\boldsymbol{x})/p_{\mathrm{U}}(\boldsymbol{x})$ is linear in θ. As a result, $\mathrm{Div}(\theta)$ is convex, and whether $\theta = \pi_{\mathrm{P}}$ is a minimizer can be checked by the first-order optimality condition $0 \in \partial\mathrm{Div}(\pi_{\mathrm{P}})$, where ∂ denotes the sub-differential. For convenience, let

$$\eta(\boldsymbol{x}) := p(y = +1 \,|\, \boldsymbol{x})$$

be the class-posterior probability for the positive class. We focus on $f(t)$ such that its minimum is attained at $t \ge 1$, though $t > 1$ is impossible when $\theta = \pi_{\mathrm{P}}$:

$$t = \pi_{\mathrm{P}} p_{\mathrm{P}}(\boldsymbol{x})/p_{\mathrm{U}}(\boldsymbol{x}) = \eta(\boldsymbol{x}) \le 1.$$

For the Kullback-Leibler divergence,

$$\operatorname*{argmin}_t f(t) = \operatorname*{argmin}_t -\ln(t) = +\infty;$$

for the Pearson divergence,

$$\operatorname*{argmin}_t f(t) = \operatorname*{argmin}_t (t-1)^2/2 = 1;$$

for the L_1-distance,

$$\operatorname*{argmin}_t f(t) = \operatorname*{argmin}_t |t-1| = 1.$$

Hence, when we constrain $t \leq 1$, $f(t)$ for the above f-divergences attains its minimum exactly at $t = 1$. Subsequently, let us consider two different cases.

In the first case, assume that $f(t)$ is differentiable so that $f(t)$ attaining its minimum at $t \geq 1$ leads to

$$\begin{cases} \partial f(t) < 0 & (t < 1), \\ \partial f(t) \leq 0 & (t = 1). \end{cases}$$

Then, differentiating (12.7) and plugging $\theta = \pi_P$ into the derivative, we can obtain

$$\partial \operatorname{Div}(\pi_P) = \int \partial f \left(\frac{\pi_P p_P(\mathbf{x})}{p_U(\mathbf{x})} \right) \frac{p_P(\mathbf{x})}{p_U(\mathbf{x})} p_U(\mathbf{x}) d\mathbf{x}$$

$$= \int \partial f(\eta(\mathbf{x})) p_P(\mathbf{x}) d\mathbf{x}.$$

To study whether $\partial \operatorname{Div}(\pi_P)$ can attain zero, we partition the support set of the positive class into

$$\mathscr{D}_1 := \{\mathbf{x} \mid \eta(\mathbf{x}) = 1 \wedge p_P(\mathbf{x}) > 0\},$$

$$\mathscr{D}_2 := \{\mathbf{x} \mid \eta(\mathbf{x}) < 1 \wedge p_P(\mathbf{x}) > 0\}.$$

In fact, $\mathscr{D}_1$ is the set of all anchor points for the positive class used in mixture proportion estimation (section 12.3.4), where the positive and negative classes do not overlap. On the other hand, the two classes are overlapped in $\mathscr{D}_2$, since

$$p_N(\mathbf{x}) = \frac{p(\mathbf{x}, y = -1)}{p(y = -1)} = \frac{p(\mathbf{x})p(y = -1 \mid \mathbf{x})}{\pi_N} = \frac{p(\mathbf{x})(1 - \eta(\mathbf{x}))}{\pi_N} > 0.$$

Therefore,

$$\partial \operatorname{Div}(\pi_P) = \int_{\mathscr{D}_1} \underbrace{\partial f(\eta(\mathbf{x})) p_P(\mathbf{x}) d\mathbf{x}}_{\leq 0} + \int_{\mathscr{D}_2} \underbrace{\partial f(\eta(\mathbf{x})) p_P(\mathbf{x}) d\mathbf{x}}_{<0}, \qquad (12.8)$$

which can attain zero only when $\mathscr{D}_2$ is empty. This indicates that $\theta = \pi_P$ cannot be a minimizer of $\mathrm{Div}(\theta)$ for all *differentiable f-divergences*, including the Kullback-Leibler and Pearson divergences, as long as the two classes are not completely separable.

12.4.3   Non-Differentiable Divergences

In the second case, assume that $f(t)$ is differentiable for $0 \le t < 1$ and non-differentiable at $t = 1$, so that $\partial f(t) < 0$ for $t < 1$ but $\partial f(t)$ is a set for $t = 1$. According to (12.8), $0 \in \partial\mathrm{Div}(\pi_P)$ is possible if and only if

$$\frac{-\int_{\mathscr{D}_2} \partial f(\eta(x)) p_P(x) dx}{\int_{\mathscr{D}_1} p_P(x) dx} \in \partial f(1). \tag{12.9}$$

This can be satisfied when either $\mathscr{D}_1$ is much larger than $\mathscr{D}_2$ (i.e., the left-hand side of (12.9) is a very small positive value), or $\partial f(1)$ has a large supremum (i.e., the right-hand side of (12.9) includes large positive values). We can see from (12.9) that the non-differentiability of $f(t)$ at $t = 1$ is essential to make $\partial f(1)$ a set and enable $0 \in \partial\mathrm{Div}(\pi_P)$. If we are lucky that (12.9) holds, $\theta = \pi_P$ can be a minimizer of $\mathrm{Div}(\theta)$ for such non-differentiable f-divergences. For example, in the case of the L_1-distance, since $\partial f(\eta(x)) = -1$ given $\eta(x) < 1$ and $\partial f(1) = [-1, +1]$, $\theta = \pi_P$ can be a minimizer of $\mathrm{Div}(\theta)$ if and only if

$$\frac{\int_{\mathscr{D}_2} p_P(x) dx}{\int_{\mathscr{D}_1} p_P(x) dx} = \frac{p_P(\mathscr{D}_2)}{p_P(\mathscr{D}_1)} \le 1.$$

This clearly illustrates that the L_1-distance is superior to all differentiable f-divergences in partial distribution matching (see section 12.4.2).

The non-differentiability of $\mathrm{Div}(p_U \| q_\theta)$ with respect to θ is also critical for other divergence measures than the f-divergence. Consider the L_2-distance:

$$\mathrm{Div}(\theta) := \int (\theta p_P(x) - p_U(x))^2 dx,$$

which is a convex function of θ. By differentiating $\mathrm{Div}(\theta)$ and plugging $\theta = \pi_P$ into the derivative, we have

$$\partial\mathrm{Div}(\pi_P) = \int_{\mathscr{D}_1} \underbrace{(\pi_P p_P(x) - p_U(x))}_{=0} p_P(x) dx$$

$$+ \int_{\mathscr{D}_2} \underbrace{(\pi_P p_P(x) - p_U(x))}_{<0} p_P(x) dx < 0.$$

This implies that $\theta = \pi_P$ cannot be a minimizer of $\mathrm{Div}(\theta)$ for the L_2-distance. Thus, the L_2-distance is also inferior to the L_1-distance in partial distribution matching, even though they look almost identical.

The implication of $\partial\mathrm{Div}(\pi_P) < 0$ for differentiable $\mathrm{Div}(\theta)$ or $0 \notin \partial\mathrm{Div}(\pi_P)$ for non-differentiable $\mathrm{Div}(\theta)$ is that π_P will be systematically overestimated:

$$\pi_P < \widehat{\pi}_P := \operatorname*{argmin}_{0 \le \theta \le 1} \mathrm{Div}(\theta).$$

The bias is positive because $\mathrm{Div}(\theta)$ is convex and $\partial\mathrm{Div}(\theta)$ is monotonically non-decreasing, so that the minimizer of $\mathrm{Div}(\theta)$ will be larger than π_P (if it has a minimizer). Note that this positive bias is not caused by *overfitting*; it exists even if we can access the true densities $p_P(x)$ and $p_U(x)$. As a result, we suffer a bias inherited from the formulation of partial distribution matching in the same way as the positive bias of κ^* in mixture proportion estimation if the irreducibility assumption is violated (see section 12.3.3).

On the other hand, the L_1-distance seems promising since $0 \in \partial\mathrm{Div}(\pi_P)$ may hold. We will further improve it in section 12.5.

12.4.4　Empirical f-Divergence Estimation

In practice, we cannot access $p_P(x)$ and $p_U(x)$, and thus $\mathrm{Div}(\theta)$ cannot be directly minimized with respect to θ. We show here how partial distribution matching can be performed based on $\mathscr{X}_P$ and $\mathscr{X}_U$ without naive density estimation. For this purpose, *empirical f-divergence estimation* suffices.

The key idea is the use of the *Fenchel-duality bounding technique* for the f-divergence (Keziou, 2003), which is based on *Fenchel's inequality*:

$$f(t) \ge tz - f^*(z), \tag{12.10}$$

where $f^*(z)$ is the *Fenchel-dual* or *convex conjugate* defined as

$$f^*(z) := \sup_t [tz - f(t)].$$

Table 12.1 lists some examples. The *biconjugate* (the convex conjugate of the convex conjugate) is defined as

$$f^{**}(t) := \sup_z [tz - f^*(z)] \le f(t),$$

where the equality $f^{**}(t) = f(t)$ holds if and only if f is convex and lower semi-continuous.

Subsequently, applying (12.10) in a pointwise manner, we obtain

$$f\left(\frac{\theta p_P(x)}{p_U(x)}\right) \ge r(x)\left(\frac{\theta p_P(x)}{p_U(x)}\right) - f^*(r(x)),$$

where $r(x)$ fulfills the role of z in (12.10). For all f-divergences, f is convex, and if we assume that f is lower semi-continuous, we have

$$f\left(\frac{\theta p_P(x)}{p_U(x)}\right) = \sup_{r(x)} r(x)\left(\frac{\theta p_P(x)}{p_U(x)}\right) - f^*(r(x)).$$

Table 12.1
Examples of f-divergences. $f^*(z)$ is the convex conjugate of $f(t)$.

Divergence	$f(t)$	$f^*(z)$
Kullback-Leibler	$-\ln(t)$	$-\ln(-z)-1$
Pearson	$\frac{1}{2}(t-1)^2$	$\frac{1}{2}z^2+z$
L_1	$\|t-1\|$	$\begin{cases} z & (-1\le z\le 1) \\ \infty & \text{(otherwise)} \end{cases}$
Penalized L_1 (with $c=\infty$)	$\begin{cases} 1-t & (t\le 1) \\ \infty & (t>1) \end{cases}$	$\max\{z,-1\}$

Multiplying both sides with $p_{\mathrm{U}}(\boldsymbol{x})$ and integrating the result over $\mathscr{X}$ give

$$\mathrm{Div}(\theta)=\sup_r L_r(\theta),$$

where

$$L_r(\theta):=\theta\int r(\boldsymbol{x})p_{\mathrm{P}}(\boldsymbol{x})\mathrm{d}\boldsymbol{x}-\int f^*(r(\boldsymbol{x}))p_{\mathrm{U}}(\boldsymbol{x})\mathrm{d}\boldsymbol{x}. \tag{12.11}$$

The two integrals are expectations over $p_{\mathrm{P}}(\boldsymbol{x})$ and $p_{\mathrm{U}}(\boldsymbol{x})$ from which we have samples $\mathscr{X}_{\mathrm{P}}$ and $\mathscr{X}_{\mathrm{U}}$, and thus $L_r(\theta)$ in (12.11) can be estimated by replacing the expectations with the sample averages:

$$\widehat{L}_r(\theta):=\frac{\theta}{n_{\mathrm{P}}}\sum_{i=1}^{n_{\mathrm{P}}}r(\boldsymbol{x}_i^{\mathrm{P}})-\frac{1}{n_{\mathrm{U}}}\sum_{i=1}^{n_{\mathrm{U}}}f^*(r(\boldsymbol{x}_i^{\mathrm{U}})). \tag{12.12}$$

Finally, define the empirical objective function as the empirical f-divergence:

$$\widehat{\mathrm{Div}}(\theta):=\sup_r \widehat{L}_r(\theta).$$

Then, the class-prior probability π_{P} can be estimated by solving the following minimax optimization problem:

$$\widehat{\pi}_{\mathrm{P}}:=\operatorname*{argmin}_{0\le\theta\le 1}\widehat{\mathrm{Div}}(\theta)=\operatorname*{argmin}_{0\le\theta\le 1}\max_r \widehat{L}_r(\theta). \tag{12.13}$$

Note that the convex conjugate $f^*(z)$ of any function $f(t)$ is convex by definition, and then $\widehat{L}_r(\theta)$ in (12.12) is concave in the function r so that the inner maximization in (12.13) is a convex optimization. If the function r is linear in parameters, the inner maximization in (12.13) is also a convex optimization with respect to the parameters of r and thus it can be easily solved. The outer minimization in (12.13) can certainly be easily solved, since it is linear in θ.

In order to derive the empirical f-divergence estimator described above, we have assumed the lower semi-continuity of f. This should indeed be satisfied by any reasonable

f-divergence. In table 12.1, we can see that the Kullback-Leibler, Pearson, and L_1 cases have continuous $f(t)$; even though $f(t)$ for the penalized L_1 case is discontinuous at $t = 1$, it is in fact lower semi-continuous (but not upper semi-continuous) at $t = 1$, since its minimum is $f(1) = 0$.

Partial distribution matching with the penalized L_1-distance will be discussed in detail in section 12.5, while empirical estimation and minimization of the penalized L_1-distance will be given in section 12.5.2.

12.5   Penalized L_1-Distance Minimization

Section 12.4.3 demonstrated that the L_1-distance is promising in partial distribution matching. In this section, we further improve and analyze it.

12.5.1   Penalized L_1-Distance

As shown in section 12.4.3, for the L_1-distance, $\theta = \pi_P$ is a minimizer of $\mathrm{Div}(\theta)$ if and only if $p_P(\mathcal{D}_1) \geq p_P(\mathcal{D}_2)$. Here,

$$\mathcal{D}_1 := \{x \mid \eta(x) = 1 \wedge p_P(x) > 0\}$$

is the anchor set for the positive class where the two classes do not overlap, and

$$\mathcal{D}_2 := \{x \mid \eta(x) < 1 \wedge p_P(x) > 0\}$$

is the set where the two classes are overlapped. However, $\mathcal{D}_1$ is usually smaller than $\mathcal{D}_2$, so that the chance of $\theta = \pi_P$ being a minimizer of $\mathrm{Div}(\theta)$ does not seem high.

The observation above motivates us to further improve the L_1-distance. For any f-divergence with $f(t)$ differentiable for $0 \leq t < 1$ and non-differentiable at $t = 1$, the optimality condition (12.9) requires that either $\mathcal{D}_1$ is much larger than $\mathcal{D}_2$ or $\partial f(1)$ has a large supremum. Since $\mathcal{D}_1$ and $\mathcal{D}_2$ are fixed and we cannot control them, we try to increase the supremum of $\partial f(1)$.

To this end, define the *penalized L_1-distance* (du Plessis et al., 2017) as

$$\mathrm{penL}_1(\theta) := \int f_c \left(\frac{\theta p_P(x)}{p_U(x)} \right) p_U(x) dx,$$

where

$$f_c(t) := \begin{cases} 1 - t & (t \leq 1), \\ c(t - 1) & (t > 1), \end{cases} \tag{12.14}$$

and $c \geq 1$ is the penalty parameter of the penalized L_1-distance. When $c = 1$, the penalized L_1-distance is reduced to the ordinary L_1-distance. Note that the sub-differential of $f_c(t)$ in

(12.14) is given by

$$\partial f_c(t) = \begin{cases} -1 & (t < 1), \\ [-1, c] & (t = 1), \\ c & (t > 1), \end{cases}$$

which indicates that the supremum of $\partial f_c(1)$ can be controlled by the penalty parameter c. Following (12.8), $\partial \mathrm{penL}_1(\pi_\mathrm{P})$ can be expressed by

$$\partial \mathrm{penL}_1(\pi_\mathrm{P}) = [-p_\mathrm{P}(\mathscr{D}_1) - p_\mathrm{P}(\mathscr{D}_2), c p_\mathrm{P}(\mathscr{D}_1) - p_\mathrm{P}(\mathscr{D}_2)],$$

and the optimality condition (12.9) is reduced to

$$c p_\mathrm{P}(\mathscr{D}_1) - p_\mathrm{P}(\mathscr{D}_2) \geq 0. \tag{12.15}$$

As a result, $0 \in \partial \mathrm{penL}_1(\pi_\mathrm{P})$ if and only if (12.15) holds.

A sufficient but not necessary condition for (12.15) is the irreducibility assumption (see section 12.3.3) and $c = \infty$. To see this, let us temporarily make the stronger anchor set assumption $p_\mathrm{P}(\mathscr{D}_1) = \epsilon > 0$ (see section 12.3.4). Then, $p_\mathrm{P}(\mathscr{D}_2) = 1 - \epsilon$, since $\mathscr{D}_1 \cap \mathscr{D}_2 = \emptyset$ and $\mathscr{D}_1 \cup \mathscr{D}_2 = \{x \mid p_\mathrm{P}(x) > 0\}$ so that $p_\mathrm{P}(\mathscr{D}_2) = 1 - p_\mathrm{P}(\mathscr{D}_1)$. In such a case, $c_\epsilon = 1/\epsilon - 1$ is enough for (12.15) to be satisfied. Since the irreducibility assumption is the limiting case of the anchor set assumption as ϵ approaches zero, we have

$$\lim_{\epsilon \to 0} [c_\epsilon p_\mathrm{P}(\mathscr{D}_1) - p_\mathrm{P}(\mathscr{D}_2)] = \lim_{\epsilon \to 0} 0 = 0,$$

and consequently (12.15) holds with $c = \lim_{\epsilon \to 0} c_\epsilon = \infty$. Nonetheless, $c = \infty$ alone is not enough, as we may encounter $c p_\mathrm{P}(\mathscr{D}_1) = \infty \cdot 0$ that is undefined.

Note that $f_c(t)$ is convex and $f_c(1) = 0$, and thus the penalized L_1-distance is still an f-divergence. In fact, $f_c(t)$ is strictly convex and $t = \theta p_\mathrm{P}(x)/p_\mathrm{U}(x)$ is linear in θ, so that $\mathrm{penL}_1(\theta)$ is strictly convex whose minimizer is unique. Therefore, if (12.15) holds and $\theta = \pi_\mathrm{P}$ is a minimizer of $\mathrm{penL}_1(\theta)$, it must be the unique minimizer of $\mathrm{penL}_1(\theta)$.

We may extend the penalization technique to general f-divergences and obtain the corresponding *penalized f-divergences* by replacing $f(t)$ with

$$\widetilde{f}(t) := \begin{cases} f(t) & (t \leq 1), \\ \infty & (t > 1). \end{cases}$$

Using the same reasoning as above, penalized f-divergences also rectify overestimation of partial distribution matching with general f-divergences. Among them, the penalized L_1-distance is of particular interest, since its minimization can be solved in linear time when $c = \infty$ (see section 12.5.2).

Let us conceptually compare mixture proportion estimation and partial distribution matching with the penalized L_1-distance ($c = \infty$).

- Without any assumption, they both have inevitable positive biases.

- Under the irreducibility assumption, the former (mixture proportion estimation) is unbiased, while solving it needs exponential time; the latter (partial distribution matching with the penalized L_1-distance) is also unbiased, and linear time is enough to solve it (no optimization is involved indeed).

- Under the anchor set assumption, the former has polynomial-time solvers, which are still practically infeasible and significantly slower than the latter.

- Under the strongest anchor point assumption, as shown in (12.5) and (12.6), the former only involves an optimization in density ratio estimation, which should be computationally efficient.

To sum up, with any of the above assumptions, mixture proportion estimation and partial distribution matching are theoretically equally superior in terms of estimation bias. On the other hand, partial distribution matching is also computationally superior in terms of time complexity.

12.5.2   Practical Implementation

Following section 12.4.4, we give a practical implementation of partial distribution matching with the penalized L_1-distance, named *penalized L_1-distance minimization*.

The convex conjugate of $f_c(t)$ in (12.14) is given by

$$
f_c^*(z) = \begin{cases} -1 & (z \le -1), \\ z & (-1 < z \le c), \\ \infty & (z > c), \end{cases}
$$

and thus we can have $f_c^*(r(x)) = r(x)$ if $-1 \le r(x) \le c$ is enforced. Moreover, $r(x)$ should be linear in parameters so that the inner maximization in (12.13) can easily be solved:

$$
r(x) := \sum_{j=1}^{b} \alpha_j \phi_j(x) - 1 = \alpha^\top \phi(x) - 1,
$$

where b denotes the number of parameters, and

$$
\alpha := (\alpha_1, \ldots, \alpha_b)^\top \in [0, \infty)^b
$$

$$
\phi(x) := (\phi_1(x), \ldots, \phi_b(x))^\top \in [0, \infty)^b
$$

are vectors of non-negative parameters and basis functions. This is a modified linear-in-parameter model to guarantee that $r(x) \ge -1$. Concerning $\phi(x)$, we may use Gaussian kernels centered at $\mathcal{X}_P \cup \mathcal{X}_U$ or a feature extractor created from a pretrained neural network as long as the activation function of the last layer is non-negative (e.g., the rectified linear unit). See section 2.1.2 for the original linear-in-parameter, kernel, and neural network models.

Subsequently, by defining

$$\boldsymbol{\beta}(\theta) := \frac{\theta}{n_{\mathrm{P}}} \sum_{i=1}^{n_{\mathrm{P}}} \boldsymbol{\phi}(x_i^{\mathrm{P}}) - \frac{1}{n_{\mathrm{U}}} \sum_{i=1}^{n_{\mathrm{U}}} \boldsymbol{\phi}(x_i^{\mathrm{U}}),$$

$$\boldsymbol{\Phi}_{\mathrm{U}} := (\boldsymbol{\phi}(x_1^{\mathrm{U}}), \dots, \boldsymbol{\phi}(x_{n_{\mathrm{U}}}^{\mathrm{U}}))^{\top},$$

the inner maximization of $\widehat{L}_r(\theta)$ in (12.13) can equivalently be solved by the following minimization problem:

$$\widehat{\boldsymbol{\alpha}} := \operatorname*{argmin}_{\boldsymbol{\alpha} \geq 0} \quad \frac{\lambda}{2} \|\boldsymbol{\alpha}\|_2^2 - \boldsymbol{\alpha}^{\top} \boldsymbol{\beta}(\theta) \tag{12.16}$$
$$\text{subject to} \quad \boldsymbol{\Phi}_{\mathrm{U}}^{\top} \boldsymbol{\alpha} \leq (c+1) \mathbf{1}_{n_{\mathrm{U}}},$$

where inequalities for vectors are applied in an element-wise manner, the regularization term $\frac{\lambda}{2} \|\boldsymbol{\alpha}\|_2^2$ with $\lambda > 0$ is included to avoid overfitting, and $\mathbf{1}_{n_{\mathrm{U}}}$ is the n_{U}-dimensional vector with all ones. Note that the constraints are only applied on unlabeled data since $f_c^*(z)$ is only applied on $r(x_i^{\mathrm{U}})$ in $\widehat{L}_r(\theta)$. This is a quadratic programming problem and thus can be solved with an off-the-shelf quadratic programming solver. With $\widehat{r}(x) := \widehat{\boldsymbol{\alpha}}^{\top} \boldsymbol{\phi}(x) - 1$, the minimax optimization (12.13) can be performed as

$$\widehat{\pi}_{\mathrm{P}} := \operatorname*{argmin}_{0 \leq \theta \leq 1} \widehat{\operatorname{penL}}_1(\theta),$$

where

$$\widehat{\operatorname{penL}}_1(\theta) := \widehat{L}_{\widehat{r}}(\theta) = \widehat{\boldsymbol{\alpha}}^{\top} \boldsymbol{\beta}(\theta) - \theta + 1.$$

Consider the special case $c = \infty$. Then, the constraint $\boldsymbol{\Phi}_{\mathrm{U}}^{\top} \boldsymbol{\alpha} \leq (c+1) \mathbf{1}_{n_{\mathrm{U}}}$ in the optimization problem (12.16) is always satisfied, and the objective function can be minimized *analytically* as

$$\widehat{\boldsymbol{\alpha}} = \frac{1}{\lambda} \max\{\mathbf{0}_{n_{\mathrm{U}}}, \boldsymbol{\beta}(\theta)\},$$

where $\mathbf{0}_{n_{\mathrm{U}}}$ is the n_{U}-dimensional vector with all zeros and the max here is the element-wise maximum. Since computing $\widehat{\boldsymbol{\alpha}}$ from $\mathscr{X}_{\mathrm{P}}$ and $\mathscr{X}_{\mathrm{U}}$ only requires linear time, it will be computationally extremely efficient to solve the minimax optimization (12.13). In practice, $\widehat{\pi}_{\mathrm{P}}$ may be found with some grid-search or line-search algorithms according to $\widehat{\operatorname{penL}}_1(\theta)$.

Finally, penalized L_1-distance minimization has a surprising relation to mixture proportion estimation. Let us remove the regularization term $\frac{\lambda}{2} \|\boldsymbol{\alpha}\|_2^2$ from the objective function of the optimization problem (12.16). Then, we have

$$\widehat{\operatorname{penL}}_1(\theta) = \begin{cases} 1 - \theta & (\boldsymbol{\beta}(\theta) \leq \mathbf{0}_b), \\ \infty & (\text{otherwise}), \end{cases}$$

whose minimizer is given by

$$\widehat{\pi}_{\mathrm{P}} = \sup\{\theta \mid \boldsymbol{\beta}(\theta) \leq \mathbf{0}_b\} = \min_{1 \leq j \leq b} \frac{\frac{1}{n_{\mathrm{U}}} \sum_{i=1}^{n_{\mathrm{U}}} \phi_j(\boldsymbol{x}_i^{\mathrm{U}})}{\frac{1}{n_{\mathrm{P}}} \sum_{i=1}^{n_{\mathrm{P}}} \phi_j(\boldsymbol{x}_i^{\mathrm{P}})}.$$

Let $b = n_{\mathrm{P}}$ and $\phi_j(\boldsymbol{x}) = K(\boldsymbol{x}, \boldsymbol{x}_j^{\mathrm{P}})$, where $K(\boldsymbol{x}, \boldsymbol{x}')$ is the Gaussian kernel. Then, we can regard $\frac{1}{n_{\mathrm{U}}} \sum_{i=1}^{n_{\mathrm{U}}} \phi_j(\boldsymbol{x}_i^{\mathrm{U}})$ as $\widehat{p}_{\mathrm{U}}(\boldsymbol{x}_j^{\mathrm{P}})$ and $\frac{1}{n_{\mathrm{P}}} \sum_{i=1}^{n_{\mathrm{P}}} \phi_j(\boldsymbol{x}_i^{\mathrm{P}})$ as $\widehat{p}_{\mathrm{P}}(\boldsymbol{x}_j^{\mathrm{P}})$ following kernel density estimation, so that

$$\widehat{\pi}_{\mathrm{P}} = \min_{\boldsymbol{x}_j^{\mathrm{P}} \in \mathscr{X}_{\mathrm{P}}} \frac{\widehat{p}_{\mathrm{U}}(\boldsymbol{x}_j^{\mathrm{P}})}{\widehat{p}_{\mathrm{P}}(\boldsymbol{x}_j^{\mathrm{P}})}.$$

This is an alternative implementation of mixture proportion estimation (12.6) under the anchor point assumption, where the density ratio estimator in (12.5) is replaced with two separate density estimators. Hence, penalized L_1-distance minimization with $c = \infty$ can be regarded as a linear-time approximate solver of mixture proportion estimation under the anchor point assumption.

12.5.3   Theoretical Analysis

Here, we theoretically analyze penalized L_1-distance minimization. It is fairly easy to find $\widehat{\pi}_{\mathrm{P}}$ given $\widehat{\mathrm{penL}}_1(\theta)$, and thus we focus on $\widehat{\mathrm{penL}}_1(\theta)$ and analyze its behavior, including

- Its consistency for estimating the penalized L_1-distance.

- Its stability for estimating the penalized L_1-distance.

- Its consistency for minimizing the penalized L_1-distance.

Below, we briefly present the main messages that can be drawn from a comprehensive theoretical analysis; for the necessary proof techniques, see section 3.1.2 and du Plessis et al. (2017).

12.5.3.1   Realizability assumption

In order to make our analysis intuitive and easy to follow, assume that $\boldsymbol{\phi}(\boldsymbol{x})$ is sufficiently powerful such that

$$\mathrm{penL}_1(\theta) = \sup_r L_r(\theta)$$

always holds. Since Fenchel's inequality (12.10) was applied in a pointwise manner, the equality is achieved by $r(\boldsymbol{x})$ such that $0 \in \partial L_r(\theta)$, leading to

$$\frac{\theta p_{\mathrm{P}}(\boldsymbol{x})}{p_{\mathrm{U}}(\boldsymbol{x})} \in \partial f_c^*(r(\boldsymbol{x})) \tag{12.17}$$

for every $\boldsymbol{x} \in \mathscr{X}$. As a consequence, $\boldsymbol{\phi}(\boldsymbol{x})$ being sufficiently powerful means that $p_{\mathrm{P}}(\boldsymbol{x})$ and $p_{\mathrm{U}}(\boldsymbol{x})$ are sufficiently regular and/or $\boldsymbol{\phi}(\boldsymbol{x})$ is sufficiently flexible to let (12.17) hold over $\mathscr{X}$.

This setting is a *realizable* or *optimistic* case in learning theory, where we do not suffer the approximation error; in contrast, in an *agnostic* or *pessimistic* setting, we need to change from the original estimation goal to an optimization goal, and there is a gap between the two targets of convergence whenever there is an approximation error (see sections 3.1.2.1 and 3.1.2.2).

12.5.3.2   Summary of main results

Under the above assumption, the first behavior of interest refers to

$$\widehat{\mathrm{penL}}_1(\theta) \to \mathrm{penL}_1(\theta) \quad \text{as} \quad n_{\mathrm{P}}, n_{\mathrm{U}} \to \infty. \tag{12.18}$$

Equation (12.18) is a minimum requirement for $\widehat{\mathrm{penL}}_1(\theta)$ to be an empirical estimator of the penalized L_1-distance.

Then, the second behavior of interest refers to that, for any finite n_{P} and n_{U}, the empirical estimate $\widehat{\mathrm{penL}}_1(\theta)$ cannot change too much with respect to the repeated sampling of $\mathcal{X}_{\mathrm{P}}$ and $\mathcal{X}_{\mathrm{U}}$. In other words, there exists $C_{n_{\mathrm{P}},n_{\mathrm{U}}} > 0$ such that $\lim_{n_{\mathrm{P}},n_{\mathrm{U}} \to \infty} C_{n_{\mathrm{P}},n_{\mathrm{U}}} = 0$, and with high probability,

$$\left| \widehat{\mathrm{penL}}_1(\theta) - \mathbb{E}_{\mathcal{X}_{\mathrm{P}},\mathcal{X}_{\mathrm{U}}}\left[\widehat{\mathrm{penL}}_1(\theta) \right] \right| \leq C_{n_{\mathrm{P}},n_{\mathrm{U}}}. \tag{12.19}$$

Note that if we let $n_{\mathrm{P}}, n_{\mathrm{U}} \to \infty$, (12.19) cannot provide more information than (12.18), since (12.18) can be rewritten as (12.19) after replacing $\mathbb{E}_{\mathcal{X}_{\mathrm{P}},\mathcal{X}_{\mathrm{U}}}[\widehat{\mathrm{penL}}_1(\theta)]$ with $\mathrm{penL}_1(\theta)$. However, $\mathbb{E}_{\mathcal{X}_{\mathrm{P}},\mathcal{X}_{\mathrm{U}}}[\widehat{\mathrm{penL}}_1(\theta)] \neq \mathrm{penL}_1(\theta)$ for $n_{\mathrm{P}}, n_{\mathrm{U}} < \infty$, and thus (12.19) is still very meaningful. While (12.18) implies that $\widehat{\mathrm{penL}}_1(\theta)$ is good in the limiting case, (12.19) implies that $\widehat{\mathrm{penL}}_1(\theta)$ is stable in the finite cases. Specifically, if (12.18) holds, $\widehat{\mathrm{penL}}_1(\theta)$ is consistent and thus asymptotically unbiased so that $\mathbb{E}_{\mathcal{X}_{\mathrm{P}},\mathcal{X}_{\mathrm{U}}}[\widehat{\mathrm{penL}}_1(\theta)]$ is close to $\mathrm{penL}_1(\theta)$; if (12.19) also holds, for any finite n_{P} and n_{U}, $\widehat{\mathrm{penL}}_1(\theta)$ is close to $\mathbb{E}_{\mathcal{X}_{\mathrm{P}},\mathcal{X}_{\mathrm{U}}}[\widehat{\mathrm{penL}}_1(\theta)]$ and thus also close to $\mathrm{penL}_1(\theta)$.

Finally, the third behavior of interest refers to

$$\mathrm{penL}_1(\widehat{\pi}_{\mathrm{P}}) \to \min_{\theta} \mathrm{penL}_1(\theta) = \mathrm{penL}_1(\pi_{\mathrm{P}}) \quad \text{as} \quad n_{\mathrm{P}}, n_{\mathrm{U}} \to \infty. \tag{12.20}$$

Equation (12.20) can be interpreted as the minimizer of $\widehat{\mathrm{penL}}_1(\theta)$ converges to the minimizer of $\mathrm{penL}_1(\theta)$, when the measure of convergence is chosen as $\mathrm{penL}_1(\theta)$. Indeed, (12.20) implies $\widehat{\pi}_{\mathrm{P}} \to \pi_{\mathrm{P}}$ as $n_{\mathrm{P}}, n_{\mathrm{U}} \to \infty$, where the convergence rate may be slower than that of $\mathrm{penL}_1(\widehat{\pi}_{\mathrm{P}}) \to \mathrm{penL}_1(\pi_{\mathrm{P}})$.

12.5.3.3   Proofs of main results

Consider the first case where $L_r(\theta)$ in (12.11) and $\widehat{L}_r(\theta)$ in (12.12) are used as they are, i.e., $f_c^*(z)$ is involved. Accordingly, define a function class $\mathscr{R}$ as

$$\mathscr{R} := \{r(\boldsymbol{x}) := \boldsymbol{\alpha}^{\top}\boldsymbol{\phi}(\boldsymbol{x}) \mid \boldsymbol{\alpha} \in \mathbb{R}^b, \boldsymbol{\phi}: \mathcal{X} \to \mathbb{R}^b, \|r\|_{\infty} \leq C_{\mathrm{r}}\}, \tag{12.21}$$

where C_r is a constant such that $1 \leq C_r \leq c$. The norm constraint is to ensure that $r(\boldsymbol{x})$ and $f_c^*(r(\boldsymbol{x}))$ are always bounded. Since $\widehat{L}_r(\theta)$ is unbiased to $L_r(\theta)$, similar to how $\widehat{R}(g)$ is unbiased to $R(g)$, its uniform deviation can be upper-bounded similar to lemma 3.7 as

$$\sup_{r \in \mathscr{R}} |\widehat{L}_r(\theta) - L_r(\theta)| = \mathscr{O}_p(1/\sqrt{n_P} + 1/\sqrt{n_U}), \tag{12.22}$$

where $\mathscr{O}_p$ denotes the order in probability. Note that (12.22) is not an asymptotic bound: It is valid for any finite n_P and n_U, but we only present the order for simplicity. Let r^* and $\widehat{r}$ be the maximizers of $L_r(\theta)$ and $\widehat{L}_r(\theta)$, respectively. Then, based on (12.22), we have

$$\widehat{L}_{\widehat{r}}(\theta) - L_{r^*}(\theta) \leq \widehat{L}_{\widehat{r}}(\theta) - L_{\widehat{r}}(\theta) = \mathscr{O}_p(1/\sqrt{n_P} + 1/\sqrt{n_U}),$$

$$L_{r^*}(\theta) - \widehat{L}_{\widehat{r}}(\theta) \leq L_{r^*}(\theta) - \widehat{L}_{r^*}(\theta) = \mathscr{O}_p(1/\sqrt{n_P} + 1/\sqrt{n_U}).$$

Note that $\widehat{L}_{\widehat{r}}(\theta) = \widehat{\mathrm{penL}}_1(\theta)$ by definition, $L_{r^*}(\theta) = \mathrm{penL}_1(\theta)$ by assumption, and consequently (12.18) is proved by combining the above inequalities:

$$\left| \widehat{\mathrm{penL}}_1(\theta) - \mathrm{penL}_1(\theta) \right| = \mathscr{O}_p(1/\sqrt{n_P} + 1/\sqrt{n_U}). \tag{12.23}$$

In the above proof, no regularizer was included in $\widehat{L}_r(\theta)$, since r already has a norm constraint—see the definition of $\mathscr{R}$ in (12.21). Nevertheless, the proof is still effective even if regularization is applied, as long as $\widehat{L}_{\widehat{r}}(\theta) \geq \widehat{L}_{r^*}(\theta)$ is still valid.

For the stability bound (12.19), there are many ways to prove it, while the easiest way would be to use the *ghost sample* $\mathscr{X}_P'$ and $\mathscr{X}_U'$. According to (12.23),

$$\left| \widehat{\mathrm{penL}}_1(\theta; \mathscr{X}_P, \mathscr{X}_U) - \widehat{\mathrm{penL}}_1(\theta; \mathscr{X}_P', \mathscr{X}_U') \right|$$

$$\leq \left| \widehat{\mathrm{penL}}_1(\theta; \mathscr{X}_P, \mathscr{X}_U) - \mathrm{penL}_1(\theta) \right| + \left| \widehat{\mathrm{penL}}_1(\theta; \mathscr{X}_P', \mathscr{X}_U') - \mathrm{penL}_1(\theta) \right|$$

$$= \mathscr{O}_p(1/\sqrt{n_P} + 1/\sqrt{n_U}).$$

Taking the expectation with respect to the ghost sample proves (12.19). This stability bound can be extended to hold uniformly over $0 \leq \theta \leq 1$, since when we prove (12.22), the underlying McDiarmid's inequality behind (12.22) also holds uniformly over $0 \leq \theta \leq 1$.

In order to prove (12.20), let us enhance (12.23) as

$$\sup_{0 \leq \theta \leq 1} \left| \widehat{\mathrm{penL}}_1(\theta) - \mathrm{penL}_1(\theta) \right| = \mathscr{O}_p(1/\sqrt{n_P} + 1/\sqrt{n_U}),$$

based on which we can obtain

$$\mathrm{penL}_1(\widehat{\pi}_P) - \mathrm{penL}_1(\pi_P)$$

$$= \left(\mathrm{penL}_1(\widehat{\pi}_P) - \widehat{\mathrm{penL}}_1(\widehat{\pi}_P) \right) + \left(\widehat{\mathrm{penL}}_1(\widehat{\pi}_P) - \widehat{\mathrm{penL}}_1(\pi_P) \right)$$

$$+ \left(\widehat{\mathrm{penL}}_1(\pi_P) - \mathrm{penL}_1(\pi_P) \right)$$

$$\leq \left| \mathrm{penL}_1(\widehat{\pi}_\mathrm{P}) - \widehat{\mathrm{penL}}_1(\widehat{\pi}_\mathrm{P}) \right| + 0 + \left| \widehat{\mathrm{penL}}_1(\pi_\mathrm{P}) - \mathrm{penL}_1(\pi_\mathrm{P}) \right|$$

$$= \mathcal{O}_\mathrm{p}(1/\sqrt{n_\mathrm{P}} + 1/\sqrt{n_\mathrm{U}}).$$

This is exactly what we were to prove as (12.20), provided that $\mathrm{penL}_1(\pi_\mathrm{P}) = \min_\theta \mathrm{penL}_1(\theta)$, which requires the optimality condition (12.15).

Finally, consider the second case where $f_c^*(z)$ is not involved in $L_r(\theta)$ and $\widehat{L}_r(\theta)$. Accordingly, define a function class $\mathscr{R}'$ as

$$\mathscr{R}' := \{ r(\boldsymbol{x}) := \boldsymbol{\alpha}^\top \boldsymbol{\phi}(\boldsymbol{x}) - 1 \mid \boldsymbol{\alpha} \in [0, \infty]^b, \boldsymbol{\phi} \colon \mathscr{X} \to [0, \infty]^b, \|r\|_\infty \leq C_\mathrm{r} \}.$$

Compared with $\mathscr{R}$ in (12.21), all r such that $-C_\mathrm{r} \leq r(\boldsymbol{x}) < -1$ may occur are excluded from $\mathscr{R}'$. Because $\mathscr{R}' \subset \mathscr{R}$, we should strengthen our realizability assumption: For any θ, there exists $r \in \mathscr{R}'$ such that (12.17) holds. Under this assumption, the above proofs and discussions are all effective, since $\mathrm{penL}_1(\theta) = \sup_{r \in \mathscr{R}'} L_r(\theta)$ is still valid.

12.5.3.4 On the convergence rate of $\widehat{\pi}_\mathrm{P}$

Given $\widehat{\pi}_\mathrm{P} \to \pi_\mathrm{P}$, the reason that we cannot establish the convergence rate of $\widehat{\pi}_\mathrm{P} \to \pi_\mathrm{P}$ is that we cannot build a *first-order growth condition* of $\mathrm{penL}_1(\theta)$ at $\theta = \pi_\mathrm{P}$: We cannot find $\gamma > 0$, such that

$$\mathrm{penL}_1(\theta) \geq \mathrm{penL}_1(\pi_\mathrm{P}) + \gamma |\theta - \pi_\mathrm{P}|.$$

Specifically, denote by $t(\theta, \boldsymbol{x}) := \theta p_\mathrm{P}(\boldsymbol{x})/p_\mathrm{U}(\boldsymbol{x})$ for convenience, and then

$$\mathrm{penL}_1(\theta) := \int f_c(t(\theta, \boldsymbol{x})) p_\mathrm{U}(\boldsymbol{x}) \mathrm{d}\boldsymbol{x}$$

$$= \int_{\mathscr{D}_1} f_c(t(\theta, \boldsymbol{x})) p_\mathrm{U}(\boldsymbol{x}) \mathrm{d}\boldsymbol{x} + \int_{\mathscr{D}_2} f_c(t(\theta, \boldsymbol{x})) p_\mathrm{U}(\boldsymbol{x}) \mathrm{d}\boldsymbol{x}$$

$$+ \int_{\mathscr{D}_3} f_c(0) p_\mathrm{U}(\boldsymbol{x}) \mathrm{d}\boldsymbol{x},$$

where $\mathscr{D}_3 := \mathscr{X} \setminus \mathscr{D}_1 \setminus \mathscr{D}_2 = \{ \boldsymbol{x} \mid p_\mathrm{P}(\boldsymbol{x}) = 0 \} = \{ \boldsymbol{x} \mid \eta(\boldsymbol{x}) = 0 \}$ is the set of all anchor points for the negative class. On $\mathscr{D}_2$, for all $\widehat{\pi}_\mathrm{P}$ sufficiently close to π_P, $t(\pi_\mathrm{P}, \boldsymbol{x}) < 1$ results in $t(\widehat{\pi}_\mathrm{P}, \boldsymbol{x}) < 1$, meaning that

$$[f_c(t(\widehat{\pi}_\mathrm{P}, \boldsymbol{x})) - f_c(t(\pi_\mathrm{P}, \boldsymbol{x}))] p_\mathrm{U}(\boldsymbol{x}) = (\pi_\mathrm{P} - \widehat{\pi}_\mathrm{P}) p_\mathrm{P}(\boldsymbol{x}).$$

However, on $\mathscr{D}_1$, for any $\widehat{\pi}_\mathrm{P} > \pi_\mathrm{P}$,

$$[f_c(t(\widehat{\pi}_\mathrm{P}, \boldsymbol{x})) - f_c(t(\pi_\mathrm{P}, \boldsymbol{x}))] p_\mathrm{U}(\boldsymbol{x}) = c(\widehat{\pi}_\mathrm{P} - \pi_\mathrm{P}) p_\mathrm{P}(\boldsymbol{x}),$$

which partially or totally cancels the change on $\mathscr{D}_2$. Especially, the minimum acceptable $c = 1/p_\mathrm{P}(\mathscr{D}_1) - 1$ increases as $p_\mathrm{P}(\mathscr{D}_1)$ decreases in order to meet the optimality condition (12.15). Moreover, on $\mathscr{D}_3$, $f_c(0)$ is constant and does not change with $\widehat{\pi}_\mathrm{P}$ at all. Thus, $\widehat{\pi}_\mathrm{P}$ may converge to π_P arbitrarily slow, if $\mathscr{D}_3$ can be arbitrarily large.

An interesting case from the theoretical aspect is when $0 \leq \theta \leq 1$ in (12.13) is replaced with $0 \leq \theta \leq \pi_{\mathrm{P}}$ and $p_{\mathrm{U}}(\mathscr{D}_3) < 1$ holds. In such a case,

$$
0 \leq \pi_{\mathrm{P}} - \widehat{\pi}_{\mathrm{P}} = \frac{\mathrm{penL}_1(\widehat{\pi}_{\mathrm{P}}) - \mathrm{penL}_1(\pi_{\mathrm{P}})}{1 - p_{\mathrm{U}}(\mathscr{D}_3)} = \mathscr{O}_{\mathrm{p}}(1/\sqrt{n_{\mathrm{P}}} + 1/\sqrt{n_{\mathrm{U}}}).
$$

The implication is that if no positive bias is ever allowed from the beginning, the convergence rate of $\widehat{\pi}_{\mathrm{P}}$ is the same as that of $\mathrm{penL}_1(\widehat{\pi}_{\mathrm{P}})$, which is the optimal parametric rate. A practical possibility to satisfy $\widehat{\pi}_{\mathrm{P}} \leq \pi_{\mathrm{P}}$ is to add a regularizer $\lambda \theta$ to $\widehat{\mathrm{penL}_1}(\theta)$ and gradually decrease λ as n_{P} and n_{U} increase. Since the minimizer of $\mathrm{penL}_1(\theta)$ may be larger than π_{P} and it is practically impossible to verify the optimality condition (12.15), it might be a good idea to add some regularization of θ to $\widehat{\mathrm{penL}_1}(\theta)$.

12.6   Class-Prior Estimation with Regrouping

In this section, we discuss how to theoretically and empirically mitigate the overestimation problem of mixture proportion estimation discussed in section 12.3 using a general technique named *regrouping*.

12.6.1   Motivation

We have shown in section 12.5 that penalized L_1-distance minimization successfully rectifies the overestimation phenomenon of partial distribution matching with the L_1-distance. The penalization technique may be applied to rectify the same issue of all general f-divergences. However, penalized f-divergences still need the irreducibility assumption (section 12.3.3), which can be seen from equation (12.15) or (12.9). There are two sources of the overestimation phenomenon:

- The lack of irreducibility;

- The mismatch between the full mixture distribution $p_{\mathrm{U}}(x)$ and the partial model $\theta p_{\mathrm{P}}(x)$.

Penalization can only eliminate the positive bias caused by the second source, while it has nothing to do with the positive bias from the first source.

Furthermore, the consistency of empirical f-divergence estimation strongly relies on the realizability assumption. To easily solve the inner maximization in (12.13), the function $r(x)$ must be linear in parameters; to ensure $\mathrm{penL}_1(\theta) = \sup_{r \in \mathscr{R}} L_r(\theta)$, the function class $\mathscr{R}$ must be sufficiently rich. Let $r(x) = \alpha^\top \phi(x)$ be a deep neural network, where $\phi(x)$ is a pre-trained basis vector and α is a trainable parameter vector. Then, the flexibility of r with respect to x would already be in a fixed pattern, and consequently the richness of the resulted $\mathscr{R}$ would be remarkably worse than the richness of a function class of r with trainable $\phi(x)$. In such a case, we would suffer an approximation error because $\mathrm{penL}_1(\theta) > \sup_{r \in \mathscr{R}} L_r(\theta)$.

To this end, we hope to improve mixture proportion estimation and make it useful without the irreducibility assumption. In section 12.3.3, we discussed an example where $\inf_{x \in \mathscr{X}} p_N(x)/p_P(x) = \delta$ for some $0 < \delta < 1$, which led to

$$\kappa^* = \pi_P + \inf_{x \in \mathscr{X}} \frac{\pi_N p_N(x)}{p_P(x)} = \pi_P + \pi_N \delta.$$

To break this bad case, consider a small set A on which $p_N(x) = \delta p_P(x)$ is achieved. Then, think of the class to which the set A belongs. Even though we may sample the points in A from both $p_P(x)$ and $p_N(x)$, the points are more likely to be sampled from $p_P(x)$ as $\delta < 1$; if $\delta < \pi_P/\pi_N$, we further have $p(y = +1 \mid x) > p(y = -1 \mid x)$ on A, so that A should be classified into the positive class. As a result, let us transport the probability mass of the set A from the negative class to the positive class, resulting in new class-conditional densities $p'_P(x)$ and $p'_N(x)$ and a new class-prior probability $\pi'_P > \pi_P$, such that

$$\begin{cases} p'_P(x) > 0 \wedge p'_N(x) = 0 & (x \in A), \\ p'_P(x) < p_P(x) \wedge p'_N(x) > p_N(x) & (x \notin A). \end{cases}$$

Then if we apply mixture proportion estimation with $\mu_H = p'_P$, we can ensure irreducibility and obtain $\widehat{\pi}_P = \pi'_P$. Since A is small, π'_P should be only slightly larger than π_P; even if A is the largest set where $p_N(x) = \delta p_P(x)$ is achieved, we can still have $\pi'_P < \kappa^*$ (see section 12.6.3.2), and there is benefit of probability mass transportation and auxiliary probability distribution creation.

Let us work on a concrete example to get the intuition. Suppose that p_P is the uniform distribution on $[\frac{1}{2}, 1]$, p_N is uniform on $[0, 1]$, and $\pi_P = \pi_N = \frac{1}{2}$. Then we have

$$p_U(x) = \begin{cases} \frac{1}{2} & (x < \frac{1}{2}), \\ \frac{3}{2} & (x \geq \frac{1}{2}). \end{cases}$$

In this case, we can confirm that $\kappa^* = \frac{3}{4}$. Then, let $\rho > 0$ be a small constant and let $A = (1 - \rho, 1]$. The mass of A in p_U from p_N is $\pi_N p_N(A) = \frac{\rho}{2}$; after we transport the mass $\frac{\rho}{2}$ from p_N to p_P, we have

$$\pi'_P(x) = \frac{1 + \rho}{2}, \quad p'_P(x) = \begin{cases} 0 & (x < \frac{1}{2}), \\ \frac{2}{1+\rho} & (\frac{1}{2} \leq x \leq 1 - \rho), \\ \frac{3}{1+\rho} & (x > 1 - \rho). \end{cases}$$

Since p_U is the same, for any $x > 1 - \rho$, $p_U(x)/p'_P(x) = \pi'_P(x)$, which can be as close to π_P as possible by controlling ρ. This is how regrouping works.

12.6.2   Practical Implementation

In practice, we have to implement the aforementioned idea based on $\mathscr{X}_P$ and $\mathscr{X}_U$. Since $\mathscr{X}_N$ is unavailable, we cannot "cut and paste" any x_i^N from $\mathscr{X}_N$ to $\mathscr{X}_P$; instead, we can "copy and paste" some x_i^U from $\mathscr{X}_U$ to $\mathscr{X}_P$. When doing so, we should select a small set of samples $\mathscr{X}_\Delta$ which look the most similar to the positive class and dissimilar to the negative class. This is why $A = (1 - \rho, 1]$ was selected in the above example: A belongs geometrically and visually to the positive class with the highest confidence among all subsets of $[0, 1]$ of size ρ.

Taking these considerations into account, we can design the following practical implementation of regrouping (Yao et al., 2022):

1. Train a binary classifier h with $\mathscr{X}_P$ and $\mathscr{X}_U$, where we treat $\mathscr{X}_U$ as the set of negative data and treat $\frac{n_P}{n_P + n_U}$ as the positive class-prior probability (see section 3.1.1).

2. Let $\rho > 0$ be a small constant. Assign each $x_i^U \in \mathscr{X}_U$ a confidence score $h(x_i^U)$, and select ρn_U unlabeled data with highest confidence scores to be $\mathscr{X}_\Delta$, i.e.,

$$\mathscr{X}_\Delta := \underset{\mathscr{X}_\Delta \subset \mathscr{X}_U, |\mathscr{X}_\Delta| \leq \rho n_U}{\operatorname{argmax}} \sum_{x_i^U \in \mathscr{X}_\Delta} h(x_i^U).$$

3. Create $\mathscr{X}_P'$ by copying $\mathscr{X}_\Delta$ to $\mathscr{X}_P$, i.e., $\mathscr{X}_P' := \mathscr{X}_P \cup \mathscr{X}_\Delta$.

4. Perform mixture proportion estimation, where we treat $\mathscr{X}_U$ and $\mathscr{X}_P'$ as the samples from μ_F and μ_H, respectively.

The constant ρ is a hyperparameter for controlling the size of $\mathscr{X}_\Delta$, which can determine the difference between the original p_P, p_N, and π_P and the new p_P', p_N', and π_P'. Theoretically, the smaller ρ is, the closer π_P' and $\widehat{\pi}_P$ will be to π_P. Empirically, ρ cannot be too small: Existing mixture proportion estimators are insensitive to tiny modifications of $\mathscr{X}_P$ (they are designed to be robust in such a way, in order to be good estimators); if the size of $\mathscr{X}_\Delta$ is only one or two, it will result in $\widehat{\pi}_P \approx \kappa^*$ rather than $\widehat{\pi}_P \approx \pi_P' \approx \pi_P$. In practice, $\rho = 10\%$ is very good and recommended to be used as the default value.

There are two fundamental concerns for copying $\mathscr{X}_\Delta$ to $\mathscr{X}_P$. First, when we have irreducibility, might regrouping make $\widehat{\pi}_P$ worse? Second, when we lack irreducibility, must regrouping make $\widehat{\pi}_P$ better? While these concerns will be formally clarified later, we give here intuitive implications of regrouping. Let us focus on $x_i^U \in \mathscr{X}_\Delta$, which is very likely to belong to the positive class.

- First, if we have irreducibility, $p_N(x_i^U)$ should be rather small (if not zero), and x_i^U should be drawn from the positive component p_P of the mixture p_U. In this case, regrouping will generally not change p_P but only increase n_P. Hence, it will not make $\widehat{\pi}_P$ worse.

- Second, if we lack irreducibility, x_i^U may be drawn from either p_P or p_N. As x_i^U is not present in $\mathscr{X}_P$, it may seem to be drawn from p_N. By regrouping, x_i^U becomes present in $\mathscr{X}_P$, which makes it now seem to be drawn from p_P. This is philosophically the same as transporting a small neighborhood of x_i^U from p_N to p_P, which will modify p_P as we expected toward irreducibility. As a consequence, regrouping will make $\widehat{\pi}_P$ better.

Since we cannot practically verify the irreducibility assumption given $\mathscr{X}_P$ and $\mathscr{X}_U$, we recommend using regrouping before mixture proportion estimation for the PU class-prior estimation in real-world applications.

Regrouping is a *meta* approach to mitigating the overestimation problem of mixture proportion estimation. It modifies the estimation problem without modifying the estimators, so that the estimators can receive a new and better problem for which the overestimation phenomenon (if present) can be significantly reduced. This is clearly advantageous over penalization of f-divergences, which modifies $f(t)$ and $f^*(z)$ in-place and is in general hard to implement (though penalized L_1-distance minimization as a special case is easy to implement).

It seems a little counterintuitive and contradictory that we can estimate π_P without the irreducibility assumption. If following the estimation goal of mixture proportion estimation (i.e., trying to estimate the true κ appeared in (12.1)), then when we lack irreducibility, any number below κ^* defined in (12.2) may be the true π_P. This is because κ^* is regarded as fixed and π_P is regarded as a dependent variable on the free μ_G. However, following the optimization goal (i.e., trying to find κ^* defined in (12.2)), π_P is fixed while κ^* is dependent on μ_G. Given the true π_P, any number above π_P may be observed as κ^*, and thus the positive bias can be arbitrarily large. Regrouping is able to reduce this arbitrarily large positive bias: If κ^* is far away from π_P, π'_P can be far away from κ^* (depending on ρ); if κ^* is close to π_P, π'_P is also close to κ^*; if κ^* is equal to π_P, π'_P is equal to them. In all three cases, we have $\pi_P \leq \pi'_P \leq \kappa^*$, where the equality is achieved under the irreducibility assumption, as shown in section 12.6.3.

12.6.3   Theoretical Justification

In the regrouping approach described above, the auxiliary class-conditional densities p'_P and p'_N are created by regrouping a small set A from p_N to p_P. Here, we analyze the properties of regrouping and theoretically justify it.

12.6.3.1   A formal definition of regrouping

In order to analyze the properties of regrouping, we need to formally define how to split, transport, and regroup A (or the mass of A). Let μ be a probability measure on a measurable space $(\mathscr{X}, \mathfrak{S})$, where $\mathfrak{S}$ is a σ-algebra on the sample space $\mathscr{X}$. Given a set $A \in \mathfrak{S}$, define measures μ^A and $\mu^{\overline{A}}$ as

$$\forall S \in \mathfrak{S}, \quad \mu^A(S) := \mu(S \cap A) \text{ and } \mu^{\overline{A}}(S) := \mu(S \setminus A),$$

where $\overline{A} = \mathscr{X} \setminus A$ denotes the relative complement of A in $\mathscr{X}$. It is clear that for any set A, $\mu^A + \mu^{\overline{A}} = \mu$ holds, since

$$\forall S \in \mathfrak{S}, \quad \mu^A(S) + \mu^{\overline{A}}(S) = \mu(S \cap A) + \mu(S \setminus A)$$

$$= \mu(S \cap A) + \mu(S \cap \overline{A})$$

$$= \mu((S \cap A) \cup (S \cap \overline{A}))$$

$$= \mu(S \cap (A \cup \overline{A}))$$

$$= \mu(S).$$

As a result, μ^A is the probability measure obtained by splitting A from μ, and $\mu^{\overline{A}}$ is the remaining probability measure after the split of A.

Now, we are ready to introduce the theory of regrouping (Yao et al., 2022). Fixing $A \in \mathfrak{S}$, we split p_N into p_N^A and $p_N^{\overline{A}}$, transport p_N^A to p_P to regroup them together, and finally renormalize p_P and p_N^A into p_P' and renormalize $p_N^{\overline{A}}$ into p_N'—namely,

$$p_U = \pi_P p_P + \pi_N p_N$$

$$= \pi_P p_P + \pi_N (\underbrace{p_N^A + p_N^{\overline{A}}}_{\text{split into two}})$$

$$= \underbrace{(\pi_P p_P + \pi_N p_N^A)}_{\text{regroup as one}} + \pi_N p_N^{\overline{A}}. \tag{12.24}$$

Without loss of generality, let $p_N(A) = \rho$, and thus $p_N(\overline{A}) = 1 - \rho$. Since we have $p_P(\mathcal{X}) = 1$, $p_N^A(\mathcal{X}) = p_N(A)$, and $p_N^{\overline{A}}(\mathcal{X}) = p_N(\overline{A})$, renormalization of (12.24) is straightforward:

$$\pi_P' := \pi_P p_P(\mathcal{X}) + \pi_N p_N^A(\mathcal{X}) = \pi_P + \rho \pi_N,$$

$$\pi_N' := \pi_N p_N^{\overline{A}}(\mathcal{X}) = \pi_N - \rho \pi_N,$$

$$p_P' := \frac{\pi_P p_P + \pi_N p_N^A}{\pi_P'} = \frac{\pi_P p_P + \pi_N p_N^A}{\pi_P + \rho \pi_N},$$

$$p_N' := \frac{\pi_N p_N^{\overline{A}}}{\pi_N'} = \frac{p_N^{\overline{A}}}{1 - \rho}.$$

Actually, by distinguishing between the cases $x \in A$ and $x \notin A$, p_P' and p_N' can be rewritten as

$$p_P'(x) = \begin{cases} \dfrac{\pi_P p_P(x) + \pi_N p_N(x)}{\pi_P + \rho \pi_N} = \dfrac{p_U(x)}{\pi_P + \rho \pi_N} > 0 & (x \in A), \\[2ex] \dfrac{\pi_P p_P(x)}{\pi_P + \rho \pi_N} = \dfrac{p_P(x)}{1 + \rho \pi_N / \pi_P} < p_P(x) & (x \notin A), \end{cases}$$

$$p_N'(x) = \begin{cases} 0 & (x \in A), \\[2ex] \dfrac{p_N(x)}{1 - \rho} > p_N(x) & (x \notin A), \end{cases}$$

where $p_U(x) > 0$ over A is from $A \in \mathfrak{S}$ (i.e., $A \subset \mathcal{X}$) and $p_U(x) > 0$ over $\mathcal{X}$.

We can see that for p_P', every $x \in A$ becomes an anchor point, and if $\rho > 0$, A itself becomes an anchor set. This means that transportation of the probability mass and creation of the auxiliary probability distribution in regrouping could implement anchor point/set creation. For p_P' and p_N' created in this way, the irreducibility or the anchor set assumption must be satisfied so that π_P' can be recovered by κ^* in (12.4) with $\mu_F = p_U$ and $\mu_H = p_P'$.

12.6.3.2   Bias reduction

Next, we clarify the previous concerns: Might regrouping result in $\pi_{\mathrm{P}}' > \pi_{\mathrm{P}}$ if the irreducibility assumption is satisfied, and must regrouping result in $\pi_{\mathrm{P}} < \pi_{\mathrm{P}}' < \kappa^*$ if the irreducibility assumption is not satisfied? We expect that A is a small set in the sample space $\mathscr{X}$ that looks most similar/dissimilar to the positive/negative class. To this end, let A^* be a set selected as

$$A^* := \underset{A \in \mathfrak{S}, p_{\mathrm{P}}(A) > 0}{\operatorname{argmin}} \frac{p_{\mathrm{N}}(A)}{p_{\mathrm{P}}(A)}, \tag{12.25}$$

where for the ease of discussion, we have assumed that the infimum is attainable (as the feasible region is an open set) so that a minimizer of the infimum exists inside, and the infimum can simply be replaced with a minimum. If it does not exist, we argue with a sequence $(A_\epsilon^*)_{\epsilon > 0, \epsilon \to 0}$ as

$$A_\epsilon^* := \underset{A \in \mathfrak{S}, p_{\mathrm{P}}(A) = \epsilon}{\operatorname{argmin}} \frac{p_{\mathrm{N}}(A)}{p_{\mathrm{P}}(A)},$$

$$\lim_{\epsilon \to 0} \frac{p_{\mathrm{N}}(A_\epsilon^*)}{p_{\mathrm{P}}(A_\epsilon^*)} = \underset{A \in \mathfrak{S}, p_{\mathrm{P}}(A) > 0}{\inf} \frac{p_{\mathrm{N}}(A)}{p_{\mathrm{P}}(A)}.$$

If the irreducibility assumption holds, A^* is the anchor set. Thus, $\rho = p_{\mathrm{N}}(A^*) = 0$, which denies the possibility of $\pi_{\mathrm{P}}' > \pi_{\mathrm{P}}$. If the assumption does not hold, it implies that

$$0 < p_{\mathrm{N}}(A^*) < \frac{p_{\mathrm{N}}(A^*)}{p_{\mathrm{P}}(A^*)}$$

due to $p_{\mathrm{P}}(A^*) < 1$. This gives $\pi_{\mathrm{P}} < \pi_{\mathrm{P}}' < \kappa^*$ since

$$\pi_{\mathrm{P}}' = \pi_{\mathrm{P}} + \pi_{\mathrm{N}} p_{\mathrm{N}}(A^*), \tag{12.26}$$

$$\kappa^* = \pi_{\mathrm{P}} + \frac{\pi_{\mathrm{N}} p_{\mathrm{N}}(A^*)}{p_{\mathrm{P}}(A^*)}.$$

At first glance, it may seem a bit strange that ρ is fixed as a hyper-parameter in the implementation but it can change in different cases in the theory. There is actually a discrepancy in the usage of ρ, which is $p_{\mathrm{U}}(A)$ in the implementation and $p_{\mathrm{N}}(A)$ in the theory. It is obvious that

$$p_{\mathrm{U}}(A^*) = \pi_{\mathrm{P}} p_{\mathrm{P}}(A^*) + \pi_{\mathrm{N}} p_{\mathrm{N}}(A^*),$$

where $p_{\mathrm{P}}(A^*) > 0$ due to (12.25), and hence it is possible that $p_{\mathrm{U}}(A^*)$ is fixed but $p_{\mathrm{N}}(A^*)$ can be zero or non-zero in different cases. That being said, for A^* defined in (12.25), $p_{\mathrm{U}}(A^*)$ is determined no matter which case it is. This is why we should tune the hyper-parameter ρ in the implementation if we pursue best estimation accuracy.

12.6.3.3   Convergence analysis

The optimal set A^* defined in (12.25) makes use of p_N, but it can be computed based on p_U and p_P, since

$$A^* = \operatorname*{argmin}_{A \in \mathfrak{S}, p_P(A) > 0} \frac{p_U(A)}{p_P(A)},$$

where

$$\frac{p_U(A)}{p_P(A)} = \pi_P + \frac{\pi_N p_N(A)}{p_P(A)}.$$

This optimization can be solved by the first mixture proportion estimator using naive density estimation proposed in Blanchard et al. (2010). Nevertheless, we will show later a computationally more efficient way for finding A^*. For this reason, we assume now that A^* is given in advance and we employ the mixture proportion estimator as it is:

$$\widehat{\pi}_P := \inf_{S \in \mathscr{S}, \widehat{p}'_P(S) > 0} \frac{\widehat{p}_U(S)}{\widehat{p}'_P(S)}. \tag{12.27}$$

Here, $\widehat{p}_U(S)$ and $\widehat{p}'_P(S)$ are the empirical estimates of $p_U(S)$ and $p'_P(S)$ based on $\mathscr{X}_U$ and $\mathscr{X}'_P$, respectively, and the σ-algebra $\mathfrak{S}$ is replaced with a set class $\mathscr{S}$ for the sake of complexity control. Let $I_S\colon \mathscr{X} \to \{0, 1\}$ be the indicator function of S such that $I_S(x) = 1$ for $x \in S$ and $I_S(x) = 0$ for $x \notin S$, Then the corresponding function class is defined as

$$\mathscr{H} := \{I_S \mid S \in \mathscr{S}\}.$$

Now we can easily see that

$$p_U(S) = \int_S p_U(x)\mathrm{d}x = \int_{\mathscr{X}} I_S(x)p_U(x)\mathrm{d}x,$$

$$p'_P(S) = \int_S p'_P(x)\mathrm{d}x = \int_{\mathscr{X}} I_S(x)p'_P(x)\mathrm{d}x,$$

and consequently

$$\widehat{p}_U(S) := \frac{1}{n_U} \sum_{i=1}^{n_U} I_S(x_i^U),$$

$$\widehat{p}'_P(S) := \frac{1}{n'_P} \sum_{x^P \in \mathscr{X}'_P} I_S(x^P),$$

where $n'_P := |\mathscr{X}'_P|$ is the size of $\mathscr{X}'_P$.

Subsequently, for $\widehat{\pi}_P$ defined in (12.27), we can derive an error bound for it as follows (for its proof, see Yao et al., 2022, theorem 4). For any $\delta > 0$, with probability at least $1 - 2\delta$, it holds that

$$|\widehat{\pi}_{\mathrm{P}} - \pi_{\mathrm{P}}| \leq \frac{\epsilon_{\delta,\mathscr{H}}(\mathscr{X}_{\mathrm{U}}) + \epsilon_{\delta,\mathscr{H}}(\mathscr{X}'_{\mathrm{P}})}{\widehat{p}'_{\mathrm{P}}(A^*) + \epsilon_{\delta,\mathscr{H}}(\mathscr{X}'_{\mathrm{P}})} + \pi_{\mathrm{N}}p_{\mathrm{N}}(A^*), \tag{12.28}$$

where the error function (with $S = \mathscr{X}_{\mathrm{U}}$ or $S = \mathscr{X}'_{\mathrm{P}}$) is defined as

$$\epsilon_{\delta,\mathscr{H}}(S) := 2\mathfrak{R}_S(\mathscr{H}) + 3\sqrt{\frac{\ln(4/\delta)}{2|S|}},$$

and $\mathfrak{R}_S(\mathscr{H})$ is the empirical Rademacher complexity of $\mathscr{H}$ given S (see section 3.1.2.3).

Equation (12.28) as an upper bound of the class-prior estimation error $|\widehat{\pi}_{\mathrm{P}} - \pi_{\mathrm{P}}|$ is in fact an excess risk bound, where the first term is an estimation error and the second term is an approximation error. In (12.28), $\epsilon_{\delta,\mathscr{H}}(\mathscr{X}_{\mathrm{U}})$ is an upper bound of the error of $\widehat{p}_{\mathrm{U}}(S)$ as an estimator of $p_{\mathrm{U}}(S)$ for all $S \in \mathscr{S}$, and so is $\epsilon_{\delta,\mathscr{H}}(\mathscr{X}'_{\mathrm{P}})$:

$$\sup_{S \in \mathscr{S}} |\widehat{p}_{\mathrm{U}}(S) - p_{\mathrm{U}}(S)| \leq \epsilon_{\delta,\mathscr{H}}(\mathscr{X}_{\mathrm{U}}),$$

$$\sup_{S \in \mathscr{S}} |\widehat{p}'_{\mathrm{P}}(S) - p'_{\mathrm{P}}(S)| \leq \epsilon_{\delta,\mathscr{H}}(\mathscr{X}'_{\mathrm{P}}),$$

where the two inequalities hold separately with probability at least $1 - \delta$. To see the convergence rate of (12.28), we need two assumptions. The *universal approximation assumption* (Scott, 2015) ensures that $\epsilon_{\delta,\mathscr{H}}(S) = \mathcal{O}_p\left(\sqrt{\frac{\ln|S|}{|S|}}\right)$, and then after assuming $\mathscr{X}'_{\mathrm{P}} \cap A^* \neq \emptyset$, the first term will be in the order of $\mathcal{O}_p\left(\sqrt{\frac{\ln\min\{n_{\mathrm{U}},n'_{\mathrm{P}}\}}{\min\{n_{\mathrm{U}},n'_{\mathrm{P}}\}}}\right)$. The second term comes from $\pi'_{\mathrm{P}} - \pi_{\mathrm{P}}$, as shown in (12.26). As a result, given A^*, $\widehat{\pi}_{\mathrm{P}}$ converges to the best possible estimate π'_{P} nearly in the optimal parametric rate.

After regrouping a small set A^* from p_{N} to p_{P}, we have $n_{\mathrm{U}} > n'_{\mathrm{P}} > n_{\mathrm{P}}$. Thus, the first term changes to $\mathcal{O}_p\left(\sqrt{\frac{\ln n'_{\mathrm{P}}}{n'_{\mathrm{P}}}}\right)$ from $\mathcal{O}_p\left(\sqrt{\frac{\ln n_{\mathrm{P}}}{n_{\mathrm{P}}}}\right)$, compared with the same mixture proportion estimator without regrouping. This suggests that regrouping can be helpful even when we have irreducibility, since besides bias reduction, it can also speed up the convergence.

Last but not least, we rely on the density-estimation-based estimator, since regrouping may only guarantee the anchor set assumption. The quantity $p_{\mathrm{U}}(S)/p'_{\mathrm{P}}(S)$ may look like a density ratio function, but it is not because it is not a function of x but defined over a set S. However, if we make the anchor point assumption, employing the density-ratio-estimation-based estimator in (12.5) and (12.6) may be better: Theoretically, with density ratio estimation, $\widehat{\pi}_{\mathrm{P}}$ converges to π'_{P} in the optimal parametric rate; empirically, $\widehat{p}'_{\mathrm{P}}$ in the denominator may cause certain instability, and density ratio estimation does not suffer from this instability (Sugiyama et al., 2012).

12.6.3.4　Computationally efficient identification of A^*

In the implementation, we can obtain $\mathcal{X}_\Delta$ by classifier training and sorting without naive density estimation that may be unstable for the denominator $p'_{\rm P}$. We explain here why equation (12.25) can be approximately solved by binary classification.

Let us define another auxiliary distribution $q(x, s)$, where $s \in \{0, 1\}$ is the positive-vs.-unlabeled label (i.e., a class label distinguishing between the positive component and the whole mixture). Specifically, $q(x, s)$ is defined as

$$q(s=1) := \frac{\pi_{\rm P}}{1 + \pi_{\rm P}}, \quad q(s=0) := \frac{1}{1 + \pi_{\rm P}},$$

$$q(x \mid s=1) := p_{\rm P}(x), \quad q(x \mid s=0) := p_{\rm U}(x).$$

Then,

$$A^* = \underset{A \in \mathfrak{S}, p_{\rm P}(A) > 0}{\arg\min} \frac{p_{\rm U}(A)}{p_{\rm P}(A)} = \underset{A \in \mathfrak{S}}{\arg\max} \frac{p_{\rm P}(A)}{p_{\rm U}(A)},$$

where the constraint $p_{\rm P}(A) > 0$ is safely removed. For the objective function,

$$\frac{p_{\rm P}(A)}{p_{\rm U}(A)} = \frac{\int_A p_{\rm P}(x)\mathrm{d}x}{\int_A p_{\rm U}(x)\mathrm{d}x} = \frac{\int_{\mathcal{X}} \mathrm{I}_A(x)p_{\rm P}(x)\mathrm{d}x}{\int_{\mathcal{X}} \mathrm{I}_A(x)p_{\rm U}(x)\mathrm{d}x} = \frac{\int \mathrm{I}_A(x)q(x \mid s=1)\mathrm{d}x}{\int \mathrm{I}_A(x)q(x \mid s=0)\mathrm{d}x},$$

where we apply the definition of the auxiliary distribution $q(x, s)$. Expressing the class-conditional densities with the class-posterior probabilities, we can obtain

$$q(x \mid s=1) = \frac{q(s=1 \mid x)q(x)}{q(s=1)} \propto q(s=1 \mid x)(\pi_{\rm P}p_{\rm P}(x) + p_{\rm U}(x)),$$

$$q(x \mid s=0) = \frac{q(s=0 \mid x)q(x)}{q(s=0)} \propto q(s=0 \mid x)(\pi_{\rm P}p_{\rm P}(x) + p_{\rm U}(x)),$$

and thus

$$A^* = \underset{A \in \mathfrak{S}}{\arg\max} \frac{\pi_{\rm P}\int \mathrm{I}_A(x)q(s=1 \mid x)p_{\rm P}(x)\mathrm{d}x + \int \mathrm{I}_A(x)q(s=1 \mid x)p_{\rm U}(x)\mathrm{d}x}{\pi_{\rm P}\int \mathrm{I}_A(x)q(s=0 \mid x)p_{\rm P}(x)\mathrm{d}x + \int \mathrm{I}_A(x)q(s=0 \mid x)p_{\rm U}(x)\mathrm{d}x}.$$

For the above optimization, its objective function has four expectations over $p_{\rm P}(x)$ and $p_{\rm U}(x)$, which can be estimated by the corresponding sample averages:

$$\mathcal{X}_\Delta = \underset{S \subset \mathcal{X}_{\rm U}}{\arg\max} \frac{\frac{\pi_{\rm P}}{n_{\rm P}}\sum_{i=1}^{n_{\rm P}} \mathrm{I}_S(x_i^{\rm P})q(s=1 \mid x_i^{\rm P}) + \frac{1}{n_{\rm U}}\sum_{i=1}^{n_{\rm U}} \mathrm{I}_S(x_i^{\rm U})q(s=1 \mid x_i^{\rm U})}{\frac{\pi_{\rm P}}{n_{\rm P}}\sum_{i=1}^{n_{\rm P}} \mathrm{I}_S(x_i^{\rm P})q(s=0 \mid x_i^{\rm P}) + \frac{1}{n_{\rm U}}\sum_{i=1}^{n_{\rm U}} \mathrm{I}_S(x_i^{\rm U})q(s=0 \mid x_i^{\rm U})},$$

where the optimization variable and feasible region $A \in \mathfrak{S}$ (which means $A \subset \mathcal{X}$ if $\mathfrak{S}$ is the largest possible σ-algebra on $\mathcal{X}$, i.e., its power set $2^{\mathcal{X}}$) is accordingly replaced with $S \subset \mathcal{X}_{\rm U}$. We assume without loss of generality that $\mathcal{X}_{\rm P} \cap \mathcal{X}_{\rm U} = \emptyset$ before regrouping takes place. Then, we have $\mathrm{I}_S(x_i^{\rm P}) = 0$ for all $x_i^{\rm P} \in \mathcal{X}_{\rm P}$ when considering $S \subset \mathcal{X}_{\rm U}$. Therefore, $\mathcal{X}_\Delta$

can be obtained by

$$\mathcal{X}_\Delta = \underset{S \subset \mathcal{X}_{\mathrm{U}}}{\mathrm{argmax}} \; \frac{\sum_{i=1}^{n_{\mathrm{U}}} \mathbb{I}_S(x_i^{\mathrm{U}}) q(s=1 \mid x_i^{\mathrm{U}})}{\sum_{i=1}^{n_{\mathrm{U}}} \mathbb{I}_S(x_i^{\mathrm{U}}) q(s=0 \mid x_i^{\mathrm{U}})}$$

$$= \underset{S \subset \mathcal{X}_{\mathrm{U}}}{\mathrm{argmax}} \; \frac{\sum_{x_i^{\mathrm{U}} \in S} q(s=1 \mid x_i^{\mathrm{U}})}{\sum_{x_i^{\mathrm{U}} \in S} q(s=0 \mid x_i^{\mathrm{U}})}. \tag{12.29}$$

Equation (12.29) is the "exact" empirical solution of (12.25) whose exactness is in the sense that there exist only approximations by replacing expectations with empirical averages.

Due to the discrete nature of $S \subset \mathcal{X}_{\mathrm{U}}$, the objective function of (12.29) may be unstable. Fortunately, it can be stabilized by moving the denominator as a constraint:

$$\mathcal{X}_\Delta = \underset{S \subset \mathcal{X}_{\mathrm{U}}}{\mathrm{argmax}} \quad \sum_{x_i^{\mathrm{U}} \in S} q(s=1 \mid x_i^{\mathrm{U}})$$

$$\text{subject to} \quad \sum_{x_i^{\mathrm{U}} \in S} q(s=0 \mid x_i^{\mathrm{U}}) \le \rho n_{\mathrm{U}},$$

where $\rho > 0$ is a small constant. Even though this optimization problem may be solved stably, obtaining a solution would be extremely time-consuming since it is a combinatorial optimization problem. As

$$q(s=1 \mid x) + q(s=0 \mid x) = 1,$$

the constraint offers no more effect than size control, so that it can be relaxed into $|S| \le \rho n_{\mathrm{U}}$.

At last, estimating the class-posterior probability $q(s=1 \mid x)$ corresponds exactly to training a binary classifier with $\mathcal{X}_{\mathrm{P}}$ and $\mathcal{X}_{\mathrm{U}}$, where $q(s=1)$ is estimated by $\widehat{q}(s=1) = \frac{n_{\mathrm{P}}}{n_{\mathrm{P}}+n_{\mathrm{U}}}$, leading to

$$\mathcal{X}_\Delta = \underset{S \subset \mathcal{X}_{\mathrm{U}}, |S| \le \rho n_{\mathrm{U}}}{\mathrm{argmax}} \sum_{x_i^{\mathrm{U}} \in S} \widehat{q}(s=1 \mid x_i^{\mathrm{U}}).$$

This is how $\mathcal{X}_\Delta$ is found in the implementation: After classifier training, sorting is enough, and no combinatorial optimization is involved, which is computationally extremely efficient. Given that $\mathcal{X}_\Delta$ is an empirical version of A^*, the technique of classifier training and sorting can also be regarded as a computationally efficient way to identify A^*.

12.6.3.5   Approximation of p'_{P} with a surrogate

The final discrepancy between the implementation and the theory of regrouping is from which distribution we sample $\mathcal{X}'_{\mathrm{P}}$. In the theory, we would like to sample the data from

$$p'_{\mathrm{P}} = \frac{\pi_{\mathrm{P}} p_{\mathrm{P}} + \pi_{\mathrm{N}} p_{\mathrm{N}}^A}{\pi_{\mathrm{P}} + \pi_{\mathrm{N}} p_{\mathrm{N}}(A)},$$

but unfortunately there is no way to split p_N^A without knowing p_N. Instead of p_P', the regrouped positive data can be regarded as sampled from

$$p_P'' = \frac{p_P + p_U^A}{1 + p_U(A)},$$

due to the definition $\mathcal{X}_P' := \mathcal{X}_P \cup \mathcal{X}_A$. Their difference can be upper-bounded uniformly over $\mathfrak{S}$ in the order of $\mathcal{O}(\rho)$, where $\rho = p_U(A)$ for convenience.

More specifically, for any $S \in \mathfrak{S}$,

$$
\begin{aligned}
p_P'(S) - p_P''(S) &= \frac{\pi_P p_P(S) + \pi_N p_N^A(S)}{\pi_P + \pi_N p_N(A)} - \frac{p_P(S) + p_U^A(S)}{1 + \rho} \\
&\leq \frac{\pi_P p_P(S) + \pi_N p_N(A)}{\pi_P + \pi_N p_N(A)} - \frac{p_P(S)}{1 + \rho} \\
&= \frac{\rho(\pi_P p_P(S) + \pi_N p_N(A)) + \pi_N p_N(A)(1 - p_P(S))}{(1 + \rho)(\pi_P + \pi_N p_N(A))} \\
&\leq \frac{\rho(p_U(S) + p_U(A)) + p_U(A)}{(1 + \rho)\pi_P} \\
&= \frac{\rho(p_U(S) + 1 + \rho)}{(1 + \rho)\pi_P} \\
&= \mathcal{O}(\rho).
\end{aligned}
$$

On the other hand,

$$
\begin{aligned}
p_P''(S) - p_P'(S) &= \frac{p_P(S) + p_U^A(S)}{1 + \rho} - \frac{\pi_P p_P(S) + \pi_N p_N^A(S)}{\pi_P + \pi_N p_N(A)} \\
&\leq \frac{p_P(S) + \rho}{1 + \rho} - \frac{\pi_P p_P(S)}{\pi_P + \pi_N p_N(A)} \\
&= \frac{\rho(\pi_P + \pi_N p_N(A) - \pi_P p_P(S)) + \pi_N p_N(A) p_P(S)}{(1 + \rho)(\pi_P + \pi_N p_N(A))} \\
&\leq \frac{\rho(\pi_P + 1 + \rho)}{(1 + \rho)\pi_P} \\
&= \mathcal{O}(\rho).
\end{aligned}
$$

Combining the two directions gives

$$\sup_{S \in \mathfrak{S}} |p_P'(S) - p_P''(S)| = \mathcal{O}(\rho).$$

Since the gap has the same order as $p_U(A)$ uniformly over $\mathfrak{S}$, it is guaranteed that whenever $p_U(A)$ is small, the gap is also small. This implies that p_P'' is a good surrogate for p_P'. As a consequence, the practical implementation of regrouping is a good approximation of the theory of regrouping, as we expected. By now, we have analyzed all of the properties of regrouping and theoretically justified all of the points in its design.

12.7 Class-Prior Estimation from Pairwise Data

Finally, we discuss how to estimate π_P in the setting of *pairwise-constraint classification* (explored previously in chapter 7). Note that the discussion here is specialized to *similar-unlabeled (SU) classification*; an extension to *dissimilar-unlabeled (DU) classification* is straightforward and hence omitted.

Assume that we have

$$\mathcal{X}_S := \{(x_i^S, x_i^{S'})\}_{i=1}^{n_S} \overset{\text{i.i.d.}}{\sim} p_S(x, x'),$$

$$\mathcal{X}_U := \{x_i^U\}_{i=1}^{n_U} \overset{\text{i.i.d.}}{\sim} p_U(x),$$

where for $\pi_S := \pi_P^2 + \pi_N^2$,

$$p_S(x, x') := \frac{\pi_P^2}{\pi_S} p_P(x) p_P(x') + \frac{\pi_N^2}{\pi_S} p_N(x) p_N(x').$$

As discussed in section 7.2.4, pairwise constraints can be treated as pointwise data. Namely, the decoupled S data

$$\widetilde{\mathcal{X}}_S := \{\widetilde{x}_i^S\}_{i=1}^{2n_S} := \{x_1^S, x_1^{S'}, \ldots, x_{n_S}^S, x_{n_S}^{S'}\}$$

can be regarded as if they were drawn independently from the pointwise density for S data defined as

$$\widetilde{p}_S(x) := \frac{\pi_P^2}{\pi_S} p_P(x) + \frac{\pi_N^2}{\pi_S} p_N(x).$$

The key ingredient is the connection among the S-, D-, and U-data densities given by

$$p_U(x) = \pi_S \widetilde{p}_S(x) + (1 - \pi_S) \widetilde{p}_D(x), \tag{12.30}$$

where $\widetilde{p}_D$ is the pointwise density for D data defined as

$$\widetilde{p}_D(x) := \frac{1}{2} p_P(x) + \frac{1}{2} p_N(x).$$

From (12.30), we can see that any class-prior estimation method for PU classification introduced in this chapter can be employed to estimate π_S from $\widetilde{\mathcal{X}}_S$ and $\mathcal{X}_U$.

Once an estimator of π_S, denoted by $\widehat{\pi}_S$, is obtained, an estimator of π_P can be obtained as

$$\widehat{\pi}_P := \begin{cases} \frac{1 + \sqrt{2\widehat{\pi}_S - 1}}{2} & (\pi_P > \frac{1}{2}), \\ \frac{1 - \sqrt{2\widehat{\pi}_S - 1}}{2} & (\pi_P < \frac{1}{2}), \end{cases}$$

which immediately follows from $\pi_S = \pi_P^2 + \pi_N^2$.

13 Conclusions and Prospects

Let us now recapitulate what we have advocated in this book and discuss future prospects of machine learning (ML), artificial intelligence (AI), and beyond.

We have explored the paradigm of *weakly supervised classification* (WSC). Our motivation for thoroughly investigating WSC was the severe difficulty of collecting fully labeled training data in many real-world applications—a situation that has significantly hindered the widespread adoption of AI and ML in society. Among various possible approaches for WSC, we have solely focused on *empirical risk minimization* (ERM). The ERM approach allowed us to systematically develop useful algorithms of WSC for diverse problem settings, such as positive-unlabeled classification, positive-confidence classification, pairwise-constraint classification, unlabeled-unlabeled classification, complementary-label classification, and partial-label classification. In the ERM approach, the classification risk is written in terms of the expectations of available data and then empirical approximation is applied to the written risk to obtain a practical learning objective. This mathematically clean nature of the ERM formulation allowed us also to systematically conduct theoretical analysis. Indeed, we have commonly used the *Rademacher complexity* to guarantee the convergence of the derived estimators.

An ultimate goal of WSC—or more generally, weakly supervised *learning* (WSL)—is to use all kinds of data. Even if it is difficult to collect a large quantity of fully labeled data, it may still be possible to obtain some kind of weakly informative data in each application. We would like to fully use such a set of precious but weak data. Actually, the WSC methods introduced in this book are basically estimating the classification risk in an unbiased fashion from available data. Therefore, in principle, linearly combining all the risk estimators (as done in chapters 5 and 7) still gives an unbiased risk estimator. With such a combined risk estimator, we can use various types of data simultaneously, including positive, negative, unlabeled, positive-confidence, negative-confidence, pairwise-similarity, pairwise-dissimilarity, and similarity-confidence data. Therefore, to further strengthen this ERM approach for WSC, it would be promising to enrich the set of WSC problems that can be solved in a similar manner.

WSC is mathematically related to *label-noise classification* (Han et al., 2020a). Indeed, as discussed in chapter 9, complementary-label classification can be regarded as a form of label-noise classification with particular noise transition probabilities. Similarly, WSC is also related to *transfer learning* (Quiñonero-Candela et al., 2009), which is aimed at recycling data used in other, related tasks. Indeed, positive-confidence classification introduced in chapter 6 implicitly performs a certain type of transfer from the positive data distribution to the unlabeled data distribution. Therefore, further exploring the interplay among WSC, label-noise classification, and transfer learning would be promising to gain more insight into these problems and to further improve the classification performance under weak supervision.

Although we have argued throughout this book that WSC is promising, it is definitely best to collect as much clean and fully supervised data as possible. As briefly discussed in chapter 1, using a *crowd-sourcing* platform to efficiently collect labels (Law and von Ahn, 2011) and conducting *active learning* to optimize the use of labeling budgets (Sugiyama and Kawanabe, 2012) are promising approaches to improve the label collection process. Another interesting direction to improve data collection is *pseudo data generation*, e.g., by using a simulator such as a car-driving game (Richter et al., 2016) or using a data generation probability distribution estimated by *generative adversarial networks* (Goodfellow et al., 2014).

WSC can be regarded as reliably training a classifier from limited information. To further pursue *reliable ML*, we still need to solve numerous challenges. *Learning from noisy data* is obviously of high importance along this direction, since data we use for training an ML system is substantially noise-prone. Therefore, developing ML methods that can cope with noise in both training input and output holds promise in further enhancing the reliability of ML systems in real-world deployment. *Robust statistics* (Huber, 1981) has been the main tool to cope with noise and outliers in general, but it would be encouraging to explore novel approaches that can take specific data collection processes into account. This is because, in the big data era, data is often collected automatically through internet-of-things devices or labeled manually by crowd-workers; furthermore, some data may even be generated artificially by a simulator. Such noisy data has peculiar characteristics, and thus incorporating the data's properties in the research of learning will be worth investigating.

In addition to robustness against uncertainty in training input and output, robustness against test input perturbation is becoming an urgent issue for the ML community to overcome. Specifically, it has been demonstrated that malicious noise in test input can arbitrarily change the classifier output, particularly in the case of deep neural networks (Goodfellow et al., 2015). Therefore, defending ML systems from such *adversarial attack* is a critical challenge for deploying ML systems in the wild. Since adversarial ML is highly related to *robust optimization* (Ben-Tal et al., 2009), it would be promising to further explore the intersection between ML and mathematical optimization.

The problem settings of WSC that we have covered have strong relevant to *data privacy*. Indeed, directly collecting labeled data can raise serious privacy concerns. Therefore, WSC, which is aimed at training a classifier only from indirect and weak data, can naturally mitigate the privacy concern, while one can still enjoy the power of ML in prediction and knowledge acquisition. So far, international tech giants have actively collected personal data to improve their services and products. On the other hand, in many countries, privacy rights and consumer protection have been enhanced, such as the *General Data Protection Regulation* (GDPR) in the European Union and, at the state level in the United States, the *California Consumer Privacy Act* (CCPA). Therefore, to use ML in a manner that is more acceptable to society, it is important to implement safe and user-controllable data-sharing systems. The *personal life repository* is one such decentralized model for private-data sharing based on a public-key cryptosystem over the cloud.[1]

Finally, the most challenging question for ML researchers would be: *What is ultimate AI?* One may think of *superhuman intelligence* as future AI. That is definitely a natural way to go. On the other hand, since AI systems will be deployed in human society, ultimate AI need not be self-intelligence but *co-intelligence* between humans and AI. For example, human designers and AI systems can teach and learn from each other to be more creative.[2] From this viewpoint, it is important to include *humans in the loop* beyond simple data labeling, and the notion of reliable ML needs to be redefined in a more human-friendly manner. We have only explored mathematical approaches to ML in this book, but toward ultimate AI, it would be important to incorporate diverse issues such as human learning (neuroscientific findings), human assistance (knowledge and creativity), and human society (culture and ethics). We do believe that the science of ML will contribute to understanding and achieving human prosperity.

Notes

Notes to Chapter 1

1. Online formulations of supervised learning and unsupervised learning also exist (Cesa-Bianchi and Lugosi, 2006).

2. Another important paradigm of statistical data analysis is *causal inference*, which tries to infer causal relation between variables beyond correlation. Note that causal relations are not necessarily reflected in probability distributions.

3. It is possible to obtain the boundary between the P and N classes, but it is not possible to know which side of the boundary is the P class or N class, since no labeled sample is available in the framework of pairwise-constraint classification.

Notes to Chapter 2

1. At a first glance, $R_{\mathrm{U,P}}(g)$ and $R_{\mathrm{U,N}}(g)$ may look a bit strange. Later, in chapters 4 and 5—positive-unlabeled (PU) classification and positive-negative-unlabeled (PNU) classification, respectively—these notations make equations short.

2. Our definition of one-versus-the-rest introduced here is a modified version with an additional $\frac{1}{c-1}$ in the second term.

Notes to Chapter 3

1. In this monograph, we solely focus on so-called discriminative classification that follows the principle of empirical risk minimization (Vapnik, 1998).

2. The estimation error can be defined for every $g \in \mathcal{G}$, but we mean the estimation error of $\widehat{g}_{\mathrm{PN}}$ or that of PN classification.

3. It seems that C_g depends on $\mathcal{G}$, but chronologically and logically, $\mathcal{G}$ should depend on C_g, in a sense that we first need a norm bound of g or its model parameters and then can define $\mathcal{G}$.

4. Here, the probability is over the repeated sampling of the data for training $\widehat{g}_{\mathrm{PN}}$.

5. Here, the probability is over the repeated sampling of the data for evaluating $\widehat{R}_{\mathrm{PN}}(g)$.

6. Note that $\mathcal{L}(\cdot,y) \circ \mathcal{G}^c := \{\mathcal{L}(\cdot,y) \circ \boldsymbol{g} \mid \boldsymbol{g} \in \mathcal{G}^c\} = \{(\boldsymbol{x},y) \mapsto \mathcal{L}(\boldsymbol{g}(\boldsymbol{x}),y) \mid \boldsymbol{g} \in \mathcal{G}^c\}$, where y is fixed, is a composed function class of scalar-valued functions but not vector-valued functions.

Notes to Chapter 4

1. For margin-based losses, the symmetric condition is reduced to $\ell(m) + \ell(-m) = 1$.

2. For margin-based losses, the linearity condition is reduced to $\ell(m) - \ell(-m) = -m$.

3. When $\alpha_{\mathrm{PU,PN}} < 1$, $\widehat{g}_{\mathrm{PU}}$ may also perform worse than $\widehat{g}_{\mathrm{PN}}$. In such a case, we should try more advanced and powerful PU classification methods before collecting more unlabeled data.

Notes to Chapter 5

1. While a *manifold* in mathematics is defined as a topological space that can be locally approximated by Euclidean space, it is just regarded as a local region or a *cluster* in semi-supervised classification.

2. For k-nearest-neighbor similarity, matrix W can be regarded as an *adjacency matrix* of a graph with n vertices—the (i, i')-entry of an adjacency matrix takes one if there exists an edge between vertices i and i'; otherwise it takes zero. In this context, matrix L is called the *graph Laplacian* (Chung, 1997).

3. In the rest of this chapter, we will assume that k is non-negative and $\lambda > \gamma c / n_{\mathrm{all}}$.

4. Note that being fixed means that g is determined before seeing the data for evaluating the empirical risk. For instance, if g is trained by some classification method, and the empirical risk is subsequently evaluated on the validation/test data, g is regarded as fixed in the evaluation.

5. https://www.openml.org/d/1460

6. The labeled data without class k means that, for example, when $\mathscr{Y} = \{1, 2, 3\}$ and $k = 2$, we have labeled data for class 1 and 3.

Notes to Chapter 6

1. $\mathfrak{R}_n(\mathscr{G}) = \mathbb{E}_{\mathscr{X}} \mathbb{E}_{\sigma_1,\ldots,\sigma_n} [\sup_{g \in \mathscr{G}} \frac{1}{n} \sum_{x_i \in \mathscr{X}} \sigma_i g(x_i)]$ where $\sigma_1, \ldots, \sigma_n$ are n Rademacher variables (see section 3.1.2.3).

2. Our demo code with synthetic data is available from https://github.com/takashiishida/pconf.

3. Negative training data was used only in the fully supervised method that was tested for performance comparison.

4. If we naively use default parameters in Sklearn (Pedregosa et al., 2011) instead, which is the usual case in the real world without negative data for validation, the classification accuracy of O-SVM is worse for all setups except D in table 6.1, which illustrates the difficulty of using O-SVM.

5. https://github.com/zalandoresearch/fashion-mnist

6. https://www.cs.toronto.edu/~kriz/cifar.html

Notes to Chapter 7

1. As you will often see, $\pi_{\mathrm{P}} \neq \frac{1}{2}$ is assumed in this chapter. This is partly because the pointwise densities $\widetilde{p}_{\mathrm{S}}$ and $\widetilde{p}_{\mathrm{D}}$ collapse with p_{U}. Hence, $\pi_{\mathrm{P}} \neq \frac{1}{2}$ is a regularity condition in pairwise-constraint classification.

2. Again, we assume $\pi_{\mathrm{P}} > \frac{1}{2}$ as in section 7.3.4. To recover the exact π_{P}, we must know $\mathrm{sign}(\pi_{\mathrm{P}} - \pi_{\mathrm{N}})$ in advance.

3. This property is essentially a consequence of $\pi_{\mathrm{S}} > \pi_{\mathrm{D}}$ and the *one-sample* case. In the two-sample case, n_{S} and n_{D} are no longer random variables, and $\pi_{\mathrm{S}}/\sqrt{2n_{\mathrm{S}}} < \pi_{\mathrm{D}}/\sqrt{2n_{\mathrm{D}}}$ holds with sufficiently large n_{S} that can be fixed independently from n_{D}.

4. https://www.openml.org/d/1460

5. In the limit of $\pi_{\mathrm{P}} \to \frac{1}{2}$, the SU risk estimator (7.8) is not defined since $\ell_{\sharp}$ diverges. Even if one implements SU classification by ignoring the diverging term $2\pi_{\mathrm{P}} - 1$ in the denominator of (7.6), we cannot obtain promising experimental results. Intuitively speaking, one cannot distinguish S data and U data because their underlying distributions in (7.3) and (7.5) are the same at $\pi_{\mathrm{P}} = \frac{1}{2}$. Hence, it becomes impossible to estimate the classification risk only with SU data. This is also the case for DU and SD classification.

6. Recall that the squared loss is U-shaped. See figure 2.4.

Notes to Chapter 8

1. Here it does not matter if either θ or θ' equals $\pi_P = p(y = +1)$ or neither of them equals it.

2. In the extremely imbalanced case where π_P is close to 0 or 1, the problem under consideration should be called *retrieval* or *detection* rather than binary classification.

3. Equation (8.5) is known as *backward correction* in classification with noisy/corrupted labels (Patrini et al., 2017).

4. http://yann.lecun.com/exdb/mnist/

5. https://github.com/zalandoresearch/fashion-mnist

6. http://ufldl.stanford.edu/housenumbers/

7. https://www.cs.toronto.edu/~kriz/cifar.html

8. https://keras.io

Notes to Chapter 9

1. There are a few exceptions such as the *Lee-Lin-Wahba multi-class model* (Lee et al., 2004) and the *error-correcting output coding* (Crammer and Singer, 2002).

2. Note that when a pattern x has already been equipped with OL y, giving CL $\bar{y}$ does not bring us any additional information (unless the OL is noisy).

3. As shown in chapter 10, CL classification is a special case of PL classification. Therefore, any PL classification method can be used for CL classification.

4. While each pattern belongs *exclusively* to one of the classes in multi-class classification, ML classification allows each pattern to belong *simultaneously* to multiple classes (Boutell et al., 2004). For example, "This image contains both a dog and a cat."

5. https://cs.nyu.edu/~roweis/data.html

6. http://archive.ics.uci.edu/ml/

7. We used $c - 1$ times more CL data than OL data since a single OL corresponds to $(c - 1)$ CLs.

Notes to Chapter 10

1. In practice, unknown class-posterior probability $p(y|x)$ may be approximated by the *softmax* of the model output $g(x) := (g_1(x), \ldots, g_c(x))^\top$ as $\widehat{p}(y|x) = \frac{\exp(g_y(x))}{\sum_{y'=1}^{c} \exp(g_{y'}(x))}$. Then we have the following risk estimator:

$$\frac{1}{n} \sum_{i=1}^{n} \sum_{y \in Y_i} \frac{\exp(g_y(x))}{\sum_{y' \in Y_i} \exp(g_{y'}(x))} \mathscr{L}\big(g(x_i), y\big).$$

Interestingly, this risk estimator coincides with that of the *progressive identification (PRODEN)* algorithm (Lv et al., 2020).

2. However, because of the generality, the second term of the upper bound derived in section 10.5.3 is twice larger than the one given in theorem 10.2.

Notes to Chapter 11

1. This chapter focuses on the overfitting incurred by the ERM-based approaches for weakly supervised classification. Nevertheless, the techniques introduced in section 11.4 are more general and could be applied to other problem settings; see, for example, Han et al. (2020b).

2. The graphs also contain *non-negative PU* (nnPU) classification, which will be introduced and explained later in Section 11.4.1.

3. By co-occurrence, we mean that both of them can be observed, or neither of them can be observed.

4. The probability measure is induced by the randomness of the two unlabeled datasets.

5. http://yann.lecun.com/exdb/mnist/

6. http://qwone.com/~jason/20Newsgroups/

7. https://www.cs.toronto.edu/~kriz/cifar.html

8. https://www.csie.ntu.edu.tw/~cjlin/libsvmtools/datasets/binary.html

9. http://yann.lecun.com/exdb/mnist/

10. https://github.com/zalandoresearch/fashion-mnist

11. https://github.com/rois-codh/kmnist

12. https://www.cs.toronto.edu/~kriz/cifar.html

13. In our experimental setup, the bUU method reduces to the balanced error (BER) minimization method (Menon et al., 2015).

14. Intuitively, as the class-priors move closer, two U sets would be more similar and thus less informative, which is significantly harder than assuming that they are sufficiently far away.

Notes to Chapter 12

1. When $p(x) > 0$ over $\mathcal{X}$, $p = q$ means that $p(x) = q(x)$ almost surely over $\mathcal{X}$, i.e., $\int (p(x) - q(x))^2 dx = 0$.

2. For example, methods based on *density ratio estimation* (Sugiyama et al., 2012) and the Fenchel-duality bounding technique (see Keziou, 2003, and also section 12.4.4) can be used for this purpose.

3. Note that we are estimating the reciprocal $p_{\mathrm{P}}(x)/p_{\mathrm{U}}(x)$, since p_{U} has a wider support than p_{P}, and then $p_{\mathrm{P}}(x)/p_{\mathrm{U}}(x)$ is smoother than the original $p_{\mathrm{U}}(x)/p_{\mathrm{P}}(x)$ (Sugiyama et al., 2012).

4. We employ PU class-prior estimation in order to perform PU classification, where collecting unlabeled data should be much cheaper than collecting the same amount of positive data.

Notes to Chapter 13

1. http://www.biscuits.work/fourth-workshop/program.html

2. https://rakutenfashionweektokyo.com/en/brands/detail/emarie/

Bibliography

Agakov, F., and D. Barber. 2006. "Kernelized Infomax Clustering." *Advances in Neural Information Processing Systems 18 (NeurIPS2005)*: 17–24.

Ali, S. M., and S. D. Silvey. 1966. "A General Class of Coefficients of Divergence of One Distribution from Another." *Journal of the Royal Statistical Society, Series B* 28 (1): 131–142.

Anderson, N., P. Hall, and D. Titterington. 1994. "Two-Sample Test Statistics for Measuring Discrepancies between Two Multivariate Probability Density Functions Using Kernel-Based Density Estimates." *Journal of Multivariate Analysis* 50 (1): 41–54.

Angluin, D., and P. Laird. 1987. "Learning from Noisy Examples." *Machine Learning* 2:343.

Baldi, P., and Y. Chauvin. 1993. "Neural Networks for Fingerprint Recognition." *Neural Computation* 5 (3): 402–418.

Bao, H., G. Niu, and M. Sugiyama. 2018. "Classification from Pairwise Similarity and Unlabeled Data." *Proceedings of 35th International Conference on Machine Learning (ICML2018)*: 452–461.

Bao, H., I. Sato, T. Shimada, M. Sugiyama, and L. Xu. 2020. "Similarity-Based Classification: Connecting Similarity Learning to Binary Classification." Technical Report 2006.06207, arXiv.

Bartlett, P. L., M. I. Jordan, and J. D. McAuliffe. 2006. "Convexity, Classification, and Risk Bounds." *Journal of the American Statistical Association* 101 (473): 138–156.

Bartlett, P. L., and S. Mendelson. 2002. "Rademacher and Gaussian Complexities: Risk Bounds and Structural Results." *Journal of Machine Learning Research* 3:463–482.

Basu, S., A. Banerjee, and R. J. Mooney. 2002. "Semi-Supervised Clustering by Seeding." *Proceedings of the 19th International Conference on Machine Learning (ICML2002)*: 27–34.

Basu, S., M. Bilenko, and R. J. Mooney. 2004. "A Probabilistic Framework for Semi-Supervised Clustering." *Proceedings of the 10th ACM SIGKDD International Conference on Knowledge Discovery and Data Mining (KDD2004)*: 59–68.

Bekker, J., and J. Davis. 2018. "Estimating the Class Prior in Positive and Unlabeled Data through Decision Tree Induction." *Proceedings of the 32nd AAAI Conference on Artificial Intelligence (AAAI2018)*: 2712–2719.

Belkin, M., P. Niyogi, and V. Sindhwani. 2006. "Manifold Regularization: A Geometric Framework for Learning from Labeled and Unlabeled Examples." *Journal of Machine Learning Research* 7 (85): 2399–2434.

Ben-Tal, A., L. E. Ghaoui, and A. Nemirovski. 2009. *Robust Optimization*. Princeton, NJ: Princeton University Press.

Bilenko, M., S. Basu, and R. J. Mooney. 2004. "Integrating Constraints and Metric Learning in Semi-Supervised Clustering." *Proceedings of the 21st International Conference on Machine Learning (ICML2004)*: 81–88.

Bishop, C. M. 1995. *Neural Networks for Pattern Recognition*. Oxford: Clarendon Press.

Bishop, C. M. 2006. *Pattern Recognition and Machine Learning*. New York: Springer.

Blanchard, G., G. Lee, and C. Scott. 2010. "Semi-Supervised Novelty Detection." *Journal of Machine Learning Research* 11 (99): 2973–3009.

Blum, A., and T. M. Mitchell. 1998. "Combining Labeled and Unlabeled Data with Co-Training." *Proceedings of the 11th Annual Conference on Computational Learning Theory (COLT1998)*: 92–100.

Boutell, M. R., C. M. Brown, J. Luo, and X. Shen. 2004. "Learning Multi-Label Scene Classification." *Pattern Recognition 37* (9): 1757–1771.

Boyd, S., and L. Vandenberghe. 2004. *Convex Optimization.* Cambridge, MA: Cambridge University Press.

Breunig, M. M., H. P. Kriegel, R. T. Ng, and J. Sander. 2000. "LOF: Identifying Density-Based Local Outliers." *Proceedings of the 2000 ACM SIGMOD International Conference on Management of Data* vol. 29: (SIGMOD2000), 93–104.

Briggs, F., X. Z. Fern, and R. Raich. 2012. "Rank-Loss Support Instance Machines for MIML Instance Annotation." *Proceedings of the 18th ACM SIGKDD International Conference on Knowledge Discovery and Data Mining (KDD2012)*: 534–542.

Brodersen, K. H., J. M. Buhmann, C. S. Ong, and K. E. Stephan. 2010. "The Balanced Accuracy and Its Posterior Distribution." *Proceedings of the 20th International Conference on Pattern Recognition (ICPR2010)*: 3121–3124.

Bromley, J., I. Guyon, Y. Le Cun, E. Säckinger, and R. Shah. 1994. "Signature Verification Using a Siamese Time Delay Neural Network." *Advances in Neural Information Processing Systems 7 (NeurIPS1994)*: 737–744.

Calandriello, D., G. Niu, and M. Sugiyama. 2014. "Semi-Supervised Information-Maximization Clustering." *Neural Networks* 57:103–111.

Candès, E. J., and B. Recht. 2009. "Exact Matrix Completion via Convex Optimization." *Foundations of Computational Mathematics* 9 (6): 717.

Cao, Y., B. An, L. Feng, G. Niu, M. Sugiyama, and Y. Xu. 2021. "Learning from Similarity-Confidence Data," Technical Report 2102.06879, arXiv.

Caruana, R. 1997. "Multitask Learning." *Machine Learning* 28: 41–75.

Cesa-Bianchi, N., and G. Lugosi. 2006. *Prediction, Learning, and Games.* New York: Cambridge University Press.

Chang, C. C., and C. J. Lin. 2011. "LIBSVM: A Library for Support Vector Machines." *ACM Transactions on Intelligent Systems and Technology* 2 (3): 1–27. Software available at http://www.csie.ntu.edu.tw/~cjlin/libsvm.

Chapelle, O., B. Schölkopf, and J. Weston. 2003. "Cluster Kernels for Semi-Supervised Learning." *Advances in Neural Information Processing Systems 15 (NeurIPS2002)*: 585–592.

Chapelle, O., B. Schölkopf, and A. Zien, eds. 2006. *Semi-Supervised Learning.* Cambridge, MA: MIT Press.

Charoenphakdee, N., J. Lee, and M. Sugiyama. 2019. "On Symmetric Losses for Learning from Corrupted Labels." *Proceedings of the 36th International Conference on Machine Learning (ICML2019)*: 961–970.

Chen, W., and G. Feng. 2012. "Spectral Clustering: A Semi-Supervised Approach." *Neurocomputing* 77 (1): 229–242.

Chiang, K. Y., I. S. Dhillon, and C. J. Hsieh. 2015. "Matrix Completion with Noisy Side Information." *Advances in Neural Information Processing Systems 28 (NeurIPS2015)*: 3447–3455.

Chung, F. R. K. 1997. *Spectral Graph Theory.* Providence, RI: American Mathematical Society.

Chung, K. L. 1968. *A Course in Probability Theory.* New York: Harcourt, Brace.

Clanuwat, T., M. Bober-Irizar, D. Ha, A. Kitamoto, A. Lamb, and K. Yamamoto. 2018. "Deep Learning for Classical Japanese Literature." In *NeurIPS Workshop on Machine Learning for Creativity and Design.* Available at: https://nips2018creativity.github.io/doc/deep_learning_for_classical_japanese_literature.pdf.

Cortes, C., and M. Mohri. 2004. "AUC Optimization vs. Error Rate Minimization." *Advances in Neural Information Processing Systems 16 (NeurIPS2003)*: 313–320.

Cour, T., B. Sapp, and B. Taskar. 2011. "Learning from Partial Labels." *Journal of Machine Learning Research* 12:1501–1536.

Cover, T. M., and J. A. Thomas. 2006. *Elements of Information Theory*, 2nd ed. Hoboken, NJ: John Wiley & Sons.

Cozman, F. G., M. C. Cirelo, and I. Cohen. 2003. "Semi-Supervised Learning of Mixture Models." *Proceedings of the 20th International Conference on Machine Learning*: 99–106.

Crammer, K., and Y. Singer. 2002. "On the Learnability and Design of Output Codes for Multiclass Problems." *Machine Learning* 47 (2): 201–233.

Csiszár, I. 1967. "Information-Type Measures of Difference of Probability Distributions and Indirect Observation." *Studia Scientiarum Mathematicarum Hungarica* 2:229–318.

Cui, Z., N. Charoenphakdee, I. Sato, and M. Sugiyama. 2020. "Classification from Triplet Comparison Data." *Neural Computation* 32 (3): 659–681.

Dan, S., H. Bao, and M. Sugiyama. 2020. "Learning from Noisy Similar and Dissimilar Data." Technical Report 2002.00995, arXiv.

Davenport, M. A., and J. Romberg. 2016. "An Overview of Low-Rank Matrix Recovery from Incomplete Observations." *IEEE Journal of Selected Topics in Signal Processing* 10 (4): 608–622.

Davis, J., I. Dhillon, P. Jain, B. Kulis, and S. Sra. 2007. "Information-Theoretic Metric Learning." *Proceedings of the 24th International Conference on Machine Learning (ICML2007)*: 209–216.

Doya, K. 2007. "Reinforcement Learning: Computational Theory and Biological Mechanisms." *HFSP Journal* 1 (1): 30–40.

Duchi, J., E. Hazan, and Y. Singer. 2011. "Adaptive Subgradient Methods for Online Learning and Stochastic Optimization." *Journal of Machine Learning Research* 12:2121–2159.

du Plessis, M. C., and M. Sugiyama. 2014a. "Class Prior Estimation from Positive and Unlabeled Data." *IEICE Transactions on Information and Systems* E97-D (5): 1358–1362.

du Plessis, M. C., and M. Sugiyama. 2014b. "Semi-Supervised Learning of Class Balance under Class-Prior Change by Distribution Matching." *Neural Networks* 50:110–119.

du Plessis, M. C., G. Niu, and M. Sugiyama. 2013. "Clustering Unclustered Data: Unsupervised Binary Labeling of Two Datasets Having Different Class Balances." *Proceedings of Conference on Technologies and Applications of Artificial Intelligence (TAAI2013)*: 1–6.

du Plessis, M. C., G. Niu, and M. Sugiyama. 2014. "Analysis of Learning from Positive and Unlabeled Data." *Advances in Neural Information Processing Systems 27 (NeurIPS2014)*: 703–711.

du Plessis, M. C., G. Niu, and M. Sugiyama. 2015. "Convex Formulation for Learning from Positive and Unlabeled Data." *Proceedings of 32nd International Conference on Machine Learning (ICML2015)*: 1386–1394.

du Plessis, M. C., G. Niu, and M. Sugiyama. 2017. "Class-Prior Estimation for Learning from Positive and Unlabeled Data." *Machine Learning* 106 (4): 463–492.

Elkan, C. 2001. "The Foundations of Cost-Sensitive Learning." *Proceedings of the Seventeenth International Joint Conference on Artificial Intelligence (IJCAI2001)*: 973–978.

Elkan, C., and K. Noto. 2008. "Learning Classifiers from Only Positive and Unlabeled Data." *Proceedings of the 14th ACM SIGKDD International Conference on Knowledge Discovery and Data Mining (KDD2008)*: 213–220.

Eric, B., N. D. Freitas, and A. Ghosh. 2008. "Active Preference Learning with Discrete Choice Data." *Advances in Neural Information Processing Systems* 21:409–416.

Fedorov, V. V. 1972. *Theory of Optimal Experiments*. New York: Academic Press.

Feldman, V., V. Guruswami, P. Raghavendra, and Y. Wu. 2012. "Agnostic Learning of Monomials by Halfspaces Is Hard." *SIAM Journal on Computing* 41 (6): 1558–1590.

Feng, L., and B. An. 2019. "Partial Label Learning with Self-Guided Retraining." *Proceedings of the 33rd AAAI Conference on Artificial Intelligence (AAAI2019)*: 3542–3549.

Feng, L., B. An, X. Geng, B. Han, J. Lv, G. Niu, M. Sugiyama, and M. Xu. 2020a. "Provably Consistent Partial-Label Learning." *Advances in Neural Information Processing Systems 33 (NeurIPS2020)*: 10948–10960.

Feng, L., B. An, B. Han, T. Kaneko, G. Niu, and M. Sugiyama. 2020b. "Learning with Multiple Complementary Labels." *Proceedings of the 37th International Conference on Machine Learning (ICML2020)*: 3072–3081.

Fisher, R. J. 1993. "Social Desirability Bias and the Validity of Indirect Questioning." *Journal of Consumer Research* 20 (2): 303–315.

Fishman, G. S. 1996. *Monte Carlo: Concepts, Algorithms, and Applications*. Berlin: Springer.

Fukushima, K. 1980. "Neocognitron: A Self-Organizing Neural Network Model for a Mechanism of Pattern Recognition Unaffected by Shift in Position." *Biological Cybernetics* 36 (4): 93–202.

Ghosh, A., H. Kumar, and P. Sastry. 2017. "Robust Loss Functions Under Label Noise for Deep Neural Networks." *Proceedings of the 31st AAAI Conference on Artificial Intelligence (AAAI2017)*: 1919–1925.

Ghosh, A., N. Manwani, and P. S. Sastry. 2015. "Making Risk Minimization Tolerant to Label Noise." *Neurocomputing* 160:93–107.

Glorot, X., and Y. Bengio. 2010. "Understanding the Difficulty of Training Deep Feedforward Neural Networks." *Proceedings of the Thirteenth International Conference on Artificial Intelligence and Statistics (AISTATS2010)*: 249–256.

Golowich, N., A. Rakhlin, and O. Shamir. 2018. "Size-Independent Sample Complexity of Neural Networks." *Proceedings of the 31st Conference On Learning Theory (COLT2018)*: 297–299. https://arxiv.org/abs/1712 .06541.

Gomes, R., A. Krause, and P. Perona. 2010. "Discriminative Clustering by Regularized Information Maximization." *Advances in Neural Information Processing Systems 23 (NeurIPS2010)*: 766–774.

Goodfellow, I., Y. Bengio, and A. Courville. 2016. *Deep Learning*. Cambridge, MA: MIT Press.

Goodfellow, I., Y. Bengio, A. Courville, M. Mirza, S. Ozair, J. Pouget-Abadie, D. Warde-Farley, and B. Xu. 2014. "Generative Adversarial Nets." *Advances in Neural Information Processing Systems 27 (NeurIPS2014)*: 2672–2680.

Goodfellow, I., J. Shlens, and C. Szegedy. 2015. "Explaining and Harnessing Adversarial Examples." *Third International Conference on Learning Representations (ICLR2015)*: 1–11.

Grandvalet, Y., and Y. Bengio. 2005. "Semi-Supervised Learning by Entropy Minimization." *Advances in Neural Information Processing Systems 17 (NeurIPS2004)*: 529–536.

Guillaumin, M., C. Schmid, and J. Verbeek. 2010. "Multiple Instance Metric Learning from Automatically Labeled Bags of Faces." *Lecture Notes in Computer Science* 63 (11): 634–647.

Hall, P., and M. P. Wand. 1988. "On Nonparametric Discrimination Using Density Differences." *Biometrika* 75 (3): 541–547.

Han, B., J. T. Kwok, T. Liu, G. Niu, M. Sugiyama, I. W. Tsang, and Q. Yao. 2020a. "A Survey of Label-Noise Representation Learning: Past, Present and Future." Technical Report 2011.04406, arXiv.

Han, B., G. Niu, M. Sugiyama, I. Tsang, M. Xu, Q. Yao, and X. Yu. 2020b. "SIGUA: Forgetting May Make Learning with Noisy Labels More Robust." *Proceedings of the 37th International Conference on Machine Learning (ICML2020)*: 4006–4016.

Härdle, W. 1990. *Applied Nonparametric Regression*. New York: Cambridge University Press.

Härdle, W., M. Müller, S. Sperlich, and A. Werwatz. 2004. *Nonparametric and Semiparametric Models*. Berlin: Springer.

Hastie, T., J. Friedman, and R. Tibshirani. 2001. *The Elements of Statistical Learning: Data Mining, Inference, and Prediction*. New York: Springer.

Hayashi, M., T. Sakai, and M. Sugiyama. 2018. "Binary Matrix Completion Using Unobserved Entries," Technical Report 1803.04663, arXiv.

He, K., S. Ren, J. Sun, and X. Zhang. 2016. "Deep Residual Learning for Image Recognition." *Proceedings of the IEEE Conference on Computer Vision and Pattern Recognition (CVPR2016)*: 770–778.

Herschtal, A., and B. Raskutti. 2004. "Optimising Area under the ROC Curve Using Gradient Descent." *Proceedings of the 21st International Conference on Machine Learning (ICML2004)*: 49.

Hinton, G. E., and R. R. Salakhutdinov. 2006. "Reducing the Dimensionality of Data with Neural Networks." *Science* 313 (5786): 504–507.

Howe, J. 2008. *Crowdsourcing: Why the Power of the Crowd Is Driving the Future of Business*. New York: Three Rivers Press.

Hsieh, C. J., I. S. Dhillon, and N. Natarajan. 2015. "PU Learning for Matrix Completion." *Proceedings of the 32nd International Conference on Machine Learning (ICML2015)*: 2445–2453.

Hu, W., E. Matsumoto, T. Miyato, M. Sugiyama, and S. Tokui. 2017. "Learning Discrete Representations via Information Maximizing Self-Augmented Training." *Proceedings of 34th International Conference on Machine Learning (ICML2017)*: 1558–1567.

Huang, G., Z. Liu, L. van der Maaten, and K. Q. Weinberger. 2017. "Densely Connected Convolutional Networks." *Proceedings of the IEEE Conference on Computer Vision and Pattern Recognition (CVPR2017)*: 4700–4708.

Huber, P. J. 1981. *Robust Statistics*. New York: Wiley.

Hüllermeier, E., and J. Beringer. 2006. "Learning from Ambiguously Labeled Examples." *Intelligent Data Analysis* 10 (5): 419–439.

Hyvärinen, A., J. Karhunen, and E. Oja. 2001. *Independent Component Analysis*. New York: Wiley.

Ioffe, S., and C. Szegedy. 2015. "Batch Normalization: Accelerating Deep Network Training by Reducing Internal Covariate Shift." *Proceedings of 32nd International Conference on Machine Learning (ICML2015)*: 448–456.

Ishida, T., W. Hu, G. Niu, and M. Sugiyama. 2017. "Learning from Complementary Labels." *Advances in Neural Information Processing Systems 30 (NeurIPS2017)*: 5644–5654.

Ishida, T., G. Niu, A. K. Menon, and M. Sugiyama. 2019. "Complementary-Label Learning for Arbitrary Losses and Models." *Proceedings of 36th International Conference on Machine Learning (ICML2019)*: 2971–2980.

Ishida, T., G. Niu, and M. Sugiyama. 2018. "Binary Classification from Positive-Confidence Data." *Advances in Neural Information Processing Systems 31 (NeurIPS2018)*: 5917–5928.

Ivanov, D. 2019. "DEDPUL: Method for Mixture Proportion Estimation and Positive-Unlabeled Classification Based on Density Estimation." Technical Report 1902.06965, arXiv.

Iyer, A., S. Nath, and S. Sarawagi. 2014. "Maximum Mean Discrepancy for Class Ratio Estimation: Convergence Bounds and Kernel Selection." *Proceedings of 31st International Conference on Machine Learning (ICML2014)*: 530–538.

Jain, S., P. Radivojac, and M. White. 2016. "Estimating the Class Prior and Posterior from Noisy Positives and Unlabeled Data." *Advances in Neural Information Processing Systems 29 (NeurIPS2016)*: 2693–2701.

Jamieson, K. G., and R. Nowak. 2011. "Active Ranking Using Pairwise Comparisons." *Advances in Neural Information Processing Systems 24 (NeurIPS2011)*: 2240–2248.

Kakade, S. M., K. Sridharan, and A. Tewari. 2009. "On the Complexity of Linear Prediction: Risk Bounds, Margin Bounds, and Regularization." *Advances in Neural Information Processing Systems 21 (NeurIPS2008)*: 793–800.

Kanamori, T., S. Hido, and M. Sugiyama. 2009. "A Least-Squares Approach to Direct Importance Estimation." *Journal of Machine Learning Research* 10 (July): 1391–1445.

Keziou, A. 2003. "Dual Representation of ϕ-divergences and Applications." *Comptes Rendus Mathématique* 336 (10): 857–862.

Kingma, D. P., and J. L. Ba. 2015. "Adam: A Method for Stochastic Optimization." *Third International Conference on Learning Representations (ICLR2015)*: 1–15.

Kiryo, R., G. Niu, M. C. du Plessis, and M. Sugiyama. 2017. "Positive-Unlabeled Learning with Non-Negative Risk Estimator." *Advances in Neural Information Processing Systems 30 (NeurIPS2017)*: 1674–1684.

Klein, D., S. D. Kamvar, and C. D. Manning. 2002. "From Instance-Level Constraints to Space-Level Constraints: Making the Most of Prior Knowledge in Data Clustering." *Proceedings of the 19th International Conference on Machine Learning (ICML2002)*: 307–314.

Koltchinskii, V. 2001. "Rademacher Penalties and Structural Risk Minimization." *IEEE Transactions on Information Theory* 47 (5): 1902–1914.

Krijthe, J. H., and M. Loog. 2017. "Robust Semi-Supervised Least Squares Classification by Implicit Constraints." *Pattern Recognition* 63: 115–126.

Krizhevsky, A. 2009. "Learning Multiple Layers of Features from Tiny Images." Technical Report, Department of Computer Science, University of Toronto.

Kullback, S., and R. A. Leibler. 1951. "On Information and Sufficiency." *The Annals of Mathematical Statistics* 22:79–86.

Lang, K. 1995. "Newsweeder: Learning to Filter Netnews." *Proceedings of the 12th International Conference on Machine Learning (ICML1995)*: 331–339.

Law, E., and L. von Ahn. 2011. *Human Computation*. San Rafael, CA: Morgan & Claypool Publishers.

LeCun, Y., Y. Bengio, L. Bottou, and P. Haffner. 1998. "Gradient-Based Learning Applied to Document Recognition." *Proceedings of the IEEE* 86 (11): 2278–2324.

Ledoux, M., and M. Talagrand. 1991. *Probability in Banach Spaces: Isoperimetry and Processes*. Berlin: Springer.

Lee, Y., Y. Lin, and G. Wahba. 2004. "Multicategory Support Vector Machines: Theory and Application to the Classification of Microarray Data and Satellite Radiance Data." *Journal of the American Statistical Association* 99 (465): 67–81.

Li, H. 2014. *Learning to Rank for Information Retrieval and Natural Language Processing*. 2nd ed. San Rafael, CA: Morgan & Claypool Publishers.

Li, Y. F., and Z. H. Zhou. 2015. "Towards Making Unlabeled Data Never Hurt." *IEEE Transactions on Pattern Analysis and Machine Intelligence* 37 (1): 175–188.

Li, Y. F., J. T. Kwok, I. W. Tsang, and Z. H. Zhou. 2009. "Tighter and Convex Maximum Margin Clustering." *Proceedings of 12th International Conference on Artificial Intelligence and Statistics (AISTATS2009)*: 344–351.

Li, Y. F., J. T. Kwok, I. W. Tsang, and Z. H. Zhou. 2013. "Convex and Scalable Weakly Labeled SVMs." *Journal of Machine Learning Research* 14 (1): 2151–2188.

Li, Z., and J. Liu. 2009. "Constrained Clustering by Spectral Kernel Learning." *2009 IEEE 12th International Conference on Computer Vision (ICCV2009)*: 421–427.

Li, Z., J. Liu, and X. Tang. 2008. "Pairwise Constraint Propagation by Semidefinite Programming for Semi-Supervised Classification." *Proceedings of the 25th International Conference on Machine Learning (ICML2008)*: 576–583.

Lichman, M. 2013. UCI Machine Learning Repository. http://archive.ics.uci.edu/ml.

Liu, L., and T. Dietterich. 2012. "A Conditional Multinomial Mixture Model for Superset Label Learning." *Advances in Neural Information Processing Systems* 25:557–565.

Liu, L.-P., and T. Dietterich. 2014. "Learnability of the Superset Label Learning Problem." *Proceedings of 31st International Conference on Machine Learning (ICML2014)*: 1629–1637.

Liu, T., and D. Tao. 2015. "Classification with Noisy Labels by Importance Reweighting." *IEEE Transactions on Pattern Analysis and Machine Intelligence* 38 (3): 447–461.

Logeswaran, L., and H. Lee. 2018. "An Efficient Framework for Learning Sentence Representations." *Proceedings of the 6th International Conference on Learning Representations (ICLR2018)*: 1–16.

Lu, N., A. K. Menon, G. Niu, and M. Sugiyama. 2019. "On the Minimal Supervision for Training Any Binary Classifier from Only Unlabeled Data." *Proceedings of the 7th International Conference on Learning Representations (ICLR2019)*: 1–18.

Lu, N., G. Niu, M. Sugiyama, and T. Zhang. 2020. "Mitigating Overfitting in Supervised Classification from Two Unlabeled Datasets: A Consistent Risk Correction Approach." *Proceedings of the 23rd International Conference on Artificial Intelligence and Statistics (AISTATS2020)*: 1115–1125.

Lv, J., L. Feng, G. Geng, X. Niu, M. Sugiyama, and M. Xu. 2020. "Progressive Identification of True Labels for Partial-Label Learning." *Proceedings of the 37th International Conference on Machine Learning (ICML2020)*: 6500–6510.

MacKay, D. J. C. 2003. *Information Theory, Inference, and Learning Algorithms*. Cambridge: Cambridge University Press.

MacQueen, J. B. 1967. "Some Methods for Classification and Analysis of Multivariate Observations." *Proceedings of the 5th Berkeley Symposium on Mathematical Statistics and Probability* 1:281–297.

Mann, G., and A. McCallum. 2007. "Simple, Robust, Scalable Semi-Supervised Learning via Expectation Regularization." *Proceedings of the 24th Annual International Conference on Machine Learning (ICML2007)*: 593–600.

Maurer, A. 2016. "A Vector-Contraction Inequality for Rademacher Complexities." *Proceedings of International Conference on Algorithmic Learning Theory (ALT2016)*: 3–17.

McDiarmid, C. 1989. "On the Method of Bounded Differences." In *Surveys in Combinatorics*, ed. J. Siemons, 148–188. Cambridge: Cambridge University Press.

McLachlan, G. J. 1975. "Iterative Reclassification Procedure for Constructing an Asymptotically Optimal Rule of Allocation in Discriminant Analysis." *Journal of the American Statistical Association* 70 (350): 365–369.

Melacci, S., and M. Belkin. 2011. "Laplacian Support Vector Machines Trained in the Primal." *Journal of Machine Learning Research* 12: 1149–1184.

Mendelson, S. 2008. "Lower Bounds for the Empirical Minimization Algorithm." *IEEE Transactions on Information Theory* 54 (8): 3797–3803.

Menon, A., C. S. Ong, B. V. Rooyen, and B. Williamson. 2015. "Learning from Corrupted Binary Labels via Class-Probability Estimation." *Proceedings of 32nd International Conference on Machine Learning (ICML2015)*: 125–134.

Mikolov, T., K. Chen, G. S. Corrado, J. Dean, and I. Sutskever. 2013. "Distributed Representations of Words and Phrases and Their Compositionality." *Advances in Neural Information Processing Systems (NeurIPS2013)*: 3111–3119.

Miyato, T., S. Ishii, M. Koyama, and S. Maeda. 2019. "Virtual Adversarial Training: A Regularization Method for Supervised and Semi-Supervised Learning." *IEEE Transactions on Pattern Analysis and Machine Intelligence* 41 (8): 1979–1993.

Mohri, M., A. Rostamizadeh, and A. Talwalkar. 2012. *Foundations of Machine Learning.* Cambridge, MA: MIT Press.

Murphy, K. P. 2012. *Machine Learning: A Probabilistic Perspective.* Cambridge, MA: MIT Press.

Nair, V., and G. E. Hinton. 2010. "Rectified Linear Units Improve Restricted Boltzmann Machines." *Proceedings of the 27th International Conference on Machine Learning (ICML2010)*: 807–814.

Nakajima, S., M. Sugiyama, and K. Watanabe. 2019. *Variational Bayesian Learning Theory.* Cambridge: Cambridge University Press.

Natarajan, N., I. Dhillon, P. Ravikumar, and A. Tewari. 2013. "Learning with Noisy Labels." *Advances in Neural Information Processing Systems 26 (NeurIPS2013)*: 1196–1204.

Netzer, Y., A. Bissacco, A. Coates, A. Y. Ng, T. Wang, and B. Wu. 2011. "Reading Digits in Natural Images with Unsupervised Feature Learning." *NIPS Workshop on Deep Learning and Unsupervised Feature Learning.* Available at: http://ufldl.stanford.edu/housenumbers/nips2011_housenumbers.pdf.

Niu, G., B. Dai, H. Hachiya, W. Jitkrittum, and M. Sugiyama. 2013. "Squared-Loss Mutual Information Regularization." *Proceedings of 30th International Conference on Machine Learning (ICML2013)*: 10–18.

Niu, G., B. Dai, M. Sugiyama, and M. Yamada. 2012. "Information-Theoretic Semi-Supervised Metric Learning via Entropy Regularization." *Proceedings of 29th International Conference on Machine Learning (ICML2012)*: 89–96.

Niu, G., Y. Ma, M. C. du Plessis, T. Sakai, and M. Sugiyama. 2016. "Theoretical Comparisons of Positive-Unlabeled Learning against Positive-Negative Learning." *Advances in Neural Information Processing Systems 29 (NeurIPS2016)*: 1199–1207.

Nocedal, J., and S. J. Wright. 1999. *Numerical Optimization.* New York: Springer.

Noroozi, M., and P. Favaro. 2016. "Unsupervised Learning of Visual Representations by Solving Jigsaw Puzzles." *Proceedings of the European Conference on Computer Vision (ECCV2016)*: 69–84.

Orr, G., and K. R. Müller, eds. 1998. *Neural Networks: Tricks of the Trade.* Vol. 1524 of *Lecture Notes in Computer Science.* Berlin: Springer.

Pan, S. J., and Q. Yang. 2010. "A Survey on Transfer Learning." *IEEE Transactions on Knowledge and Data Engineering* 22 (10): 1345–1359.

Parzen, E. 1962. "On Estimation of a Probability Density Function and Mode." *The Annals of Mathematical Statistics* 33 (3): 1065–1076.

Paszke, A., L. Antiga, J. Bai, J. Bradbury, G. Chanan, S. Chintala, S. Chilamkurthy, A. Desmaison, Z. DeVito, L. Fang, N. Gimelshein, S. Gross, T. Killeen, A. Kopf, A. Lerer, Z. Lin, F. Massa, M. Raison, B. Steiner, A. Tejani, and E. Yang. 2019. "PyTorch: An Imperative Style, High-Performance Deep Learning Library." *Advances in Neural Information Processing Systems 32 (NeurIPS2019)*: 8024–8035.

Patrini, G., A. Rozza, A. K. Menon, R. Nock, and L. Qu. 2017. "Making Deep Neural Networks Robust to Label Noise: A Loss Correction Approach." *Proceedings of the IEEE Conference on Computer Vision and Pattern Recognition (CVPR2017)*: 1944–1952.

Pearson, K. 1900. "On the Criterion That a Given System of Deviations from the Probable in the Case of a Correlated System of Variables Is Such That It Can Be Reasonably Supposed to Have Arisen from Random Sampling." *Philosophical Magazine Series 5* 50 (302): 157–175.

Pedregosa, F., M. Blondel, M. Brucher, D. Cournapeau, V. Dubourg, E. Duchesnay, A. Gramfort, O. Grisel, V. Michel, A. Passos, M. Perrot, P. Prettenhofer, B. Thirion, J. Vanderplas, G. Varoquaux, and R. Weiss. 2011. "Scikit-learn: Machine Learning in Python." *Journal of Machine Learning Research* 12 (85): 2825–2830.

Pennington, J., C. D. Manning, and R. Socher. 2014. "GloVe: Global Vectors for Word Representation." *Proceedings of the Conference on Empirical Methods in Natural Language Processing (EMNLP2014)*: 1532–1543.

Pukelsheim, F. 1993. *Optimal Design of Experiments*. New York: Wiley.

Quiñonero-Candela, J., N. Lawrence, A. Schwaighofer, and M. Sugiyama, eds. 2009. *Dataset Shift in Machine Learning*. Cambridge, MA: MIT Press.

Ramaswamy, H., C. Scott, and A. Tewari. 2016. "Mixture Proportion Estimation via Kernel Embeddings of Distributions." *Proceedings of 33rd International Conference on Machine Learning (ICML2016)*: 2052–2060.

Ratner, A., S. H. Bach, H. R. Ehrenberg, J. A. Fries, C. Ré, and S. Wu. 2017. "Snorkel: Rapid Training Data Creation with Weak Supervision." *Proceedings of the VLDB Endowment* 11 (3): 269–282.

Richter, S. R., V. Koltun, S. Roth, and V. Vineet. 2016. "Playing for Data: Ground Truth from Computer Games." *Proceedings of the 14th European Conference on Computer Vision (ECCV2016)*: 102–118.

Robbins, H., and S. Munro. 1951. "A Stochastic Approximation Method." *The Annals of Mathematical Statistics* 22: 400–407.

Rosenblatt, M. 1956. "Remarks on Some Nonparametric Estimates of a Density Function." *The Annals of Mathematical Statistics* 27 (3): 832–837.

Rumelhart, D. E., J. L. McClelland, and the PDP Research Group. 1986. *Parallel Distributed Processing: Explorations in the Microstructure of Cognition: Foundations*. Vol. 1. Cambridge, MA: MIT Press.

Russell, S. J., and P. Norvig. 2010. *Artificial Intelligence: A Modern Approach*. 3rd ed. Upper Saddle River, NJ: Prentice Hall.

Saerens, M., C. Decaestecker, and P. Latinne. 2002. "Adjusting the Outputs of a Classifier to New a Priori Probabilities: A Simple Procedure." *Neural Computation* 14 (1): 21–41.

Sakai, T., G. Niu, and M. Sugiyama. 2018. "Semi-Supervised AUC Optimization Based on Positive-Unlabeled Learning." *Machine Learning* 107 (4): 767–794.

Sakai, T., G. Niu, M. C. du Plessis, and M. Sugiyama. 2017. "Semi-Supervised Classification Based on Classification from Positive and Unlabeled Data." *Proceedings of 34th International Conference on Machine Learning (ICML2017)*: 2998–3006.

Saunshi, N., S. Arora, H. Khandeparkar, M. Khodak, and O. Plevrakis. 2019. "A Theoretical Analysis of Contrastive Unsupervised Representation Learning." *Proceedings of the 36th International Conference on Machine Learning (ICML2019)*: 5628–5637.

Schölkopf, B., and A. J. Smola. 2002. *Learning with Kernels*. Cambridge, MA: MIT Press.

Schölkopf, B., J. C. Platt, J. Shawe-Taylor, A. J. Smola, and R. C. Williamson. 2001. "Estimating the Support of a High-Dimensional Distribution." *Neural Computation* 13 (7): 1443–1471.

Scott, C. 2015. "A Rate of Convergence for Mixture Proportion Estimation with Application to Learning from Noisy Labels." *Proceedings of the 17th International Conference on Artificial Intelligence and Statistics (AISTATS2015)*: 838–846.

Scott, C., G. Blanchard, and G. Handy. 2013. "Classification with Asymmetric Label Noise: Consistency and Maximal Denoising." *Proceedings of the 26th Annual Conference on Learning Theory (COLT2013)*: 489–511.

Settles, B. 2009. "Active Learning Literature Survey." Computer Sciences Technical Report 1648, University of Wisconsin–Madison.

Shalev-Shwartz, S., and S. Ben-David. 2014. *Understanding Machine Learning: From Theory to Algorithms*. Cambridge: Cambridge University Press.

Shannon, C. 1948. "A Mathematical Theory of Communication." *Bell Systems Technical Journal* 27: 379–423.

Shi, J., and J. Malik. 2000. "Normalized Cuts and Image Segmentation." *IEEE Transactions on Pattern Analysis and Machine Intelligence* 22 (8): 888–905.

Shimada, T., H. Bao, I. Sato, and M. Sugiyama. 2021. "Classification from Pairwise Similarities/Dissimilarities and Unlabeled Data via Empirical Risk Minimization." *Neural Computation* 33 (5): 1234–1268.

Shinoda, K., H. Kaji, and M. Sugiyama. 2021. "Binary Classification from Positive Data with Skewed Confidence." *Proceedings of the Twenty-Ninth International Joint Conference on Artificial Intelligence (IJCAI2020)*: 3328–3334.

Silverman, B. W. 1986. *Density Estimation for Statistics and Data Analysis*. London: Chapman & Hall.

Sohn, K., D. Berthelot, N. Carlini, E. D. Cubuk, A. Kurakin, C.-L. Li, C. A. Raffel, H. Zhang, and Z. Zhang. 2020. "FixMatch: Simplifying Semi-Supervised Learning with Consistency and Confidence." *Advances in Neural Information Processing Systems 33 (NeurIPS2020)*: 596–608.

Sokolovska, N., O. Cappé, and F. Yvon. 2008. "The Asymptotics of Semi-Supervised Learning in Discriminative Probabilistic Models." *Proceedings of the 25th International Conference on Machine Learning (ICML2008)*: 984–991.

Springenberg, J. T., T. Brox, A. Dosovitskiy, and M. Riedmiller. 2015. "Striving for Simplicity: The All Convolutional Net." *Third International Conference on Learning Representations (ICLR2015)*: 1–14.

Srivastava, N., G. Hinton, A. Krizhevsky, R. Salakhutdinov, and I. Sutskever. 2014. "Dropout: A Simple Way to Prevent Neural Networks from Overfitting." *Journal of Machine Learning Research* 15: 1929–1958.

Sugiyama, M. 2010. "Superfast-Trainable Multi-Class Probabilistic Classifier by Least-Squares Posterior Fitting." *IEICE Transactions on Information and Systems* E93-D (10): 2690–2701.

Sugiyama, M. 2013. "Machine Learning with Squared-Loss Mutual Information." *Entropy* 15 (1): 80–112.

Sugiyama, M. 2015a. *Introduction to Statistical Machine Learning*. Amsterdam: Morgan Kaufmann.

Sugiyama, M. 2015b. *Statistical Reinforcement Learning: Modern Machine Learning Approaches*. Boca Raton, FL: Chapman and Hall/CRC.

Sugiyama, M., T. Kanamori, S. Liu, M. C. du Plessis, T. Suzuki, and I. Takeuchi. 2013. "Density-Difference Estimation." *Neural Computation* 25 (10): 2734–2775.

Sugiyama, M., T. Kanamori, and T. Suzuki. 2012. *Density Ratio Estimation in Machine Learning*. Cambridge: Cambridge University Press.

Sugiyama, M., and M. Kawanabe. 2012. *Machine Learning in Non-Stationary Environments: Introduction to Covariate Shift Adaptation*. Cambridge, MA: MIT Press.

Sugiyama, M., H. Hachiya, M. Kimura, G. Niu, and M. Yamada. 2014. "Information-Maximization Clustering Based on Squared-Loss Mutual Information." *Neural Computation* 26 (1): 84–131.

Sutton, R. S., and G. A. Barto. 1998. *Reinforcement Learning: An Introduction*. Cambridge, MA: MIT Press.

Suzuki, T., T. Kanamori, J. Sese, and M. Sugiyama. 2009. "Mutual Information Estimation Reveals Global Associations between Stimuli and Biological Processes." *BMC Bioinformatics* 10 (1): 52–12.

Tang, J., Q. Mei, M. Qu, M. Wang, J. Yan, and M. Zhang. 2015. "LINE: Large-Scale Information Network Embedding." *Proceedings of the 24th International Conference on World Wide Web (WWW2015)*: 1067–1077.

Tax, D. M. J., and R. P. W. Duin. 2004. "Support Vector Data Description." *Machine Learning* 54 (1): 45–66.

Tewari, A., and P. L. Bartlett. 2007. "On the Consistency of Multiclass Classification Methods." *Journal of Machine Learning Research* 8: 1007–1025.

Thurstone, L. L. 1927. "A Law of Comparative Judgment." *Psychological Review* 34 (4): 273.

Tokui, S., J. Clayton, S. Hido, and K. Oono. 2015. Chainer: "A Next-Generation Open Source Framework for Deep Learning." *NIPS Machine Learning Systems Workshop*. Available at: http://learningsys.org/papers/LearningSys _2015_paper_33.pdf

Torralba, A., R. Fergus, and W. T. Freeman. 2008. "80 Million Tiny Images: A Large Data Set for Nonparametric Object and Scene Recognition." *IEEE Transactions on Pattern Analysis and Machine Intelligence* 30:1958–1970.

Tsuchiya, T., N. Charoenphakdee, I. Sato, and M. Sugiyama. 2019. "Semi-Supervised Ordinal Regression Based on Empirical Risk Minimization." Technical Report 1901.11351, arXiv.

Valizadegan, H., and R. Jin. 2007. "Generalized Maximum Margin Clustering and Unsupervised Kernel Learning." *Advances in Neural Information Processing Systems 19 (NeurIPS2006)*: 1417–1424.

Vapnik, V. N. 1998. *Statistical Learning Theory*. New York: Wiley.

Wagstaff, K., C. Cardie, S. Rogers, and S. Schrödl. 2001. "Constrained k-Means Clustering with Background Knowledge." *Proceedings of the 18th International Conference on Machine Learning (ICML2001)*: 577–584.

Wang, X., and A. Gupta. 2015. "Unsupervised Learning of Visual Representations Using Videos." *Proceedings of the IEEE International Conference on Computer Vision (ICCV2015)*: 2794–2802.

Weinberger, K. Q., J. Blitzer, and L. K. Saul. 2005. "Distance Metric Learning for Large Margin Nearest Neighbor Classification." *Advances in Neural Information Processing Systems (NeurIPS2005)*: 1473–1480.

Weston, J., A. Bordes, and S. Chopra. 2015. "Memory Networks." *Third International Conference on Learning Representations (ICLR2015)*: 1–15.

Wu, Z., and M. Sugiyama. 2021. "Learning with Proper Partial Labels." *Technical Report 2112.12303, arXiv*.

Xiao, H., K. Rasul, and R. Vollgraf. 2017. "Fashion-MNIST: A Novel Image Dataset for Benchmarking Machine Learning Algorithms." *Technical Report 1708.07747, arXiv*.

Xing, E. P., M. I. Jordan, A. Y. Ng, and S. Russell. 2002. "Distance Metric Learning, with Application to Clustering with Side-Information." *Advances in Neural Information Processing Systems (NeurIPS2002)* 15: 521–528.

Xu, L., J. Honda, G. Niu, and M. Sugiyama. 2019. "Uncoupled Regression from Pairwise Comparison Data." *Advances in Neural Information Processing Systems 32 (NeurIPS2019)*: 3994–4004.

Xu, L., B. Larson, J. Neufeld, and D. Schuurmans. 2005. "Maximum Margin Clustering." *Advances in Neural Information Processing Systems 17 (NeurIPS2004)*: 1537–1544.

Xu, Y., D. Tao, C. Xu, and C. Xu. 2017. "Multi-Positive and Unlabeled Learning." *Proceedings of the 26th International Joint Conference on Artificial Intelligence (IJCAI2017)*: 3182–3188.

Yamada, M., J. Simm, M. Sugiyama, and G. Wichern. 2011. "Improving the Accuracy of Least-Squares Probabilistic Classifiers." *IEICE Transactions on Information and Systems* E94-D (6): 1337–1340.

Yan, R., A. G. Hauptmann, J. Yang, and J. Zhang. 2006. "A Discriminative Learning Framework with Pairwise Constraints for Video Object Classification." *IEEE Transactions on Pattern Analysis and Machine Intelligence* 28 (4): 578–593.

Yao, Y., T. Liu, B. Han, M. Gong, G. Niu, M. Sugiyama, and D. Tao, D. 2022. "Rethinking Class-Prior Estimation for Positive-Unlabeled Learning." Tenth International Conference on Learning Representations (ICLR2022): 1–11.

Yi, J., A. Jain, R. Jin, Q. Qian, and L. Zhang. 2013. "Semi-Supervised Clustering by Input Pattern Assisted Pairwise Similarity Matrix Completion." *Proceedings of the 30th International Conference on Machine Learning (ICML2013)*: 1400–1408.

Yu, F. X., S. F. Chang, T. Jebara, D. Liu, and S. Kumar. 2013. "$\propto$SVM for Learning with Label Proportions." *Proceedings of 30th International Conference on Machine Learning (ICML2013)*: 504–512.

Yu, X., M. Gong, T. Liu, and D. Tao. 2018. "Learning with Biased Complementary Labels." *Proceedings of the 15th European Conference on Computer Vision (ECCV2018)*, vol. pt. I: 68–83.

Yuan, G. X., C. H. Ho, and C. J. Lin. 2012. "An Improved GLMNET for l1-Regularized Logistic Regression." *Journal of Machine Learning Research* 13 (64): 1999–2030.

Yuille, A. L., and A. Rangarajan. 2003. "The Concave-Convex Procedure." *Neural Computation* 15 (4): 915–936.

Zeiberg, D., S. Jain, and P. Radivojac. 2020. "Fast Nonparametric Estimation of Class Proportions in the Positive-Unlabeled Classification Setting." *Proceedings of the 34th AAAI Conference on Artificial Intelligence (AAAI2020)*: 2304–2313.

Zelnik-Manor, L., and P. Perona. 2005. "Self-Tuning Spectral Clustering." *Advances in Neural Information Processing Systems 17 (NeurIPS2004)*: 1601–1608.

Zeng, Z.-N., T.-H. Chan, S.-H. Gao, K. Jia, Y. Ma, S.-J. Xiao, and D. Xu. 2013. "Learning by Associating Ambiguously Labeled Images." *Proceedings of the IEEE Conference on Computer Vision and Pattern Recognition (CVPR2013)*: 708–715.

Zhang, J., and R. Yan. 2007. "On the Value of Pairwise Constraints in Classification and Consistency." *Proceedings of the 24th International Conference on Machine Learning (ICML2007)*: 1111–1118.

Zhang, M.-L., C.-Z. Tang, and F. Yu. 2017. "Disambiguation-Free Partial Label Learning." *IEEE Transactions on Knowledge and Data Engineering* 29 (10): 2155–2167.

Zhang, M.-L., and F. Yu. 2015. "Solving the Partial Label Learning Problem: An Instance-Based Approach." *Proceedings of the 24th International Joint Conference on Artificial Intelligence (IJCAI2015)*: 4048–4054.

Zhang, T. 2004. "Statistical Analysis of Some Multi-Category Large Margin Classification Methods." *Journal of Machine Learning Research* 5: 1225–1251.

Zhou, Z.-H. 2018. "A Brief Introduction to Weakly Supervised Learning." *National Science Review* 5 (1): 44–53.

Index

Adaptive Computation and Machine Learning

Francis Bach, Editor

Christopher Bishop, David Heckerman, Michael Jordan, and Michael Kearns, Associate Editors

Bioinformatics: The Machine Learning Approach, Pierre Baldi and Søren Brunak

Reinforcement Learning: An Introduction, Richard S. Sutton and Andrew G. Barto

Graphical Models for Machine Learning and Digital Communication, Brendan J. Frey

Learning in Graphical Models, Michael I. Jordan

Causation, Prediction, and Search, second edition, Peter Spirtes, Clark Glymour, and Richard Scheines

Principles of Data Mining, David Hand, Heikki Mannila, and Padhraic Smyth

Bioinformatics: The Machine Learning Approach, second edition, Pierre Baldi and Søren Brunak

Learning Kernel Classifiers: Theory and Algorithms, Ralf Herbrich

Learning with Kernels: Support Vector Machines, Regularization, Optimization, and Beyond, Bernhard Schölkopf and Alexander J. Smola

Introduction to Machine Learning, Ethem Alpaydin

Gaussian Processes for Machine Learning, Carl Edward Rasmussen and Christopher K. I. Williams

Semi-Supervised Learning, Olivier Chapelle, Bernhard Schölkopf, and Alexander Zien, Eds.

The Minimum Description Length Principle, Peter D. Grünwald

Introduction to Statistical Relational Learning, Lise Getoor and Ben Taskar, Eds.

Probabilistic Graphical Models: Principles and Techniques, Daphne Koller and Nir Friedman

Introduction to Machine Learning, second edition, Ethem Alpaydin

Machine Learning in Non-Stationary Environments: Introduction to Covariate Shift Adaptation, Masashi Sugiyama and Motoaki Kawanabe

Boosting: Foundations and Algorithms, Robert E. Schapire and Yoav Freund

Machine Learning: A Probabilistic Perspective, Kevin P. Murphy

Foundations of Machine Learning, Mehryar Mohri, Afshin Rostami, and Ameet Talwalker

Introduction to Machine Learning, third edition, Ethem Alpaydin

Deep Learning, Ian Goodfellow, Yoshua Bengio, and Aaron Courville